Bistatic Radar

Bistatic Radar

Nicholas J. Willis
Technology Service Corporation

SciTech Publishing, Inc
Raleigh, NC
www.scitechpub.com

SciTech
PUBLISHING, INC.

Published by SciTech Publishing Inc.
Raleigh, NC
www.scitechpub.com

SciTech President: Dudley R. Kay
Page Composition: Integrated Book Technology, Troy, NY
Cover Design: Brent Beckley

This is the SciTech corrected and reprinted version of the Second Edition published by Technology Service Corporation in 1995 and published in the First Edition by Artech House.

10 9 8 7 6 5 4 3 2 1

This book is available at special quantity discounts to use as premiums and sales promotions, or for use in corporate training programs. For more information and quotes, please contact:

Director of Special Sales
SciTech Publishing, Inc.
911 Paverstone Dr. — Ste. B
Raleigh, NC 27613
Phone: (919)847-2434
Fax: (919)847-2568
E-mail: sales@scitechpub.com
http://www.scitechpub.com

ISBN 1-891121-45-6
ISBN 13: 978-1-891121-45-6

To my mother, who complained when I didn't study,
and
To my wife, who didn't complain when I did.

CONTENTS

Preface   xi

Acknowledgments   xv

Chapter 1   Introduction and Overview   1
   1.1   Introduction   1
   1.2   Definitions   2
   1.3   Requirements   6
   1.4   Applications and Issues   7

Chapter 2   History   15
   2.1   Early History   16
      2.1.1   Developments in the United States   22
      2.1.2   Developments in the United Kingdom   25
      2.1.3   Developments in France   30
      2.1.4   Developments in the Soviet Union   32
      2.1.5   Developments in Japan   33
      2.1.6   Developments in Germany   34
      2.1.7   Developments in Italy   35
   2.2   First Resurgence   35
      2.2.1   Theory and Measurements   35
      2.2.2   Semiactive Homing Missiles   36
      2.2.3   Hitchhiking   36
      2.2.4   Forward-Scatter Fences   37
      2.2.5   Multistatic Radars   38
   2.3   Second Resurgence   42
      2.3.1   Air Defense   43
      2.3.2   Clutter Tuning   46
      2.3.3   Hitchhiking   49
      2.3.4   Forward-Scatter Fences   53
      2.3.5   Multistatic Radars   55
   2.4   Nonmilitary Applications   57

Chapter 3   Coordinate Systems, Geometry, and Equations   59

Chapter 4   Range Relationships   67
  4.1   Range Equation   67
  4.2   Pattern Propagation Factors   68
  4.3   Ovals of Cassini   70
  4.4   Operating Regions   72
  4.5   Target Path Dynamic Range   73
  4.6   Isorange Ellipsoids, Isorange Contours, and Range Cells   76
  4.7   Operating Limits and Instantaneous S/N Dynamic Range   80
    4.7.1   Operating Limits   81
    4.7.2   Instantaneous S/N Dynamic Range   81

Chapter 5   Location and Area Relationships   85
  5.1   Target Location   86
  5.2   Measurement and Location Errors   90
  5.3   Coordinate Conversion   100
  5.4   Display Correction   104
  5.5   Coverage   105
    5.5.1   Detection-Constrained Coverage   106
    5.5.2   Line-of-Sight-Constrained Coverage   108
  5.6   Clutter Cell Area   111
    5.6.1   Beamwidth-Limited Clutter Cell Area   112
    5.6.2   Range-Limited Clutter Cell Area   112
    5.6.3   Doppler-Limited Clutter Cell Area   114
  5.7   Maximum Unambiguous Range and PRF   114

Chapter 6   Doppler Relationships   119
  6.1   Target Doppler   119
  6.2   Isodoppler Contours   122
  6.3   Clutter Doppler Spread   124
    6.3.1   Clutter Doppler Spread over an Isorange Contour   125
    6.3.2   Clutter Doppler Spread over a Range Cell   127
    6.3.3   Clutter Doppler Spread through Sidelobes   128
  6.4   Doppler Beat Frequency   129

Chapter 7   Target Resolution   131
  7.1   Range Resolution   131
  7.2   Doppler Resolution   133
  7.3   Angle Resolution   135
  7.4   Synthetic Aperture Radar Isorange Resolution   138

Chapter 8   Target Cross Section   145
  8.1   Pseudomonostatic RCS Region   145
  8.2   Bistatic RCS Region   147

8.3	Glint Reduction in the Bistatic RCS Region	148
8.4	Forward-Scatter RCS Region	150
Chapter 9	**Clutter**	**157**
9.1	Surface Clutter	157
	9.1.1 In-Plane Ground Clutter	160
	9.1.2 In-Plane Sea Clutter	164
	9.1.3 Out-of-Plane Ground Clutter	167
	9.1.4 Out-of-Plane Sea Clutter	170
9.2	Chaff	171
Chapter 10	**Electronic Countermeasures and Counter-Countermeasures**	**175**
10.1	Noise Jamming	176
10.2	Deception Jamming	185
Chapter 11	**Multistatic Radars**	**191**
11.1	Receiving Aperture Characteristics	192
11.2	System S/N	193
11.3	Implementation Requirements	195
Chapter 12	**Special Concepts and Applications**	**197**
12.1	Hitchhiking	199
	12.1.1 Range Extension	202
	12.1.2 Passive Situation Awareness	206
	12.1.3 Monitoring	209
	12.1.4 Launch Alert	212
12.2	Forward-Scatter Fences	218
	12.2.1 Single Forward-Scatter Fence	219
	12.2.2 Netted Forward-Scatter Fences	222
12.3	Hybrid Radars	224
12.4	Clutter Tuning	234
	12.4.1 Synthetic Aperture Radar	234
	12.4.2 Moving Target Indication	238
Chapter 13	**Special Problems and Requirements**	**245**
13.1	Beam Scan-on-Scan	245
13.2	Pulse Chasing	254
13.3	Sidelobe Clutter	257
13.4	Time Synchronization	258
13.5	Phase Synchronization and Stability	260
Appendix A	**Early Publications of Bistatic Radar Phenomenology**	**265**
A.1	Introduction	265
A.2	1934 NRL Patent	265
A.3	1933 BTL Paper	271

A.4   1932 GPO Paper   273

Appendix B   Width of a Bistatic Range Cell   277
  B.1   Introduction   277
  B.2   Development of the Approximate Expression for the Width of a
        Bistatic Range Cell, $\Delta R_B'$ 279
  B.3   Development of an Exact Expression for the Width of a Bistatic
        Range Cell, ΔR_B 280
        B.3.1   Geometry and Definitions   280
        B.3.2   Development   281
        B.3.3   Discussion   283
  B.4   Error Analysis and Examples   284

Appendix C   Approximation to the Location Equation   291
  C.1   Introduction   291
  C.2   Development   291
  C.3   Error Analysis, Discussion, and Examples   293

Appendix D   Area within a Maximum Range Oval of Cassini   295

Appendix E   Relationships between Parameters in Target Location and
             Clutter Doppler Spread Equations   299
  E.1   Introduction   299
  E.2   Target Location   299
  E.3   Parameter Ratios   301
  E.4   Angular Rates   302
  E.5   Pseudobeamwidth and Beam Pointing Angle   303
  E.6   Approximations to Pseudobeamwidth and Beam Pointing Angle   304

Appendix F   Orthogonal Conic Section Theorems   311
  F.1   Introduction   311
  F.2   Orthogonal Bisector—Tangent Theorem   311
  F.3   Orthogonal Ellipse—Hyperbola Theorem   314

Bibliography   317

Index   327

PREFACE

Bistatic radars have been on or lurking near the radar scene for almost 60 years. During this time about 200 bistatic radars became operational, with two known to be in operation as of this writing. They are the Space Surveillance System and the Multistatic Measurement System. As might be expected, bistatic radar literature reflects this modest level of activity: a chapter or section on bistatic radars in a radar textbook or handbook, occasional papers or letters in professional journals, and on rare occasions a special journal issue devoted to bistatic radars or, as they are referred to when in their special netted configuration, multistatic radars. The last such issue was published by the Institute of Electrical Engineers (IEE) in December 1986 (Vol. 133, Pt. F., No. 7). One textbook on bistatic radars, *Dispersed Radar Stations and Systems,* was authored by V. Ya. Aver'yanov and published by Nauka i Tekhnika, Minsk, in 1978. It was machine translated into English in 1980, but never distributed outside of U.S. government agencies. Thus, it is unfortunately lost to the bistatic radar community at large. Perhaps someday Aver'yanov's work will become accessible and his contributions properly recognized. Until then he retains phantom recognition for the first bistatic radar textbook.

This book is a major extension of a chapter on bistatic radar I wrote for Merrill Skolnik in his *Radar Handbook,* Second Edition, published by McGraw-Hill, New York, in 1990. Everything in that chapter appears in this book—replicated, revised, or expanded—along with many details and topics that did not appear in the chapter, due to space limitations or simply because they had not yet been written. The format of the chapter and this book are similar, and like the chapter, the book assumes that the reader is familiar with the theory and workings of a monostatic radar. I would like to thank Merrill Skolnik for his encouragement in pursuing the book version and acknowledge McGraw-Hill's permission to use all material in the *Radar Handbook* for this work.

Chapter 1 provides a definition and subjective overview of bistatic radars—subjective in the sense that I decided to include the "bistatic radar issue" and my assessment of bistatic radar developments in the context of this issue. (In essence, the issue is that bistatic radars are special-purpose devices and must be treated as such.) Chapter 2 and Appendix A summarize virtually everything I could find on bistatic radar history, from concepts through hardware, with emphasis on hardware that has been tested or deployed. Two principal sources of information concerning early—circa World War II—bistatic radars are Guerlac's *Radar in World War II,* published in 1987 but written much earlier, and Swords's *Technical History of the Beginnings of RADAR,* 1986. Both Guerlac and Swords so thoroughly researched their subjects that I could do no better than extract verbatim the material I needed for the bistatic radar story. My contribution was to weave a bistatic thread through this and associated material.

Developing and analyzing bistatic radars require large doses of geometry, which in turn requires that a coordinate system be defined. There are almost as many coordinate systems as there are published bistatic radar papers. I settled on one basic coordinate system—Jackson's North-referenced coordinate system—for the book. Chapter 3 defines Jackson's system and equations that describe ellipses, hyperboles, and other strange bistatic relationships that we either learned and soon forgot in our school years, or had no earthly need to learn in the first place. Incidentally, Jackson's paper, "The Geometry of Bistatic Radar Systems," published by the IEE in 1986, is in my opinion one of a half dozen seminal papers on bistatic radars; it significantly influenced the writing of this book and the earlier handbook chapter. Because I do not wish on any reader the onerous task of redeveloping any of the equations and relationships summarized in Chapter 3, I provide their development and proofs in Appendices E and F. They are not readily available from other sources.

While Jackson's North-referenced coordinate system is a fine general-purpose system for defining and analyzing bistatic operation, I was forced to invoke other coordinate systems for special applications. For example, ovals of Cassini are more easily defined (and drawn) in a polar coordinate system; some geometry proofs are more easily developed in a rectangular coordinate system; and arcane bistatic clutter measurements have always been referenced to a variant of the spherical coordinate system. I introduce these systems as the need arises.

Chapter 4 develops geometry-dependent, bistatic range relationships. It starts with the well-defined bistatic range equation using Blake's methodology, and then derives range-related characteristics that are unique to bistatic radars. These characteristics include the now well-known ovals of Cassini (all self-proclaimed experts in bistatic radar speak in terms of Cassini's ovals); various kinds of dynamic range variations; and the tricky isorange contour, where the separation between isorange contours defines a geometry-dependent bistatic range cell. Appendix B wrestles with the geometry to quantify this separation. The bistatic range equation is also a convenient method for defining bistatic operating regions and limits, which are summarized in Chapter 4.

Chapter 5 details a similar development for target location and the nearly always distorted area relationships. As with the proliferation of coordinate systems, almost as many ways exist to solve the bistatic triangle (formed by the transmitter, receiver, and target) and thus locate the target. The most useful ones are covered in this chapter, including a simplified location equation for "over-the-shoulder" operation, which is developed in Appendix C. Converting "normal" radar measurements to the bistatic plane defined by the bistatic triangle and treating errors associated with these radar measurements are particularly messy processes. As far as I know, no one has had the patience (or need) to publish them in the open literature; they are here. Correcting raw bistatic data for display on a plan position indicator scope is another tricky process, as is calculating bistatic clutter cell areas, even in two dimensions. I have provided the simple forms here with an admonition to use a computer for three dimensions. Bistatic radar coverage comes in two packages: detection-limited and line of sight-limited coverage. Detection-limited coverage is conveniently defined as the area within the oval of Cassini. Contrary to occasional assertions that the calculation is a simple problem in first-year calculus, I found the problem to be less than simple and furthermore not readily available. Appendix D provides the calculation, which is then summarized in this chapter.

Unlike monostatic doppler equations, which are fairly well behaved, bistatic doppler equations aren't. Bistatic target doppler becomes tractable when everything is referenced to the bisector of the bistatic angle, the "bistatic bisector." (The bistatic angle is shown on the cover of this book, and if one were to choose a parameter that best characterizes bistatic radar operation, it is this angle.) Isodoppler contours (with both the transmitter and receiver in motion) are almost tractable in two dimensions. A computer program is required for three dimensions. Clutter doppler spread is simply ugly. I first struggled with various approximations to simplify the process, which only made things worse. Then by blind luck I found that, at least in two dimensions, an exact solution could be had by invoking the isorange contour (or ellipse) eccentricity parameter, e. (Our bistatic experts might now be accused of eccentric speaking. Sorry!) The development is given in Appendix E and summarized in Chapter 6.

Chapter 7 is again unique to bistatic radars. In monostatic radar textbooks, target resolution is straightforward and requires only a cursory discussion. Yet once more geometry rears its head to distort the bistatic process. I had originally concentrated on range and doppler resolution, assuming that at least angle resolution was unchanged; but, of course, it's not. I included synthetic aperture radar isorange resolution here simply because the name is the same. It could just as well have been in Chapter 6, because it requires clutter doppler spread for operation.

Chapters 8 and 9 summarize open literature data on target radar cross section (RCS) and clutter scattering coefficients. For target RCS I found Kell's paper published by the IEEE in 1965 most useful—despite its modest title: "On the Derivation of Bistatic RCS from Monostatic Measurements." Based on Kell's work I defined three bistatic RCS regions in terms of the bistatic angle: pseudomonostatic, bistatic, and forward scatter. Each has unique characteristics that are useful for modeling bistatic operation. The forward scatter region, where the bistatic RCS can become enormous, defies intuition. My intuition was satisfied by invoking Babinet's principle, which is described in Chapter 8.

Chapter 9 separates the bistatic clutter scattering coefficient into two regions: in-plane and out-of-plane, which are cast in the spherical coordinate system variant. Probably the best summary and analysis of this subject was given by Weiner in a 1981 report, "Multistatic Radar Phenomenology Terrain and Sea Scatter." Unfortunately the unclassified report is restricted to U.S. government agencies, so I was not able to reference it. His references and private discussions, however, proved invaluable. The semiempirical models for in-plane clutter data were developed from initial work by Barton and discussions with Nathanson. Out-of-plane clutter measurements are both sparse and controversial. I attempted to resolve or at least mitigate the controversy.

Chapter 10 is an extension of David K. Barton's recently published chapter on ECM and ECCM, in his *Modern Radar System Analysis,* published by Artech House, Norwood, MA, in 1988. As in Barton's, my chapter uses Blake's power spectral density method for calculating noise jammer-limited radar performance. The deception ECM section was developed from many sources, but principally from unpublished notes and memoranda by L.E. Davies, who was my ECM mentor at Stanford Research Institute many years ago.

Multistatic radars—basically special netted configurations of bistatic radars—receive a cryptic but obligatory treatment in Chapter 11. They are, however, clearly more important than the relative page allocation in this book because the two currently operational bistatic radars are in fact multistatic radars. Fortunately, nearly everything that is said about bistatic radars also applies to multistatic radars; only the geometry is messier. I concentrated on characteristics unique to some types of multistatic radars, specifically the coherent type, in which the multistatic radar operates as a thinned phased-array antenna.

After reading the first part of my manuscript, my technical reviewer, Ciro Pinto-Coelho, sent me a note saying he was eager to read more about the bistatic radar concepts that I had briefly discussed in Chapters 1 and 2, such as hitchhiking, forward-scatter fences, and hybrid radars. So I was spurred on to write about them in Chapter 12. Initially, I planned simply to describe these techniques, but then realized that a system analysis, or feasibility analysis, would be more useful. The chapter rapidly grew, finally to the point where I was forced to shut down the process. ("Pencils up," as the professor says in a timed test.) I attempted to cover important aspects of bistatic radar operation and performance in the chapter, but was deficient in hardware and software implementation. Thus, the chapter covers necessary but not sufficient conditions for bistatic radar feasibility.

Finally, Chapter 13 identifies things that a bistatic radar must do that monostatic radars either can do more easily or don't have to do at all. These are the penalties for even thinking about a bistatic solution. Some can be solved with a penalty in performance; others can be solved by brute force but at a penalty in complexity and cost. In all cases, they must be considered to achieve a feasible bistatic radar design.

PREFACE TO SECOND EDITION

Minor changes have been made to the book: inevitable typographic errors were corrected; some figures were enlarged; the ISLR analysis (Section 13.5) was modified; and the bistatic radar cost discussion (Section 1.4) was tempered to account for performance advances and attendant cost reductions in digital signal processing and local oscillators.

Sadly, I must report that one of the objectives in writing the book—to prevent reliving bistatic lessons of history—remains unfulfilled. In the roughly five years between publication of the first and second editions, I have encountered about as many goofy bistatic ideas as in the previous five years. Possibly Cervantes had the right outlook:

"Del dicho al hecho hay gran trecho."
It's a far cry from speech to deed.

So we must be patient.

ACKNOWLEDGMENTS

I would like to thank Ciro Pinto-Coelho (Westinghouse) for his review of the book. Ciro so thoroughly reviewed the book that the publication date was extended by six months in order to incorporate his suggestions and constructive criticisms. (Yet it took 60 years to publish the first English language book on bistatic radar; so, an additional six months is "in the noise.") I leaned heavily on the advice and publications of many radar experts, including Merrill Skolnik (NRL), Alexander Mac-Mullen, Dick DiDomizio, and Tony Andrews (TSC), Lamont Blake and Theo Cheston (TSC Consultants), Fred Nathanson (formerly TSC), David Barton (ANRO), Melvin Weiner (MITRE), M.C. Jackson (GEC-Marconi), Archie Gold (SAIC), and Joe Henry and Ray Martin (Westinghouse). My thanks for their insights and guidance. I am also compelled to recognize the almost daily counsel of TSC's radar engineering principal, Wayne Rivers; he kept me honest. The engineering world is divided into two parts: those who are computer literate and those who are not. I fall into the latter category. Thus, the computer expertise of Ciro Pinto-Coelho (again), Mary Heimer and Houng Pham (TSC), and Greg Willis (free-lance graphics) is much appreciated. My thanks to Cori Beckman, Barbara- Anne Ward and Lisa Miller for typing the (left-handwritten) manuscript and its many revisions, and to Phyllis Herron, East Sussex, England, who counseled me on English geography and unsplit my infinitives. Once again, to my wife, Carlaine, who was incredibly patient and supportive, thank you.

Chapter 1
INTRODUCTION AND OVERVIEW

1.1 INTRODUCTION

A useful way to introduce a bistatic radar is to define it, outline its operation, and summarize its utility. The basic definition is straightforward: a radar operating with separated transmitting and receiving antennas. How far the antennas are separated is sometimes included in the definition and can be ambiguous. Variations of a bistatic radar have been developed, the most significant of which is the multistatic radar. In its basic form, a multistatic radar uses two or more separated receiving antennas operating with a single transmitting antenna. These topics are discussed in Section 1.2. While most of the material in this book is about bistatic radars, multistatic radars can be considered extensions of bistatic radar and are treated as such, usually in special sections and in one case as a separate chapter of the book (Chapter 11).

The basic requirements for bistatic radar operation are outlined in Section 1.3. A convenient approach is to relate bistatic radar requirements to those of monostatic radar (collocated transmitting and receiving antennas) because most readers are familiar with the ubiquitous monostatic radar. In fact, bistatic and monostatic radar requirements, operation, and performance are compared throughout the book to provide a sense of continuity between the two types of radars. This comparison also provides a "sanity check" for all bistatic radar equations because in the limit a bistatic equation must collapse to the monostatic equation.

The utility of bistatic radars is summarized in Section 1.4. Nearly 200 bistatic radars were built and deployed for air defense operations prior to and during World War II and about ten were built and deployed after the war. However, very few are operational as of this writing, primarily because the single site monostatic radar is much easier to operate and usually—but not always—performs better. Thus, after they were developed in the late 1930s, monostatic radars simply replaced the earlier bistatic radars. The "but not always" caveat is the reason bistatic radars survive. The special niches that a bistatic radar has filled or can fill—when properly analyzed and developed—are summarized in this section and indeed are the motivation for this book.

1.2 DEFINITIONS

Bistatic radar is defined as a radar that uses antennas at different locations for transmission and reception [2]. As shown in Figure 1.1, a transmitting antenna is placed at one site and a receiving antenna is placed at a second site, separated by a distance L, called the baseline range or simply baseline. The target is located at a third site. Any of the sites can be on the earth, airborne, or in space, and may be stationary or moving with respect to the earth.

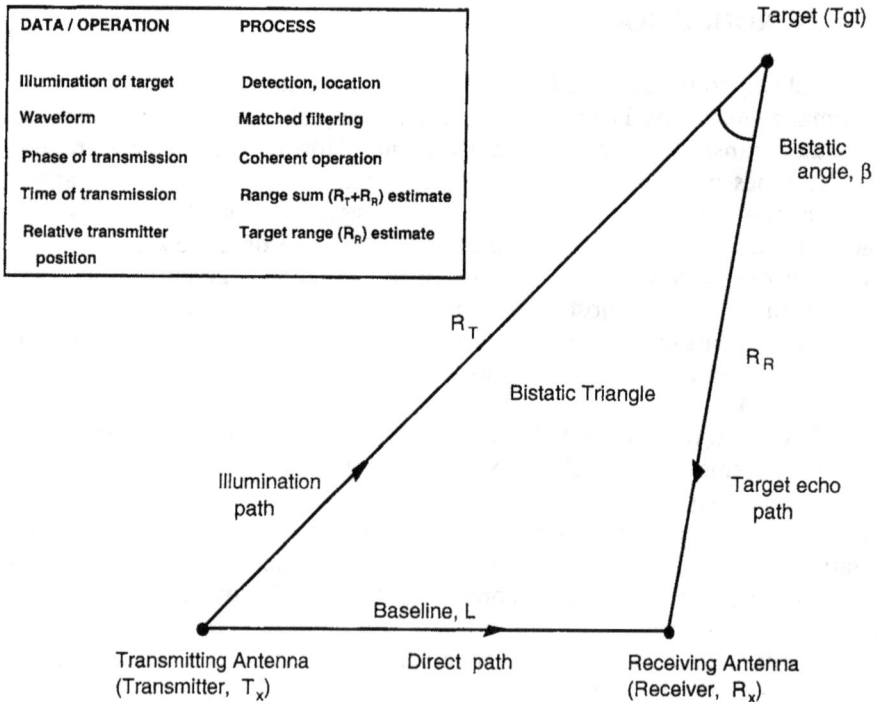

Figure 1.1 Bistatic geometry and typical requirements for bistatic radar operation.

While the necessary condition in the bistatic radar definition is that the antennas be at different locations, the entire transmitting subsystem is almost always located at one site and the entire receiving subsystem is located at a second site.

For convenience, this case will be assumed throughout the book, with the understanding that this assumption is a sufficient but not a necessary condition for bistatic radar operation.

The angle between the transmitter and receiver with the vertex at the target is the bistatic angle, β (Figure 1.1). It is also called the cut angle or the scattering angle. Although not part of the bistatic radar definition, this parameter is unique to bistatic radars and appears in many bistatic equations. This angle is also used to characterize the difference between monostatic and bistatic performance.

The IEEE definition [2] of a bistatic radar does not specify how far the transmitting and receiving sites must be separated. Attempts have been made to quantify this separation. For example, Skolnik [1] recently defined the separation simply as "a considerable distance." In his earlier, seminal paper on bistatic radar [15], he defined considerable distance as "comparable with the target distance," giving as examples a few miles to several hundred miles for aircraft targets and hundreds to thousands of miles for satellite targets. This definition applies principally to bistatic forward-scatter fences, which were used in the World War II era and occasionally in the postwar era. In the same paper he also defined the separation such that "... the echo signal does not travel over the same [total] *path* as the transmitted signal."

Blake [136] takes this last definition a step further by defining two conditions for separation:

Either: "the *directions* of the transmitter and receiver [from the target] differ by an angle that is comparable to or greater than either beamwidth,"

or: "... the *distances* from the target to the transmitter [R_T] and receiver [R_R] differ by an amount that is a significant fraction of either distance."

Both Blake's direction-distance and Skolnik's path separation criteria apply to all bistatic configurations, including forward-scatter fences, although quantification of terms such as "comparable" and "significant fraction" might cause arguments in some quarters. An alternative approach is to specify any separation, as long as that separation is *selected* to achieve a specific operational objective. Reasons for designing a radar to operate bistatically are outlined in Section 1.4 and detailed in subsequent chapters. Under any of the three criteria, separation can range from meters and tens of meters for high resolution, industrial imaging, and intrusion detection to tens and hundreds of kilometers for aircraft, missile, and satellite detection.

The purpose for establishing these separation criteria is simply to distinguish special types of radars, such as continuous wave (CW) and over-the-horizon-backscatter (OTH-B) radars, and one implementation of an air traffic control (ATC) radar from bistatic radars. The CW and some OTH-B radars often use site separation for receiver isolation [1, 213]. Other OTH-B radars [1, 190] and an experimental air height surveillance ATC [191] use site separation for convenience: the transmitting and receiving antennas are differently configured and thus are separately located. In some cases OTH-B site separation can be 100–200 km, and these configurations are sometimes called bistatic [213]. However, while each of these radar types must know the separate transmitting and receiving site locations, provide an external transmitting signal reference to the receiver, and occasionally correct for parallax in space and time dimensions, they otherwise operate as monostatic radars. Also, because they usually do not satisfy any of the separation criteria given previously (the OTH-B radar typically operates between 1000–4000 km) they are not, strictly speaking, bistatic radars. In any case, when bistatic radars use site separation to achieve some operational objective, they gain receiver isolation and separate transmitting and receiving antenna configurations as a side benefit.

When two or more receiving sites with common (or overlapping) spatial coverage are employed, and data from targets in the common coverage area are combined at a central location, the system is called a *multistatic radar*. In the conventional form of multistatic radar operation, independent target measurements, such as time-of-arrival (TOA, or range sum), time-difference-of-arrival (TDOA, or range difference), doppler (or velocity) or angle measurements, from each site are combined to form a target state estimate. The data combining is usually characterized as noncoherent processing. Examples are the Multistatic Measurement System [13, 14], which combines TOA, TDOA, and doppler measurements from two or three sites, and the Space Surveillance System [1, 7, 184, 185] which combines angle measurements from two sites to form a target state estimate.

When the phase and amplitude of target echoes from each receiving site are combined coherently, a large distributed receiving antenna is formed. Thinned, random, distorted, and distributed arrays [3–6] and the Radio Camera [11, 12] are examples and are usually considered a subset of multistatic radars. Multiple transmitters can be used with any of these multistatic configurations. They can be located at separate sites or colocated with the receiving sites, and their transmissions can also be phased to form a large transmitting aperture.

A multistatic radar is a special case of a *radar net;* both process target data from multiple sites at a central location. A radar net typically consists of a set of monostatic radars, with target data processed noncoherently at the central location. The net can be configured in one of two ways, depending on site separation. When the sites are widely spaced, the net can join target data from each site so that an extended target track is established over the coverage of the net, thereby expanding spatial coverage. This coverage is the union of each site's coverage area. When the sites are closely spaced, the net can combine target data from sites having common spatial coverage to improve the quality of target state estimates. This coverage is the intersection of two or more site coverage areas. Obviously the common coverage area increases as the site separations decrease, as might be the case for ships steaming in formation.

A multistatic radar is also configured to combine target data within the common coverage area to improve the quality of target state estimates. If the multistatic radar combines data coherently, quality can be further improved. Conceptually, we should be able to combine monostatic radar data coherently within a net. In such a case, the operation and implementation of a "coherent radar net" and a coherent multistatic radar system become similar, if not identical. When a bistatic radar uses a monostatic radar as the transmitter and data from the monostatic and bistatic common coverage areas are combined, the configuration is again multistatic. This special, and potentially useful, configuration is also called a *hybrid radar*.

Bistatic radars can also be netted, typically to expand their spatial coverage, which again is the union of each receiving site's coverage. Of course, when a bistatic radar net combines target data within its common spatial coverage, it becomes a multistatic radar. In some cases, multiple bistatic receivers can be serviced by a single transmitter. However, unless the receivers are netted in some way, the configuration remains bistatic, with each transmitting-receiving pair operating independently and with a unique geometry.

A *trilateration radar* is another special case of a radar net. It combines range-only data from three sites on targets within the common coverage area, again to improve the quality of target state estimates. Early trilateration concepts defined the sites using range-only monostatic radars [16]. When three or more sites are used, the concept is sometimes called a *multilateration radar* [203, 204]. The concept can be expanded readily to include a multistatic radar using TDOA measurement techniques for a hyperbolic solution to target location [204]. In this case a multilateration radar becomes a special case of a multistatic radar. Figure 1.2 shows coverage comparisons between various radar nets, a multistatic radar, and a multilateration radar. In general, an inverse relationship exists between coverage and target state estimation accuracy.

Passive receiving systems, or *electronic support measure* (ESM) systems, often use two or more receiving sites to detect, identify, and locate transmitters such as monostatic radars. They are also called *emitter locators*. Target location is by means of combined angle measurements from each site (e.g., triangulation), TDOA, or differential doppler measurements between sites. They usually are not designed to detect and process the echoes from targets illuminated by the transmitter. They can, however, be used with a bistatic or multistatic radar to identify and locate a suitable transmitter to initialize radar operations. Thus, while they have many requirements and characteristics common to multistatic radars, they are not radars and hence are not considered here.

The preceding definitions are broad and traditional [1, 15, 16], but they are by no means uniformly established in the literature. Terms such as quasibistatic, quasimonostatic, pseudomonostatic, tristatic, polystatic, real multistatic, and multibistatic have also been used [17–20]. They are usually special cases of the broad definitions given above. In subsequent chapters the term *pseudomonostatic* is used to characterize special bistatic geometries that approximate monostatic operation.

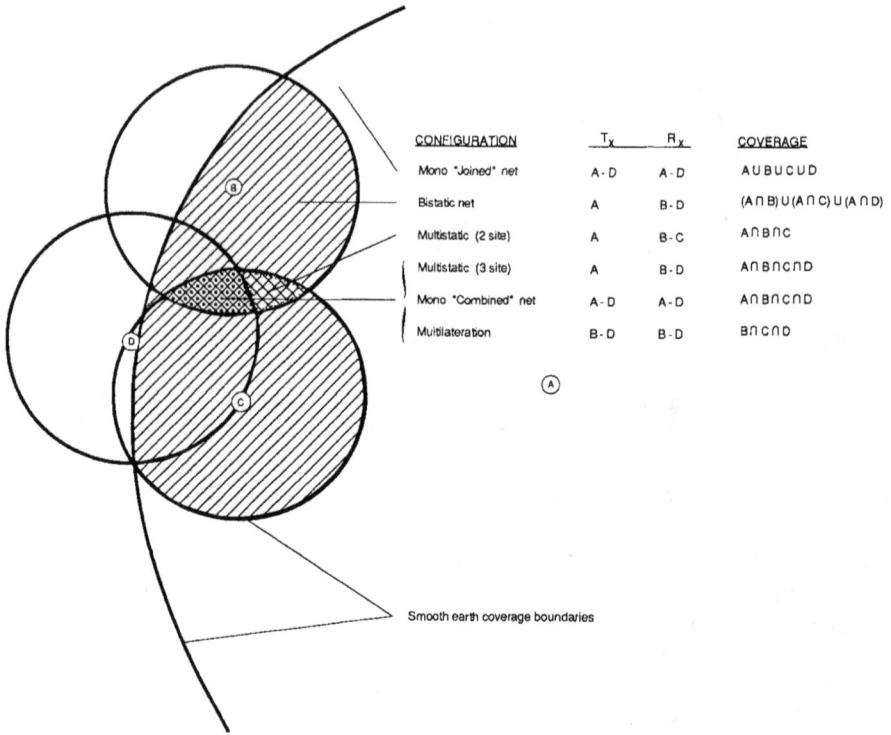

Figure 1.2 Coverage comparison between radar nets, multistatic radars, and multilateration radars.

1.3 REQUIREMENTS.

Bistatic target detection and location are similar to, but more complicated than, those of a monostatic radar. For target detection, the target is illuminated by the transmitter and target echoes are detected and processed by the receiver, as with a monostatic radar. For matched filter operation, the transmitted waveform must be available to the receiver. For coherent receiver operation, the phase of the transmitted waveform must also be available to the receiver.

For target location, the bistatic receiver usually estimates the range and measures angles of arrival (AOA) of the target echo, as does a monostatic radar. Bistatic AOA measurements are typically made in an azimuth and elevation plane centered on the receiver site. Bistatic range estimates are derived from measurements of total signal propagation time from transmitter to target to receiver. Thus, the time of signal transmission must be available to the receiver. This time measurement is converted to a range sum estimate, $(R_T + R_R)$, where R_T is the transmitter-to-target range and R_R is the target-to-receiver range. In a monostatic radar, $R_T = R_R$; in a bistatic radar, $R_T \neq R_R$ in nearly all cases; a typical example is shown in Figure 1.1. Therefore, to obtain an estimate for R_T or R_R, the bistatic radar must solve the transmitter-target-receiver triangle, called the *bistatic triangle*. This solution usually requires an estimate of the transmitter position with respect to the receiver. These general requirements are summarized in Figure 1.1.

The receiver can obtain the required transmission time and phase data in two ways: (1) directly, by reception and demodulation of the transmitted signal over the baseline, or "direct path," if an adequate line-of-sight (LOS) is available; and (2) indirectly, by the use of identical stable clocks in the transmitter and receiver that have been synchronized before the start of operations. Transmitter waveform and position data can be obtained via a data link or from an intercept receiver and emitter location system, usually colocated with the bistatic receiver.

Sometimes the direct path signal from the transmitter is sufficiently strong to saturate the bistatic receiver, particularly when the main beam from the transmitting antenna points at the receiving site. In these cases the receiver must take measures to attenuate the transmitting signal. These measures can include (1) time gating, (2) doppler filtering, and (3) spatial nulling, for example by terrain masking or sidelobe cancellation. The first measure operates on a pulsed signal; the second and third measures operate on pulsed or CW signals.

Like monostatic radars, bistatic radars need not perform all of the functions listed in Figure 1.1. The required functions are obviously set by a specific application. In all cases, however, the distinguishing features of a bistatic radar are (1) the required data coupling between transmitter and receiver, (2) the geometry, defined by the bistatic triangle, and (3) the measures taken by a bistatic radar to either mitigate the adverse effects of the geometry or occasionally capitalize on the beneficial effects of the geometry.

1.4 APPLICATIONS AND ISSUES*

Bistatic radars have been proposed, analyzed, developed, and tested for many military applications. They are reported to have been deployed for a few applications, which are, in descending order of fielded numbers:

Author's Note: I wrote two versions of this section. The first version was an objective summary of bistatic radar military applications, which are detailed in subsequent chapters, particularly Chapters 2 and 12. The second version extended the first with a subjective assessment of the "bistatic radar issue." Because "issue" has the connotation of a debate or dispute, it did not appear to have a place in an engineering text. However, upon reflection, I concluded that all aspects of the bistatic radar saga—spanning about 70 years—should be aired, especially if we are to avoid reliving the lessons of history. Thus, I have included the second version, warts and all. In this version I have omitted references when they are used in later chapters; otherwise they are included.

1. semiactive homing missiles;
2. forward-scatter fences;
3. multistatic radars;
4. hitchhikers.

A semiactive homing missile is configured like a bistatic radar; the transmitter is usually located on the launch platform and the receiver is located on the missile. Many thousands of semiactive homing missiles have been deployed, with a separate community evolving to pursue this technology. The missile and radar communities continue to go their separate ways, with separate programs and publications. This book recognizes the link between these two communities and their occasional technical interchanges, and then presses on with the radar side.

A forward-scatter fence uses widely spaced transmitting and receiving sites—Skolnik's considerable distance criterion—often with fixed transmitting and receiving beams oriented along or near the baseline to detect moving targets near the baseline. In this geometry the bistatic angle is large, approaching 180° as the target approaches the baseline. The target's radar cross section (RCS) is significantly enhanced in this region. Almost 200 of these fences were deployed by Japan, France, and the Soviet Union for aircraft detection before and during World War II, and one, the U.S. AN/FPS-23, was deployed for a short time after World War II. All were eventually replaced by monostatic radars. Also, cryptic reports [213, 214] exist of three forward-scatter OTH fences deployed in the mid-1960s to detect missiles launched from the Soviet Union. Their current status is not known.

About ten multistatic radars were deployed by the United States in the years following World War II to enhance detection and location of aircraft, missiles, and satellites. Two remain in operation: the Space Surveillance System (SPASUR), a multistatic, fixed-beam (but not forward-scatter) fence for estimating satellite ephemerides, and the Multistatic Measurement System (MMS), a cued multistatic system for tracking ballistic missiles.

A bistatic hitchhiker uses a transmitter of opportunity, usually another radar, to detect and locate targets near the transmitting or receiving site. The transmitter can be *friendly* (cooperative) or *hostile* (noncooperative). The bistatic hitchhiker has also been called a parasitic radar. One bistatic hitchhiker system, the German Klein Heidelberg, was deployed in Denmark for a short time during World War II. It hitchhiked off the British Chain Home radar transmitters to warn of allied bombing raids when they were over the English Channel. A second Klein Heidelberg apparently was installed in Oostvorne, the Netherlands, to cover the more southern bombing routes. [234]. Thus, ignoring the missile connection, two multistatic radars and possibly three forward-scatter OTH fences survive in the military community as of this writing. Why?

In 1977, Skolnik made the following observation [32]:

The idea of bistatic radar, as with similar techniques of marginal value, is resurrected on a regular basis, but usually goes away after careful examination. It might be very well used for special purposes, but it is not likely to be widely deployed in the near future.

This observation is the "bistatic radar issue." On first reading, one is struck by the term "marginal value," and, in view of the number of surviving bistatic radars compared to the thousands of deployed monostatic radars, the term, while somewhat captious, characterizes the situation. Yet the key to Skolnik's observation is "special purposes." A bistatic radar must offer some cost, technical, or operational advantage—value added to the system—that outweighs its inherent disadvantages for it to be accepted by the current radar community. And because of the pragmatic monostatic mindset of this community, that advantage must be significant. Jackson [73] characterizes this situation as follows:

. . .bistatic radars have a number of properties that differ significantly from the monostatic equivalents. Some of these produce operational advantages that provide a motivation for deployment, while others bring [only] trouble and expense.

Thus for SPASUR, a fixed-beam CW multistatic fence provided acceptable satellite location performance at the lowest cost. For MMS, an estimate of the ballistic missile cross-range position and velocity was significantly improved by using a set of multistatic receivers that were cued, or pointed, at the target by a monostatic radar, and probably at a lower cost than if multiple monostatic radars, each with an expensive high powered transmitter, were used. For the forward-scatter OTH fences operating at very low frequencies (3-30 MHz), the coverage was such that a high percentage of missile launches could be detected over a very large area with a minimum number of transmitting and receiving sites.

In contrast, forward-scatter fences operating at any frequency were found to have very limited utility for air defense. While aircraft penetration of a single fence could be detected, the aircraft rapidly flew out of fence coverage, and in any event, could not be located and tracked—a critical air defense requirement. Only when adjacent or crossing fences were deployed could a rough position and velocity estimate be made—at considerable increase in system complexity. These estimates were adequate for cueing and fighter vectoring, but not adequate to support a fire control solution. Consequently, the early forward-scatter fences were replaced by monostatic radars with better spatial coverage and location accuracy.

A bistatic hitchhiker operates at the pleasure of the transmitter, which is usually a monostatic radar. In the Klein Heidelberg case, the fixed site British Chain Home radars operated nearly continuously over a fixed coverage area, which is ideal for the hitchhiking concept. In modern air defense scenarios, the monostatic radars are usually mobile or airborne, and because of threats, such as the antiradiation missile (ARM), may not transmit continuously. Thus, unless a friendly transmitter can be dedicated to the task, a bistatic hitchhiker becomes a sensor of opportunity, which limits its utility. Further, the argument arises that if a monostatic radar is dedicated to support the bistatic hitchhiker, the system might as well use the monostatic radar data in the first place. Thus, these types of bistatic radar systems faded from the inventory.

To the forgoing list of bistatic radars that have been deployed can be added a list of bistatic concepts that have been analyzed, developed, and occasionally tested, but apparently never deployed:

5. counter-ARM;
6. counter-retrodirective jammers;
7. counter-stealth;
8. clutter tuning;
9. hitchhiking variants.

ARMs are typically designed to attack ground-based air surveillance radars. The effectiveness of an ARM attack can be reduced by moving the radar's transmitter away from the battle area into a "sanctuary," possibly a satellite-based sanctuary. This bistatic "counter-ARM" concept, appropriately called Sanctuary, was field tested and showed that the messy bistatic ground clutter could be suppressed and a modest air surveillance capability, in terms of range and azimuth coverage, could be generated. It was not deployed. Decoys, which preserve the monostatic configuration without modification, were deployed to protect these radars.

A retrodirective jammer, or retrojammer, uses an AOA intercept receiver to point a high-gain jamming antenna at a monostatic radar. Retrojammer effectiveness can be reduced by selecting the geometry in a bistatic radar so that the bistatic angle is greater than the jammer's beamwidth. Thus the jamming-to-signal ratio (JSR) at the bistatic receiver is reduced by the ratio of the retrojammer's mainlobe-to-sidelobe gain, where the main lobe is pointed at the transmitter and the side lobes are pointed at the receiver. No reports of this concept are available.

The counter-stealth bistatic concept resurrects the early forward-scatter fences, but with the important observation that the RCS of any type target is significantly increased for targets near the baseline. Specifically, the forward-scatter RCS of a stealth target is the same as that of an equisized conventional target in this geometry [15, 91]. This RCS enhancement was inherent in the early fences, but it was not known at the time. (The enhancement may have contributed to an extraordinarily long fence range of about 700 km for one of the Japanese Type-A radars in World War II.) Laboratory measurements have confirmed this RCS enhancement for both conventional and coated (stealth-like) objects. No bistatic counter-stealth deployments have been reported, possibly due to security.

Clutter tuning is implemented when both transmitter and receiver are moving, for example, airborne. In this configuration the separate velocity vectors of the transmitter and receiver and their relative geometry can be changed to control, or "tune," the ground clutter's doppler shift and doppler spread. In this way, a synthetic aperture radar (SAR) map can be generated directly on the velocity vector of the receiving platform—an impossible task for a monostatic SAR. In addition, moving target indication (MTI) operation is possible broadside to the receiving platform's velocity vector—a difficult task for a monostatic radar. Often, the transmitter is a monostatic radar that cues the bistatic receiver to a particular ground location. Two successful field tests of a bistatic SAR, one called Tactical Bistatic Radar Demonstration (TBIRD) and the other reported by Auterman [53], have been conducted. Both operated in a broadside rather than a forward-looking geometry and at modest ranges. Again reports of deployment have not been made.

Variants on the German hitchhiking concept abound. In one field test configuration called Bistatic Alerting and Cueing (BAC), cooperative (but nondedicated) monostatic airborne surveillance radars, such as AWACS, were used as a transmitter and a small bistatic receiver on the ground was used to alert and cue ground forces to aircraft and moving ground targets at short ranges from the receiver. The concept has the advantage of eliminating registration and time-delay problems inherent in sending data from a remote radar to mobile ground units. The bistatic receiver is also immune to ARM attack. Again no reports of deployment are available.

In a second hitchhiking test, a commercial television transmitter and modified TV receiver were used for short-range aircraft detection. The *negative* test report by Griffiths, *et al.*, [55] is unique in the annals of bistatic radar literature. Occasionally targets were detected, but not consistently, principally due to inadequate clutter suppression. The obvious (but unstated) problem was limited funding, which forced "corner-cutting" in the experiment.

In a third hitchhiking test, CW communication satellite transmitters and ground-based receivers were used to detect aircraft. Because the effective radiated power level of the satellites was modest and the transmitter-to-target ranges were large, receiver-to-target detection ranges were small, < 4 km, unless a very large receiving aperture was used.

In a fourth hitchhiking test, very low frequency (VLF) signals generated by lightning were used as the illuminator and a *Scordes* (sferics correlation detection system) receiver detected and located ionospheric disturbances at transatlantic distances. No reports of deployments are available.

The sun and radio stars have been studied and possibly tested as transmitter sources, with the conclusion that they are useful only when other RF sources are not available and modest detection ranges are acceptable.

Another hitchhiking concept uses a small bistatic receiver on an aircraft and transmitters of opportunity (typically radars) to detect targets near the aircraft or near the transmitting site. In these two regions the bistatic signal-to-noise ratio (SNR or *S/N*), characterized by ovals of Cassini, is often adequate for target detection. In the former case, target detections can warn the pilot of an intruding aircraft or missile, while the pilot's primary sensor, the monostatic radar, is off. The concept is called situation awareness, or passive situation awareness (PSA). In the latter case, the pilot can monitor activity near a threat radar, including possible missile launches from platforms at or near the threat radar. Details of either configuration apparently have not been published.

None of these bistatic radar concepts appears to have reached the military inventory. Some of the concepts simply did not cross the special-purpose, or value-added, threshold established by the bistatic radar issue. Others appear to have approached but not crossed the threshold. Although details are not available, an "insufficiency list" can be created for these cases. In rough order of importance, the list is as follows:

1. excessive complexity;
2. cost credibility (no bistatic radar cost history);
3. other solutions competing for limited funding;
4. degraded performance:
   - reduced coverage (range, angle),
   - limited engagement capability,
   - degraded resolution and accuracy (range, isorange, doppler, and angle),
5. limited field tests and test data;
6. threats not of significant concern;
7. monostatic mindset.

Each of the undeployed bistatic radar concepts probably suffered from two or more of these insufficiencies.

Three final points are offered. First, Skolnik established the bistatic radar issue in 1977. Much of the open literature on bistatic radar systems and tests contained in this book was published after that time. Thus, Skolnik probably established the issue at least in part on unpublished data—data that may never be published. In short, the mistakes will remain buried. Consequently, a modest objective of this book is to document what is known in an attempt to prevent future bistatic radar mistakes and to guide future bistatic radar developments.

Second, an element common to all bistatic configurations is their unique spatial coverage patterns—the regions over which a target can be detected. They are markedly different from typical monostatic circular (or sector) patterns, in terms of shape, size, and location. In fact, Section 5.5 shows that bistatic spatial coverage is always smaller than that of a monostatic radar for equivalent radar parameters. Thus, a successful bistatic application requires mapping these unique patterns into a useful operational configuration. Semiactive homing missiles are a classic example: the seeker is cued to a moving space point by a monostatic radar, and needs to search only a small region about the space point for target acquisition. The MMS and some hitchhiking and clutter tuning concepts are other examples of monostatic radar cueing. SPASUR, a fixed-beam fence, is a variant of the theme. The target (satellite) must fly through the fence, and SPASUR waits for the event to happen. The Klein Heidelberg is another example of the variant; it was effective as long as the bombing raids were routed near Denmark or the Netherlands.

In contrast, the limited spatial coverage (and, of course, location accuracy) of a forward-scatter fence was eventually found to be inadequate for air defense tracking and engagement. However, for the time, before monostatic radars, it was the only solution available. It might become part of the counter-stealth solution by operating as the initial detection sensor and cueing other sensors for tracking and engagement. Limited spatial coverage also plagued the counter-ARM sanctuary concept, where coverage (i.e., detection range) was sacrificed by moving the transmitter away from the battle area.

However, in cases where cueing or constrained operations can be invoked, the bistatic spatial coverage limitation can often be mitigated. As outlined above, examples are (1) cueing by a monostatic radar, (2) a bistatic fence constrained to cue other sensors, and (3) special targets with constrained trajectories or flight paths, such as satellites and ballistic missiles. In all these cases, other bistatic implementation issues, as outlined in Chapter 13, must be accommodated before the concept becomes feasible.

Third, bistatic radar performance is sometimes compared to monostatic radar performance. Objectives of such a comparison are usually (1) to argue for replacement of monostatic radars, (2) to demonstrate a bistatic advantage, and (3) to show that in the limit all bistatic radar equations simplify to monostatic equations.

The first "replacement" objective is specious. It violates the mandate of Skolnik's bistatic radar issue and probably contributed to the issue in the first place. While replacement is not acceptable, modification might be acceptable in some cases. Modern monostatic radars are often designed for multimode operation, due in large part to the development of high-speed digital signal processing (and whenever the supporting software can keep pace). Bistatic operation might be considered as one mode in such a system. Clutter tuning, the hybrid radar, and some multistatic radar concepts can be designed to operate this way, whenever the value-added threshold is sufficiently crossed.

The second "advantage" objective has been exercised many times [1, 15, 19, 20, 73, 138-140, 209, 210] and is a useful exercise—if fairly done. Geometry, characterized by monostatic and bistatic radar site locations, is the critical factor in demonstrating a fair bistatic advantage of relative performance. Jackson [73] identifies these geometry factors as follows:

> Apart from the improved isolation of the receiver from the transmitter, and its relative immunity to some forms of attack, most of the advantages (and problems) of bistatic radars stem from the general geometrical properties of the system. There are two radiation paths involved which differ significantly in length and direction unless the ranges concerned are much greater than the station separation. At very long range, defined in this relative way using the baseline as the yardstick, the performance closely resembles that of a monostatic radar. Bistatic effects appear as the range is reduced and some of them eventually become embarrassingly large, especially near the baseline, where the [bistatic] angle between the sightlines is obtuse, and near the receiver.

In general, whenever one bistatic site can be positioned near a target, the bistatic performance, in terms of coverage, accuracy, and countermeasures to some types of jamming, can be superior to that of a monostatic radar with equivalent parameters and located near the other bistatic site (farther from the target) [19, 209]. As might be expected, site location is also critical to results for netted monostatic and netted bistatic (or multistatic) relative performance [19, 140, 209]. However, in nearly all comparisons when monostatic radars with equivalent parameters are located at each bistatic receiving site, netted monostatic performance is superior to netted bistatic performance. For example, Section 12.1.1 details the potential coverage advantage of bistatic range extension, but observes that if the monostatic radar can be located at the forward bistatic receiving site, the coverage advantage swings to the monostatic side. Thus, bistatic range extension would be an advantage only in special cases when passive operation at a site is needed or when the forward-based, monostatic radar has significantly less performance. In a rare exception to this rule, when monostatic air defense coverage is degraded by multiple, high-powered stand-off retrodirective jammers, a properly sited bistatic receiver can recover a significant fraction of the lost monostatic coverage. This concept is detailed in Section 12.3.

The third "simplification" objective is a useful "sanity check" in any bistatic radar analysis. In all bistatic radar equations involving target ranges R_T and R_R, the baseline L, or the bistatic angle β, the expression must reduce to the monostatic case in which $R_T = R_R$, $L = 0$, and $\beta = 0$; otherwise, the bistatic equation is simply wrong. While this rule is a necessary condition to validate bistatic radar equations, it is not a sufficient condition. One can derive a bistatic expression based on erroneous assumptions about the geometry—especially geometry involving ellipses or ovals of Cassini—and then develop a false sense of security when the expression reduces to the monostatic case, only to discover the error later (or worse, to *have* the error discovered later.) The author suffered this affliction more than once. Nevertheless, this process is useful and appears throughout the book for two purposes: as the sanity check and to show how bistatic radar operation and performance deviate from that of a monostatic radar.

Chapter 2
HISTORY

With the possible exception of the first radar demonstration in 1904, all early radar experiments were of the bistatic type. They were conducted nearly simultaneously and totally independently by the United States, United Kingdom, France, the Soviet Union, Japan, Germany, and Italy. Japan, France, and the Soviet Union actually deployed bistatic forward-scatter fences, and Germany deployed a bistatic hitchhiker, all for aircraft detection in World War II. Even the British Chain Home monostatic radars had a reversionary bistatic mode. As is well documented[1, 25, 159], development of the monostatic radar, with its fundamental, single-site operational advantage, followed the early bistatic radar experiments, and by the end of World War II all bistatic radar work had stopped. Since then bistatic radars have had periodic but modest resurgences when a specific bistatic application was found attractive, or when the concept was simply rediscovered; the resurgence cycle appears to be about 15 to 20 years [1].

The first resurgence occurred in the 1950s, when bistatic radars were developed and deployed again as forward-scatter fences, as semiactive homing missiles, and as precision test range instrumentation and satellite tracking systems. These last systems were configured as multistatic radars. The term "bistatic" first appeared in 1952 [34]. Then, in the 1970s, bistatic radars were developed in response to the new antiradiation missile threat [220]. Experimental systems were tested, but not deployed. The bistatic concepts of pulse chasing and clutter tuning were also tested; the hitchhiking concept was retested. A new multistatic instrumentation system for precision ballistic missile tracking became operational in 1980 [14]. Most recently, bistatic radars have been tested as security fences and have been proposed and analyzed in a forward-scatter configuration as a counterstealth measure [54, 101, 141, 142, 211, 212]. Nonmilitary bistatic radar concepts, including high-resolution imaging, collision avoidance, environmental measurements, and police aircraft speed-trap warning, were developed principally in the 1970s and 1980s. The last concept was commercially produced in 1989 in the United States.

This chapter details bistatic radar history in an attempt to illuminate (pun intended) special, potentially worthwhile bistatic applications, as well as bistatic "dead ends," and to ease the process of rediscovery in the next resurgence cycle. Surprisingly, much of the early bistatic radar history has only recently been published, and because this period was the heyday of bistatic radar testing and deployment, this chapter concentrates on that period.

2.1 EARLY HISTORY

The first radar was developed in 1904 by a German engineer, Christian Hulsmeyer, to detect ships. The radar, called a "telemobiloscope" [25, 163], used a spark-gap transmitter, which operated on a 40- to 50-cm wavelength—relatively short for that time. For the first demonstration the transmitter was located on a "high tower" [172] and the receiver was located on the Hohenzoellern Bridge in Cologne. The receiver, shown in Figure 2.1, rang a bell, also shown in the figure, when a ship on the river was detected. Unfortunately, transmitter-to-receiver and ship-to-receiver ranges were not documented, but in all likelihood the transmitter-receiver separation was designed only for receiver isolation. Hulsmeyer later demonstrated ship detection at ranges up to 3 km from the receiver in Rotterdam, but with no documentation of transmitter location [163].

In any case, his 1904 British patent (No. 13,170) shows the receiving antenna located directly above the transmitting antenna on a ship's mast, in a monostatic configuration. Hulsmeyer's new machine failed to capture support from the tradi-

Figure 2.1 Christian Hulsmeyer's equipment of 1904 on display in the Deutsches Museum, Munich. (Courtesy of IEE [159].)

tion-bound naval authorities and public companies, and his experiments ended in 1904 [25, 163].

The concept of a bistatic radar was first documented in the August 1917 edition of *The Electrical Experimenter*, when its editor, Hugo Gernsbach, interviewed Nikola Tesla on methods of "subjecting [submerged] enemy submarines" [163]. As reported by S.S. Swords in his excellent account of the early history of radar [159], Tesla is quoted as follows:

... consider that a concentrated ray from a searchlight is thrown on a balloon at night. When the spot of light strikes the balloon, the latter at once becomes visible from many different angles. The same effect would be created with the electric ray if properly applied. When the ray struck the rough hull of a submarine it would be reflected, but not in a concentrated beam—it would spread out; which is just what we want. Suppose several vessels are steaming along in company; it thus becomes evident that several of them will intercept the reflected ray and accordingly be warned of the presence of the submarine or submarines.

Swords observes that Tesla appears not to have considered the attenuation suffered by radio waves propagating through water. Nevertheless, Tesla's idea clearly defines the concept of both bistatic and multistatic radar operation.

The first, unambiguous evidence of bistatic detection phenomenon was obtained by U.S. Navy civilian engineers in 1922. It is summarized by Swords [159] from accounts by Guerlac [25] and Allison [164] as follows:

In September 1922, Dr. Albert Hoyt Taylor and his assistant Leo Clifford Young ..., who were stationed at the United States Naval Aircraft laboratory, Anacostia, DC, were carrying out VHF propagation experiments at 60 MHz. They employed a superheterodyne receiver and a 50 W transmitter amplitude-modulated at 500 Hz. Initially tests were carried out in the grounds of the Naval Air Station and audible maxima and minima caused by reflections from steel buildings were observed. The receiver was placed in a car which was driven a few miles from the Station to Haines Point across the Potomac River. These reflection phenomena were again observed from trees and other objects and in particular from the wooden steamer Dorchester which passed down the river.

On the 27th September a memorandum drawn up by Taylor was sent to the Bureau of Engineering. It gave details of the experiments and, among several suggestions, proposed that radio beams at about this frequency could be used for the radio detection of objects such as, for instance, enemy vessels passing between two destroyers.

"If it is possible to detect, with stations one half mile apart, the passage of a wooden vessel, it is believed that with suitable parabolic reflectors a transmitter and receiver, using a concentrated instead of a diffused beam, the passage of vessels, particularly of steel vessels (warships) could be noted at much greater distances. Possibly an arrangement could be worked out whereby destroyers located on a line a number of miles apart could be immediately aware of the passage of an enemy vessel between any two destroyers in the line, irrespective of fog, darkness or smoke screen. It is impossible to say whether this idea is a practical one at the present stage of the work, but it seems worthy of investigation."

Guerlac adds the following details of the Dorchester detection [25].

While these [Haines Point] experiments were in progress and unobstructed signals were being received from the other side, the steamer Dorchester, a wooden vessel of no great size, passed down the channel. Fifty feet before the bow of the steamer crossed the line of vision between transmitter and receiver, the signals jumped to nearly twice the previous intensity. When the steamer actually passed across this line they dropped to half the normal value. Again when the stern of the vessel had passed 50 ft further down-stream, the signals rose to normal intensity, then up to about twice normal, and then dropped back down again.

Guerlac [25] reports that "No approval or encouragement, however, was forthcoming from the Navy Department and the 5 meter work was dropped." Subsequently Taylor and Young became involved in the ionospheric sounding experiments of Breit and Tuve in 1925, which used high-frequency transmissions from the Naval Research Laboratory's (NRL) station NKF.

The ionospheric sounding experiments conducted in the 1920s, principally in the United States and United Kingdom, can be considered a precursor of modern radar experiments, including bistatic radar. The purpose of these experiments was to confirm the existence of the Heavyside-Kennelly layer, the lower layer of the ionized region in the upper atmosphere that creates a sky wave propagation path for long-range communications. The 1925 Breit and Tuve experiments used 1-ms pulsed transmiussions [25] from station NKF, with the receiver separated about eight miles, as shown in Figure 2.2. The target was, of course, the ionosphere. Both the ground wave and the sky wave reflected from the ionosphere were detected and made visible via a rotating mirror in an amplitude versus time display—similar to an A-scope [159]. The ground wave is identical to the baseline or direct path reception of the transmitted signal. The sky wave is identical to the transmitter-target-receiver path shown in Figure 1.1. The time separation (≈ 0.5 ms) between the two paths was a critical factor in confirming the presence of the sky wave.

In slightly earlier ionospheric experiments, 1924–1925, Appleton and Barnett of the United Kingdom used CW transmissions from the BBC Bournemouth station on about 770 kHz with the receiver located at Oxford, about 100 miles away. They measured interference (i.e., the beat frequency) between the ground and sky

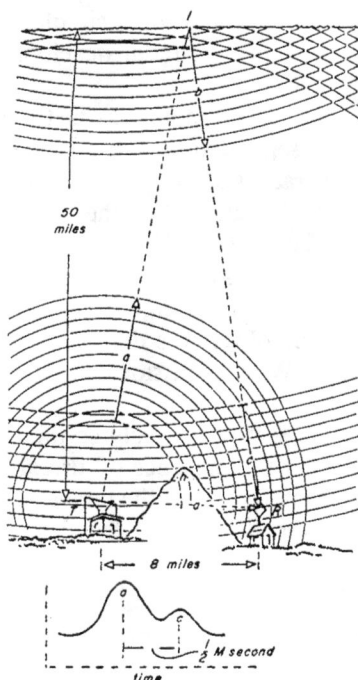

Figure 2.2 Geometry for the 1925 Breit and Tuve ionosphere experiments, where T is the NRL transmitter, R is the receiver at the Department of Terrestrial Magnetism, which was partially shielded from the transmitter by a hill, h, and I is the ionosphere. (Courtesy of Anchor Books [162].)

wave by uniformly changing the wavelength 10 m over 10–30 s, similar to an FM-CW ranging system. Thus, the difference between the sky wave path and the ground wave path could be calculated [159]. In the bistatic radar analogue this difference corresponds to $(R_T + R_R) - L$, as shown in Figure 1.1. The Appleton and Barnett experiments are similar to the early bistatic radar implementations, which also used the interference or beat frequency phenomenon, but with the important difference that the interference was generated by doppler from a moving target.

These and other ionospheric sounding experiments, along with geodetic surveying and radio-altimetry experiments, laid the ground work for radar development in the early 1930s. The development was conducted nearly simultaneously and totally independently by the United States, United Kingdom, France, the Soviet Union, Japan, Germany, and Italy. The first widely circulated documentation of bistatic detection phenomenon (consisting of one paragraph and one figure) was by Bell Telephone Laboratory (BTL) researchers Englund, Crawford, and Mumford in 1933 [156]. (An earlier report [169] by British Post Office engineers in 1932 was as significant, but not as widely circulated as the BTL report.) The BTL report apparently spurred French radar developments, which, like all other work, continued in total secrecy.

Most of the early radar development was of the bistatic type, in which the transmitter and receiver were separated by a distance comparable to the target distance [21–26]. These bistatic radars used CW transmitters and detected a beat frequency between the direct path signal from the transmitter and the doppler-frequency shifted signal scattered by a moving target [25, 156, 162]. This effect is shown in Figure 2.3(a) and (b). It has also been called sky wave interference [21], wave interference [161], CW wave interference [1], and flutter [159]. Virtually all of the early bistatic radar technology was derived from existing communication technology—separated sites, CW transmissions, and frequencies ranging from 25–80 MHz [27, 156].

These early bistatic radars were typically configured as fixed, ground-based, forward-scatter fences to detect the presence of aircraft, a major, emerging threat in the 1930s. In this fence geometry, when the target position was near the baseline joining transmitter and receiver, its echo exhibited large fluctuations, with excur-

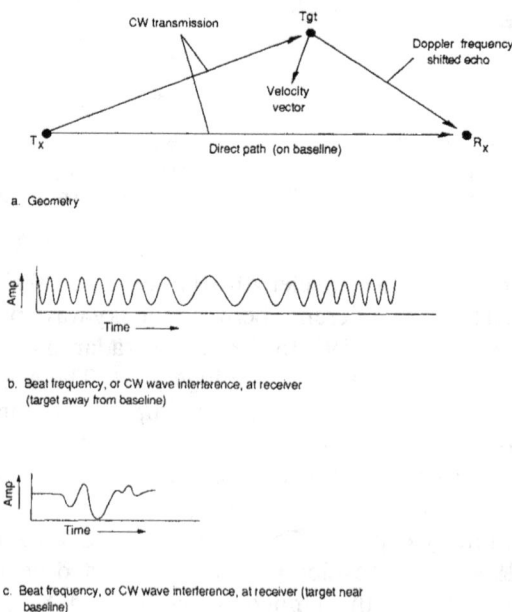

Figure 2.3 Typical early bistatic radar fence operation: (a) geometry, (b) beat frequency, or CW interference, at receiver (target away from baseline) [21] and (c) beat frequency, or CW interference, at receiver (target near baseline)[21].

sions significantly greater and less than those from target positions elsewhere, Figure 2.3(c). The 1922 NRL observation of the Dorchester fluctuations is an early example. At the operating frequency of these fences, target RCS was near the Mie or resonance region, which, depending on target aspect angle, can both enhance the target RCS and cause it to fluctuate. (This "half-wave dipole" resonance effect was known at the time, and exploited by the United Kingdom in the 1935 Wilkins and Watson-Watt Daventry experiment; see Section 2.1.2.) Furthermore, when the target was near the baseline, RCS enhancement and fluctuation can be caused by the forward-scatter effect; see Section 12.2. Although this phenomenon was not known at the time, these early bistatic radars often exploited the inherent RCS enhancement for their operation, sometimes achieving extraordinary detection ranges. The problem of extracting target *position* information, specifically range, from such radars could not readily be solved with techniques available at that time [1]. Indeed, the extraction of target position information remains a difficult task when the target is near the baseline, and an impossible task for range (and velocity) information when the target is directly on the baseline.

Because of this limitation, some radar historians, most notably Guerlac [25, p. 163] and Watson-Watt [170], chose not to call these bistatic fence configurations "radars," which after all is an acronym for radio detection *and* ranging. In fact, Guerlac limited his radar definition to systems using "pulses of radar energy," which is clearly too restrictive. In contrast, Skolnik ignored this distinction and called the bistatic fence a bistatic radar [15, 30]. With occasional exceptions, one of which is documented in the description of the U.S. SPASUR system (Section 2.2.5), this latter viewpoint prevails.

As an aside, the beat frequency phenomenon is observed in commercial television transmissions when an aircraft crosses the TV transmitter-to-home receiver baseline, close to either the transmitter or receiver. In this geometry the received picture will often flutter [1, 55]. Coincidentally, the TV operates with CW transmissions and near the same frequencies used by the early experimental radars. Griffiths, *et al.*, [55] give criteria for the TV flutter effect to occur:

> . . . (i) the delay of the reflected signal must be small compared with the line length and picture features, so that similar signals are beating together; (ii) the rate of change of the delay must be low, so that the effect is not masked by the persistence of screen phosphor and viewer's vision; (iii) the reflected signal strength must be a significant fraction of the direct; and (iv) the whole effect must last long enough to be noticed.

All these conditions are met when an aircraft crosses the bistatic baseline close to either the transmitter or receiver. Exactly on the baseline, the bistatic range is zero and the Doppler shift is zero whatever the direction and speed of motion. At such grazing incidence angles (forward scatter), the bistatic cross-section also assumes rather higher values than usual. . . . By going close to one end of the baseline the $r_1 r_2$ [$R_T R_R$] product can be made low, giving strong reflected signals, and if the track crosses the line obliquely then even high-speed targets pass slowly through the zero Doppler null. Aircraft flutter is, therefore, a short-range phenomenon. . . .

2.1.1 Developments in the United States

In the United States, bistatic radar detections again occurred by accident in the summer of 1930 when Leo Young and Lawrence Hyland, with the new NRL, were testing the directional properties of an aircraft antenna system. Swords [159] summarizes events from accounts by Guerlac [25], Allison [164], Howeth [165], and Gebhard [166]:

> Hyland was at the aircraft which was parked on the compass rose [Generally a special concrete circular area at airfields on which an aircraft can be turned when "swinging" or calibrating the aircraft's magnetic compass] of the Air Station at Bolling field and about two miles distant from a ground beacon at the Naval Research Laboratory that transmitted on 32.8 MHz using horizontal polarisation. The behavior of a 15 ft fore-and-aft wire antenna on the aircraft [an OU2] was being observed. . . . The polar pattern of the wire pattern antenna would ideally be a figure-of-eight, with two maxima and two minima. The beacon's signal was received in the aircraft, and as the aircraft turned on the compass rose, Hyland observed the bearings at which the maxima and minima of the signal occurred. He noticed that a perfect minimum position could be seriously disturbed by aircraft passing overhead and along a line between his position and that of the beacon station. He was deeply impressed by this and reported it. The next day the phenomenon was again observed. . . .

This time NRL was not to be denied. Guerlac continues [25]:

> A small portable field receiver with a single-wire antenna was hurriedly thrown together and the experiments were repeated at distances of 4, 6, and 10 miles from the transmitter. In all cases the presence of planes was indicated by the periodic variation (beats) of the signal from NRL.
>
> This observation was very soon interpreted in terms of the earlier Taylor and Young observation of 1922. In a memorandum . . . the phenomenon was described by Hyland as resulting from interference between a directly transmitted wave and a wave reflected or reradiated . . . from the airplane. . . .
>
> The surprising fact was not that radio waves were reflected from aircraft—the laws of physics predicted this—but that the effect was large enough to be observed. In their discussions of the 1922 observations Taylor and Young had considered the possibility of plane detection, but thought that the energy from a plane would be too small.

Similar observations were made during the late summer and early fall with some interesting variations. It was determined that the effect could be observed from airplanes flying as high as 8000 ft. The phenomenon was also studied using a vertical beam antenna which confined its energy to a narrow vertical cone. An airplane flew across this cone from various directions and at various altitudes. An extremely weak ground signal—attenuation was necessary if the ground wave was to beat against the sky wave—was obtained from this beam at a distance of 10 miles. When the plane flew into the cone the usual change in the ground signal was observed. Successful tests were made on a frequency of 65 MHz and the same variations were heard from a receiver placed in a car. It was experimentally demonstrated that the effect could not be obtained when the ground wave was not present. A few calculations were also made to ascertain just what the periodicity of the variations should be for a plane flying at a given speed at certain altitudes between the transmitter and the receiver. The results obtained appeared to agree, as well as could be ascertained without recording apparatus, with the results observed in the tests.

Although these calculations appear to have been lost, BTL researchers Englund, Crawford, and Mumford published results of similar calculations in 1933 [156], which are given in Appendix A.

While the U.S. Navy approved further "research on the use of very high-frequency radio waves to detect the presence of enemy vessels or aircraft" [25] in January 1931, little funds and encouragement were forthcoming. Guerlac [25] reports that NRL continued tests after working hours:

In the autumn [1931] some tests were made on the airship Akron which had been asked to circle the Naval Research Laboratory in order that high-frequency direction finders might be calibrated. The reflection tests were made with personal equipment belonging to Hyland and Young and at Hyland's own expense, and consisted in picking up energy reflected by the airship from three Washington broadcasting stations. . . .

These tests appear to be the first experimental bistatic radar to use a cooperative (but nondedicated) transmitter, as defined in Section 4.4. Guerlac continues with NRL developments [25]:

During 1932 attempts were made to determine how well the velocity of aircraft could be determined from the beat frequency of the interference effects. Some scattered observations were made between ship and ship, and between ship and shore, in the course of which it was shown that a tug could be detected a mile away, when it passed between the two stations. Nevertheless comparatively little work was done on such slowly moving objects, for fast moving aircraft showed the phenomenon much more readily.

Page adds details of these aircraft experiments [162]:

> With transmitter and receiver separation of 3 miles, and careful screening of the receiver from the transmitter by intervening hills and the "blind spot" in the receiving antenna, and with 500 watts of steady tone being radiated from the transmitter on a frequency of 29 megacycles, the apparatus could detect the presence of an airplane in flight at distances out to 40 miles. Neither the distance nor the direction to the airplane was determined by this method. The pilot of the airplane had to give the experimenters his location.

This NRL work remained classified. Then, in March 1933, Englund, Crawford, and Mumford published a brief account of their accidental detection of an airplane during propagation experiments in New Jersey [156]. NRL quickly declassified their results and filed a patent application on June 13, 1933, which was granted to Taylor, Young, and Hyland in 1934 [21]. Appendix A reviews the NRL patent [21] and the BTL paper [156] that precipitated the patent. It is significant to note that both the NRL "sky wave interference with the ground wave" and the BTL "beat frequencies" can be correlated directly with bistatic target doppler, Equation (6.4b) of Chapter 6. In fact, all the reported bistatic phenomena in the 1930s—beat frequencies [25, 156], sky wave interference [21], wave interference [162], CW interference [1], and the flutter effect [159]—appear to be manifestations of this effect.

Guerlac [25] reports that NRL's bistatic radar ("beat equipment") experiments continued at 60 MHz into 1934 and, in fact, for several years after development of the pulse method was undertaken. Guerlac continues with an account of pulse origins:

> It is generally conceded at the Naval Research Laboratory, where so many persons had a share in bringing pulse radar into existence, that Leo Young, who had been actively interested in both previous radio detection attempts, was the first to hit upon the idea of using reflected pulses of radio energy for the detection of aircraft and other targets. Young has said that the idea occurred to him at the time of the revival of interest in radio detection resulting from the Hyland observation; he recalls having mentioned the possibility to Hyland late in 1930. . . . Be this as it may, nothing was done about the idea for some time because the difficulties seemed almost insurmountable. Extremely short pulses—much shorter than anything being used in ionospheric work—would be required. It would be difficult to produce these short pulses and to build a receiver that would detect them, and still more difficult to design an indicator, or electronic time-base (as the British rather aptly prefer to call it) to display the transmitted and received pulses and to measure the extremely short intervals of time. Before a program was actually launched some calculations were made and the possibility was pretty thoroughly discussed, with Taylor, the Superintendent of the Radio Division, with Ross Gunn, the Technical Advisor to the Director and others. . . .

Finally [25],

It is an important fact that during the early phases of its development at NRL, pulse radar encountered a great deal of skepticism not only in the Bureaus, where in certain quarters it amounted to outright antagonism, but in NRL's own Radio Division, and that this did not evaporate until the feasibility of the equipment had been demonstrated beyond a doubt. The vicissitudes and obvious weaknesses of the beat method were to some degree responsible for this attitude; in fact, the earlier experience with radio detection may be said to have made it harder, rather than easier, to recognize in embryo the vast importance and the essential simplicity of the new idea.

In any case, with the invention of the duplexer in 1936 by Taylor and Page, all subsequent work focused on the pulsed, monostatic radar, with its vastly improved operational flexibility.

Following observations of the early NRL experiments, the U.S. Army Signal Corps at Fort Monmouth, New Jersey, initated radar experiments that were in a bistatic configuration, principally for transmitter-receiver isolation [25]. In 1934 they used the beat frequency method to detect a Ford truck at 250 ft as it moved down the perpendicular bisector of the baseline, $L = 40$ ft. The transmitter was a 0.5-W split-anode magnetron operating at a 9- to 10-cm wavelength (≈ 3 GHz) [25, 159, 161]. Later, small ships (500 T) were detected at about 3000 ft. Subsequent tests in 1936 used pulse transmissions from a 110-MHz 75-W triode transmitter. The transmitter, using a Yagi-Uda antenna, and receiver, using a dipole antenna, were separated by one mile for isolation. Airplanes were detected at ranges out to seven miles [159]. With a convergence of data from pulsed and duplexing experiments provided by NRL and U.S. industrial laboratories, further U.S. Signal Corps research focused on monostatic radars.

2.1.2 Developments in the United Kingdom

In the United Kingdom, bistatic radar detections were observed by British Post Office engineers Nancarrow, Mumford, Carter, and Mitchell, in 1932 [169]. The conditions and geometry of their observations were very similar to those of BTL [156] made a year later. In this case they experienced "nuisance" interference from aircraft, in terms of a beat frequency, up to $2\frac{1}{2}$ miles from the receiver. They concluded that, "The only feasible explanation of the phenomena seems to be that interference is set up between the directly received waves and those reradiated from the airplane. . . ." They, too, made calculations to predict the beat frequencies. Appendix A shows that these predictions are again correlated with the bistatic doppler equation, Equation (6.4b). As Watson-Watt [170] reports, these observations were generally overlooked at the time.

In January 1935, following the first meeting of the British Committee for the Scientific Survey of Air Defense, Arnold Wilkins suggested to Watson-Watt that radio waves might be used for the detection of aircraft. Wilkins was undoubtedly aware of the earlier British Post Office observations [159, 171]. Watson-Watt immediately drafted a memorandum incorporating Wilkins's findings, entitled "Detection and location of aircraft by radio methods," which is reproduced in [159].

On February 26, 1935, they conducted what became known as the Daventry experiment as an initial demonstration of their aircraft detection concept. The Daventry experiment was configured as a forward-scatter fence. It used Appleton's receiver equipment, which was designed to measure the angle of incidence at the ground of downcoming waves from the ionosphere at Slough [159, 173]. The basic receiving system consisted of a two dipole element radio polarimeter, similar to an interferometer, oriented to measure elevation AOA, which was then displayed on a cathode-ray tube (CRT) (Figure 2.4). The receiver was modified so that there was no deflection on the CRT from the direct path or ground wave from the transmitter. Only signals arriving either from other AOAs or with doppler modulation, such as from an aircraft, would cause a deflection.

Swords reports on the selection of the transmitter and receiver sites and the target (Figure 2.5) [171] and summarizes results of the experiments as follows [159]:

> Again Wilkins' experience in propagation work proved an asset. Watson-Watt's initial response was to operate the Slough ionospheric transmitter at 6 MHz but to so modify it that it would operate on short pulses and with a considerably increased peak pulse power beyond its then maximum of 1 kW.

Figure 2.4(a) Front view of the receiver used in the 1934 Daventry experiment. (Courtesy of IEE [159].)

Figure 2.4(b) Back view of the receiver used in the 1934 Daventry experiment. (Courtesy of Science Museum, London.)

However, Wilkins considered this impossible in the time available and recollected that the BBC Empire short-wave station at Daventry (call sign GSA) operated on 49.8 metres and beamed its 10 kW of power in a southerly direction. The Daventry antenna consisted of an array of horizontal dipoles which produced an azimuth beamwidth of some 60° with a main vertical lobe at an elevation of 10°.

Wilkins, in calculating the effect of re-radiation from an aircraft, as postulated in the . . . memorandum, had considered a monoplane bomber with

Figure 2.5 BBC transmitter (right), Handley Page Heyford target (top left), and the "traveling laboratory" receiver (bottom left) used in the 1934 Daventry experiment [171].

a typical wing span of about 75 ft or 25 metres and had assumed this effectively equivalent to a horizontal half-wave dipole operating at 6 MHz. This now was the frequency of the Daventry transmitter. Thus the plan for the demonstration was to position a receiver at a suitable distance away from the Daventry transmitter and in its main beam, and to fly a Heyford bomber up and down the beam noting any fluctuations in the received signal caused by interference between the direct signal and the signal reflected from the aircraft.

The equipment was set up on 25th February inside a van in a field near Weedon, Northamptonshire, and the experiment was successfully carried out the following morning between 09.45 and 10.00 hrs.

Watson-Watt [170] reports a transmitter-receiver separation of 10 miles, whereas Skolink [1] reports a 5.5-mile separation, and Johnson [171] reports a 1-mile separation. (Weedon is about 4 miles from the town of Daventry.) In any case, the separation was selected such that the direct path signal was greatly attenuated [25]. Johnson [171] provides details of the tests:

Down in the van, Wilkins, Watson Watt and Rowe [Air Ministry observer] watched the tiny green spot on the tube of the oscilloscope which represented the direct signal from the BBC transmitter. If the aircraft reflected the signal, it would cause the spot to move vertically.

The stationary spot glowed in the darkened van. Then the occupants heard the faint hum of the lumbering Heyford approaching at a stately 90 mph. On the face of the cathode-ray tube the spot began to move slowly up the screen. As the bomber flew over, "well to one side," Wilkins remembers, the spot moved up and down, oscillating faster and faster. . . . The varying output from the receiver caused the spot to oscillate, which it continued to do until, when it was about eight miles away, the Heyford flew out of range and the three delighted observers were left looking at the stationary green spot.

Following this initial success, work commenced in May 1935 at Orfordness on a pulsed system with transmitting and receiving sites separated sufficiently (≈ 300 m) only for receiver isolation. Initial objectives were to shorten the pulse width to about 15 μs and to increase the peak power to about 50 kW. The work culminated in the Chain Home (CH) early warning radar system operating between 22 and 50 MHz.

The Chain Home radars operated with separate transmitting and receiving sites, but again with separation (≈ 1 km) only for isolation. However, they had a standard, reversionary mode which, in the presence of ECM or a transmitter failure, a receiving site could operate with a transmitter at an adjacent site about 40 km distant, hence becoming bistatic [28].

In fact, just such an event occurred in 1936, according to Bowen [157], when a prototype of the Chain Home system (10- to 13-m wavelength), located at Bawdsey Manor, England (Figure 2.6), was first demonstrated in an air defense exercise for the commander-in-chief of the British Fighter Command. A new transmitter installed on a 240-ft tower at Bawdsey Manor failed on the first day. The old transmitter located on 75-ft masts at Orfordness, about 18 km away, was quickly recommissioned. This bistatic configuration generated ". . . something comparable with the original performance" of about 100-km detection and tracking of aircraft [157], although according to Guerlac [25] and Swords [159], the British Fighter Command was less than impressed.

Figure 2.6 Bawdsey Manor at the mouth of the River Deben. The headquarters of radar research in the United Kingdom from 1936 until the outbreak of war in 1939. (Courtesy of Cambridge University Press [157]).

The first British airborne radar experiments in 1937 were in a bistatic configuration. Swords [159] reports that:

Ground radar was initially known as RDF1 and the airborne radar development programme was then referred to as RDF2; RDF1 1/2 referred to the bistatic system of a ground transmitter illuminating a target aircraft with only a receiver in the interception aircraft.

The ground transmitter was located at Bawdsey, using a 9-ft dipole antenna and silica envelope transmitter tubes operating in a push-pull mode. They generated about 40-kW peak power with 3-μs pulses at 6.8-m wavelength [25, 172]. Swords [159] describes the receiver configuration and test results:

> The first experiments were carried out in June 1937. A Heyford bomber fitted with a receiver operating at 6.8 metres and with an indicator, and with a half-wave dipole wire antenna, connected between the undercarriage spats, circled at a few thousand feet above Bawdsey. Aircraft were picked up at ranges of 8 to 10 miles. Two reasons which prompted the use of the Heyford were the generous space available in the fuselage and the fact that its Kestrel engines were well screened against ignition noise. The wavelength of 6.8 metres was chosen because there was readily available an EMI television receiver of the TRF type tuned to this wavelength which was perfectly suited for the task. A car ignition system was used to provide the HT for the cathode-ray tube. The total weight of the installation was some 50 lbs. . . . This RDF1 1/2 system, which gave excellent ranges even at such an early stage of development, was strongly advocated by Bowen, but Watson-Watt was equally strongly against it, and the latter's viewpoint prevailed.

Subsequent airborne radar development was in the RDF2 monostatic configuration, with separate transmitting and receiving antennas on the aircraft.

2.1.3 Developments in France

In France, bistatic radar research was begun by Pierre David in 1933 after reading the Englund, Crawford, and Mumford paper [156]. This paper confirmed his 1928 predictions, which were not tested at that time [159]. The first successful tests of a forward-scatter fence were obtained in 1934 using the beat frequency method. A 50-W transmitter operating at 75 MHz with the receiver separated by 5 km detected aircraft up to a maximum height of 5 km.

The transmitter-receiver baseline was soon extended to 15–21 km, with similar results. In 1935 construction of three transmitters and six receivers was begun on a David *"barrage électromagnétique"* (electromagnetic barrier) system, that ultimately was forseen to stretch between the continent and Corsica [159]. Development continued, with Guerlac [25] reporting that, "About 20 sets built by the Army, the Thomson Company, and the Sadir Company were incorporated into a network that was tested during aerial maneuvers in July and August 1938." He continues with details of the system:

> The equipment was extremely simple and light and could be transported by only a few men. The antennas were nondirective dipoles and the transmitter power . . . was still very low. In order to set up a warning system that would convey some idea of the location of the approaching planes it was necessary to combine the stations into a fixed chain. . . . The chains were so laid out that a single observer compared the receivers of two adjacent elements of the

chain, the information from the observers being fed into a filter station. The system was expected to serve both for early warning and for ground control of interception. A chain of this sort operating on 30 MHz was set up near the city of Rheims in the summer of 1938 and tested during the maneuvers. It was used to direct night fighters to intercept "enemy" bombers. The data [automatically recorded on a moving tape] from the various observers was telephoned at intervals to the filter room which in turn directed the night fighting planes to the enemy by means of ultrahigh-frequency radio telephone.

The same system was studied by the Naval Laboratory in Toulon, and a few stations were set up along the Mediterranean coast and on one or two small ships. In 1939, a chain protecting the Naval bases along the British Channel, the Atlantic Ocean, and the Mediterranean Sea was put into service. It was composed of stations operating at 30 MHz, with each receiver-transmitter pair separated by a distance of from 20 to 60 miles.

Swords [159] reports that during the Rheims tests:

Three configurations of stations were used, the most complex of which was referred to as *"maille en Z"* and allowed the speed, direction and altitude of an aircraft to be determined to within 10%, 10° to 20° and 1000 metres, respectively.

Details of the *maille en Z* configuration apparently are not available. Because one translation of *maille* is "mesh," the configuration probably consisted of a mesh of fences in the form of a Z. However, a three-fence mesh, one fence per leg of the Z, does not provide a unique solution for the speed and direction of an aircraft, for example, crossing the Z from top to bottom at various offsets and oblique angles. A four-fence configuration in the form of an Σ, or "double Z," will provide a unique solution for all trajectories, as long as the aircraft (1) maintains a constant velocity vector and (2) crosses all four legs at different times. This double-Z configuration might have been implemented as follows:

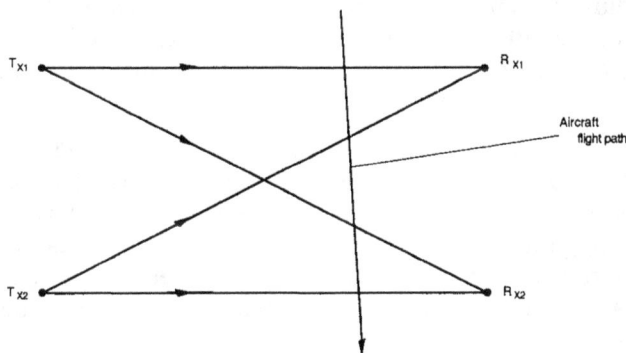

Each transmitter services both receivers, generating four "trip wire fences" in a true multistatic configuration. The time of aircraft detection by each trip wire is measured, which is sufficient to determine aircraft speed, direction, and point of fence penetration. This configuration is analyzed in Section 12.2.2. Aircraft altitude might have been estimated by using vertical fan beams and measuring the time duration of detections in each fan, ignoring sidelobe detections. Then with an estimate of the constant aircraft trajectory, altitude becomes proportional to the time duration.

Monostatic pulsed radar development was also pursued during the prewar period—with considerable urging by David—and following the fall of France in 1940, French work concentrated almost exclusively on this technique.

2.1.4   Developments in the Soviet Union

In the Soviet Union, radar experiments began in 1934 when the Air Defense Command or PVO (Voiska Protivo-vozdushnoi Oborony) signed an agreement with the Leningrad Electro-Physics Institute or LEFI (Leningradskii Elektrofizicheskii Institute). Pavel Oshchepkov of the PVO was a principal investigator of these experiments. The first tests of a forward-scatter fence were conducted in July and August 1934 and are summarized by Swords [159] from accounts by Oshchepkov and others, as follows:

> The equipment used, designated RAPID, was a bistatic continuous-wave system; the transmitter gave an output of 150–200 W on a wavelength of 4.7 m, and with the carrier frequency modulated at 1000 Hz. The receiver was of the super-regenerative type and employed a small horizontal dipole antenna, which . . . was mounted on a short vertical rod that could be held in the operator's hand and rotated. The presence of an aircraft was indicated by a beat note in the receiver headphones. Tests were carried out the 10th July 1934 with a separation between transmitter and receiver of 3 km, and on the following day with a separation of 11 km. Aircraft up to heights of 1000 m and within a radius of 3 km from the receiver were detected. Further tests were carried out on the 9th and 10th August 1934 to help establish the ultimate range of the equipment, and it was discovered that maximum ranges of 75 km were obtainable.

On the basis of these and other successful tests, the PVO placed an order for five factory-produced experimental sets called REVEN (RHUBARB) in October 1934, with the goal of producing a fence up to 10 km high along the Soviet frontier.

However, opposition from the Red Army Signals Command to the whole radar venture started in 1935 and greatly impeded Soviet radar development [159]. Oshchepkov was arrested and imprisoned for ten years in the great purge of 1937–

1938. Ironically, an organization of the Army Signals Command took over Osh-chepkov's laboratory and apparently what they saw changed their attitude toward REVEN. It went into production as RUS-1 and was tested in 1938 [213]. Some 45 of these sets were produced and a number of them were installed in the Soviet Far East and the Trans-Caucasus starting in 1939 [213]. They operated on about 75 MHz, with a 35-km separation between the transmitter and receiver [24, 159]. Subsequent Soviet radar work concentrated on pulsed, monostatic systems, probably for the same basic reason all other developers did—operational utility.

2.1.5 Developments in Japan

In Japan radar work was started by Professors Okabe and Yagi in about 1936, when they demonstrated that aircraft and other moving objects could be detected by a CW doppler technique [26, 159]. Details of the demonstration, including transmitter-receiver separation, are not available, although Price [26] asserts that the configuration was a typical forward-scatter fence using beat frequency detection. In any case, CW (and pulsed) radar developments continued, and in 1937 the Navy's research laboratory obtained ranges of up to 5 km from ships using an FM-CW radar. Because of dissatisfaction with results by the military, the experiments were not continued.

At about this time the Japanese army appears to have taken over CW radar research, culminating in the deployment of about 100 forward-scatter fences, called Type-A, starting in 1941 [26]. Swords [159] summarizes their operation from accounts by Wilkinson [168] as follows:

Before the Japanese attack on Pearl Harbour, an Air Defense System had been organized by the Army to protect the Japanese mainland and this comprised three districts with an information centre in each district. These centres were at Tokyo, Osaka and Fukuoka. The centres received information from a network of radars around the coast and from observers. The first radars to be used were Type-A systems, bistatic continuous-wave installations. These remained in operation throughout the war. It was the intention to replace them eventually with monostatic pulse radars [Type-B] except in certain mountainous regions. With a transmitter at one point and a receiver at another distant point, a fence was created. An operator would report to the Information Centre that an aircraft had passed through the beam or fence. On the display board in the Centre, each link of the Type-A Chain was denoted by a line which could be illuminated in red whenever an aircraft had passed through the beam. The frequencies of these Type-A sets lay in the band 40 MHz to 80 MHz. Typical transmitter power outputs were 3 W, 10 W and 100 W (Price [26] reports 400 W also) with distances between transmitters and corresponding receivers varying from about 40 miles to 150 miles. . . .

The longest Type-A used was not in Japan itself, but between Taiwan (Formosa) and Shanghai, a distance of over 400 miles.

The Navy operated, completely independently of the Army, its own network of stations, all Type-B sets, although towards the end of hostilities they were contemplating the installation of some Type-A systems. The coverage of their system in many cases duplicated that of the Army, but they were principally concerned with protecting their own installations.

2.1.6 Developments in Germany

In Germany, the beat frequency phenomenon had been observed in 1934 when a ship passed through a decimeter radio beam. The phenomenon was called the "studied ship effect." However, it was not pursued in a forward-scatter fence configuration, but rather in a monostatic configuration using two antennas separated only enough for isolation. Successful tests were conducted in 1934 using 630-MHz equipment with a transmitter-receiver separation of 200 m. A ship was detected at 12 km and an aircraft at less than 1 km. Further German research concentrated on pulsed operation for improved signal isolation and range measurement [158, 159].

However, during World War II the Germans developed a bistatic hitchhiker system, known as the Klein Heidelberg, that used a British Chain Home radar as the transmitter [17, 158]. The receiver was designed to give passive warning of the onset of Allied bombing raids when they were over the English Channel. According to Price [158]:

... the Klein Heidelberg receiver was locked to the direct transmissions coming from the British radar station but in addition it picked up echoes from the aircraft in the area. The receiver was connected to an indicator tube which therefore displayed "two blips". . . .

The first blip is a measure of the baseline range, L; the second blip is a measure of the range sum, $R_T + R_R$. Then with a third measure of a "directional bearing on the echo signals," the bistatic triangle is solved, and the receiver-to-aircraft range R_R is calculated [via Equation (5.1) of Chapter 5]. Price continues [158]:

Under ideal circumstances the aircraft's position could be determined to within six miles and the Klein Heidelberg erected by the German air force on the island of Roms, off the west Coast of Denmark, could plot aircraft moving up to [$R_R =$] 280 miles away [near the English coast]. However, by the time the device was fully operational there, [the R.A.F.] Bomber Command had virtually ceased to route its aircraft over Southern Denmark, and this unusual device had little effect on night battles.

The Klein Heidelberg appears to be the first operational bistatic radar to use a noncooperative transmitter, as defined in Section 4.4.

2.1.7 Developments in Italy

In Italy, Guglielmo Marconi observed the beat frequency phenomenon in 1933. In the course of testing his microwave (90-cm wavelength) telephone link between the Vatican and the Pope's summer residence at Castel Gandolfo, he observed rhythmic variations in a modulated tone, which was transmitted to adjust the system's signal level. In this case the transmitter and receiver were at adjacent sites, but with undocumented separation. Marconi noticed that the variations occurred each time a steam roller, which was working on the road in front of the antenna, moved [159].

Marconi ran controlled experiments and demonstrated the concept to the Italian military authorities in 1933. He suggested that research should begin on a bistatic cw apparatus with the objective of proceeding to a monostatic pulse system when difficulties in pulsing high-emission valves had been overcome [167]. Limited research continued, and in 1934 Marconi detected automobiles at a distance of about 1.5 miles. Further bistatic radar work under limited funding was not successful, and in 1941 pulsed monostatic radar was vigorously pursued, partially based on information provided by Germany [159].

In virtually every case, when a method was found to use common transmitting and receiving antennas, or even adequate isolation between adjacent transmitting and receiving antennas, the radar community abandoned these early bistatic radar developments and concentrated on the single-site monostatic radar. Pulsed waveforms and the duplexer were, of course, the major developments leading to monostatic radar, and its greater utility for aircraft, ships, and mobile ground units.

2.2 FIRST RESURGENCE

Not until the 1950s was interest in bistatic radars revived. In retrospect, the revival was not unexpected because monostatic radar theory, techniques, technology, measurements, and resulting systems had continued to expand since World War II; thus, the time had come to revisit, or "reinvent" [1] bistatic radars in light of these monostatic radar developments. The term "reinvent" is appropriate because little of the bistatic radar work conducted through World War II had been published or, indeed, had survived. The reinvention (and invention) process focused principally on theory and measurements, forward-scatter fences, semiactive homing missiles, and multistatic radars. Naturally occurring sources of illumination were also evaluated and tested.

2.2.1 Theory and Measurements

Bistatic RCS theory, including forward-scatter RCS theory, was developed and corresponding measurements were taken between 1955 and 1965 [33–41]. Subsequently bistatic clutter measurements were taken [42, 43]. These and later theories and data are summarized in Chapters 8 and 9. Bistatic radar system theory was codified by Skolnik in 1961 [15]. Skolnik also summarized bistatic fence configurations, compared bistatic to monostatic operations, and developed bistatic location techniques in [15]. These topics are summarized in this book. The name "bistatic radar" was coined by K.M. Siegel and R.E. Machol in 1952 [34].

2.2.2 Semiactive Homing Missiles

A major development at this time was the semiactive homing missile seeker, in which the large, heavy, and costly transmitter could be off-loaded from the small, expendable missile onto the launch platform. While these seekers are clearly a bistatic radar configuration, missile engineers have developed a different lexicon to describe their technology and operation; for example, semiactive versus bistatic, illuminator versus transmitter, seeker versus receiver, rear reference signal versus direct path signal, *et cetera*. The missile and radar communities continue to go their separate ways, particularly with separate publications. Three classic references [187] on semiactive seerks are by A.S. Locke [188], D.A. James [189], and A. Ivanov [216].

A small contribution the bistatic radar community may have recently made to the missile community is end-game angle glint reduction. Bistatic RCS measurements [54] show that when the bistatic angle is increased, so that the semiactive homing missile is not on the illuminator-to-target LOS, the target's angle glint is reduced. These measurements are summarized in Section 8.3.

2.2.3 Hitchhiking

Bistatic radars were also analyzed and tested when hitchhiking off naturally occurring sources of illumination. Caspers [16] summarizes two such configurations, one using lightning [178, 179] and the other using radio stars [181]:

Scordes (sferics correlation detection system) is bistatic VLF radar using sferics, i.e., VLF signals generated by lightning, as the illuminating source. . . . This system . . . is capable of detecting abnormal or disturbed characteristics of the ionospheric D and E regions. The sferics signal is propagated via great-circle paths to the region of ionospheric disturbance (target) as well as directly to the Scordes receiving site. Signals scattered by the ionospheric disturbance are also received at the Scordes receiving site. Range-sum and angle measurements are then used to locate the scattering region. . . . In the VLF experiment, location measurements of transatlantic sferic sources for 150 cases showed an absolute deviation from the mean of only 31 nmi.

"Stellar radar, . . . using the sun or certain radio stars as transmitters, has been studied for possible use in detecting or tracking objects in space. . . . It was concluded the "stellar or solar radars are apparently only useful in situations which absolutely preclude active RF emanation and where modest detection ranges are valuable."

2.2.4 Forward-Scatter Fences

Unfortunately, little information is available about these second-generation forward-scatter fences, which were designed for aircraft and missile detection. The United States built and deployed the AN/FPS-23 (Fluttar) in the 1950s for use with the distant early warning (DEW) air defense line in the Arctic [1, 24, 30–32, 54, 184]. The DEW line employs 31 AN/FPS-19 and AN/FPS-30 monostatic surveillance radars spaced about 160 km apart. Fluttar was a CW, fixed-beam bistatic radar fence designed as a low-altitude gap filler between the monostatic radars. The Fluttar geometry appears to be similar to forward-scatter geometry, although forward-scatter RCS enhancement had not been identified at the time of Fluttar's development. Fluttar was operational for about five years and then was removed. Reasons for its removal have not been documented. The Canadians also developed a bistatic radar for their McGill Fence [29], but apparently it was not deployed.

The United States also built and deployed three forward-scatter OTH fences to detect ballistic missile launches from the Soviet Union. According to Greenwood [214]:

Unlike conventional radar, over-the-horizon, or OTH, radar is not restricted in its range by the curvature of the earth. By reflection from the ionosphere OTH radar can penetrate to great distance, making possible the detection of missiles soon after they are launched. The currently deployed "forward scatter" OTH radar detects the disturbances in the ionosphere caused by the ionized jet of gas emanating from a rocket's motor. Since each type of missile disturbs the ionosphere somewhat differently, a detected missile can be identified by its characteristic OTH signature. In the currently operational system three transmitters are deployed in Taiwan, Japan and the Philippines. These transmitters are matched with corresponding receivers in Italy, Germany and another European country. Although the system was originally intended as an early-warning system for a massive missile attack, it has detected a high percentage of the known single events [missile launches] since 1968. All long-range missiles fired from test sites in the U.S.S.R. are detectable.

Kolosov [213] identifies the third receiving site located in "another European country" as the United Kingdom. He also identified the problem of "maintaining synchronous signal reception" between the transmitter and receiver:

Because the distance between the transmitting and receiving sites may exceed 10,000 km, it is difficult to achieve signal synchronization with communication lines. The receiver, therefore, includes a reference oscillator synchronized with the help of unique time signals . . . which reduces the radar's reliability.

The current status of these forward-scatter OTH fences has not been documented.

2.2.5 Multistatic Radars

Multistatic radars were designed, developed, and a surprising number were deployed during this time to improve detection, location, and tracking of ballistic missiles and satellites. The U.S. Plato and Ordir ballistic missile detection systems were designed as the first multistatic radars; they combined range sum and doppler information from each receiver site to estimate target position. However, they were not deployed [24, 32].

The Azuza, Mistram, and Udop interferometer radars, a variant of multistatic radars, were installed at the U.S. Eastern Test Range for precision measurement of target trajectories. They used a single cw transmitter, multiple receivers at separate, precisely located sites, and cooperative beacon transponders on the target [9, 10]. Caspers [16] summarizes details of the Azuza, Mistram, and Udop systems:

Azuza is the oldest CW radar system at the Eastern Test Range. It is a C-band, short-baseline (500 m), interferometer system consisting of nine receivers and one transmitter located along two crossed baselines. Each baseline consists of three antenna pairs spaced at 5, 50, and 500 m. The longest baseline provides the accuracy, and the two shorter baselines resolve ambiguities. The Azuza system measures range by phase measurement of sideband frequencies modulating the carrier, coherent range [sic.] by doppler count, two direction cosines, and two cosine rates. Accuracies of less than 10 ft in range and 20 ppm in direction cosine are obtainable.

Mistram (missile trajectory measurement) is a CW interferometer system employing ground antennas located along two mutually perpendicular baselines spaced at 10,000 and 100,000 ft. It provides range, four range differences, range rate, and four range-difference rates. Range accuracy is 2.4 ft.

Udop (UHF doppler) transmits to the target on 450 MHz, and five receiving sites receive a 900-MHz signal from the target transponder. Baseline lengths are 25 to 75 miles. The five receiving sites yield slant-range rate. To compute range or position, an initial position is required from some other tracking system. The random error is 0.2 ft but with a systematic error of 9 ft plus the initial error. Udop is of relatively low cost compared with other high-accuracy systems.

This multistatic, interferometer technology was extended to the satellite skin detection and tracking fence SPASUR. It was first deployed in 1958 and 1959 along a great circle at 33.5° N latitude from Fort Stewart, Georgia, to San Diego, California, about 3500 km apart [7]. Three transmitting and four receiving sites were initially deployed along the great circle in the following configuration:

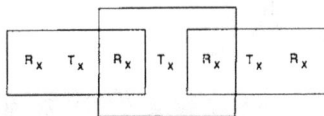

The two interior receiving sites operate with adjacent transmitters yielding three multistatic sets. Each set consists of a CW transmitter and two receivers, one on each side of the transmitter and separated from the transmitter by 400 to 500 km. Each site generates a stationary, vertical, fan beam oriented along the great circle. All beams are coplanar. The AOA of a satellite echo is measured at each receiving site by an interferometer antenna system, with the satellite's position established by the intersection of two AOA measurements [7]. The geometry is shown in Figure 2.7, with only the portions of the fan beams intersecting the satellite shown on the figure. In this geometry the maximum bistatic angle is approximately 90° for a minimum satellite altitude of 100 km. Thus the SPASUR fence does not operate in the forward-scatter RCS enhancement mode. In the mid-1960s two additional receiving sites were added to the eastern part of the network to improve the accuracy of low-altitude satellite position fixing [185].

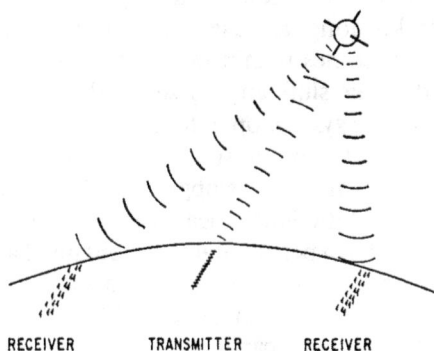

RECEIVER   TRANSMITTER   RECEIVER

Figure 2.7 SPASUR geometry, showing a transmitting linear array and two receiving interferometers, each with three linear arrays [7]; actually, each receiving site used up to nine linear arrays. (© 1960 IRE [185].)

The transmitters, one of which is shown in Figure 2.8, initially operated at 108 MHz, a frequency that was outside the military band, but authorized for use by the U.S. Federal Communications Commission (FCC) on a temporary basis. It was changed to 216 MHz in the mid-1960s. The largest transmitter is located at Kickapoo Lake, Texas, in the middle of the fence, and operates with about 1 MW of CW transmitter power from a linear array about 3.3 km long that is oriented in a north-south direction [1, 185].

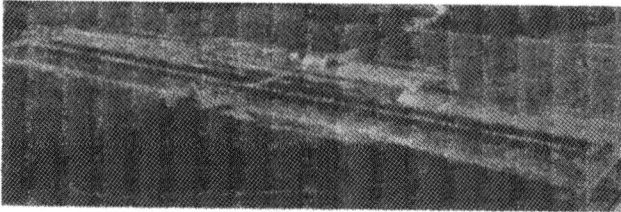

Figure 2.8 SPASUR transmitting site at Jordan Lake, Alabama. (© 1960 IRE [7].)

SPASUR receiver operation is based on the Minitrack (108-MHz) satellite beacon tracking system [8], which was in turn based on the earlier Azusa missile tracking system [16]. The receiving interferometer antenna system is configured for either high-altitude operation (two of the six sites) or medium-altitude operation (the remaining four sites). For high-altitude operation the antenna system consists of seven parallel, phase-measuring linear arrays, again oriented north-south. Effective array length is 0.74 km, with a total array separation of 0.37 km. Two additional linear arrays, 1.48 km long, are used as alerting antennas. These antennas detect the satellite before it comes within range of the phase-measuring arrays to measure the satellite's doppler shift and to select the appropriate narrow band receiver. The medium-altitude system operates in a similar manner, but with eight smaller phase-measuring arrays and one smaller alerting array [185].

In 1965, SPASUR performance was upgraded when construction was started on a second, experimental bistatic fence located in southern Texas about 800 km south of the existing fence. The experimental fence obtained a second satellite measurement to improve the ephemeris estimate from a single pass. This system consists of one transmitter and one receiver separated by about 160 km. The transmitter uses an unspecified modulated carrier, so that satellite position is determined by bistatic range and an AOA measurement, Equation (5.1). The transmitting and receiving antenna configurations are similar to those of the SPASUR system [185].

Data from each SPASUR receiving site are sent via land line [185] to a Space Surveillance Operations Center at Dahlgren, Virginia, where orbital computations are made. In 1960, these computations were made on a new, high-speed (15,000 operations per second) digital computer with a 20,000-word magnetic core storage capacity [7]. Gumble [184] reports that SPASUR "... is currently [*circa* 1985]being upgraded to double its range from 7,500 to 15,000 nautical miles."

Calibration of the SPASUR interferometer system was reported [7] to use satellites with known ephemerides, which are calculated with data from the Minitrack stations. Reported Minitrack angular accuracies range from 0.1 to 1.0 milliradians, depending on ionospheric conditions. (Minitrack calibrated its interferometer antennas against radio stars, and used radio station WWV for time synchronization, with an accuracy of approximately 1 ms [8]). Thus, SPASUR would be calibrated against these standards, though with diminished accuracy [7].

Ironically, Easton and Fleming in their 1960 SPASUR paper [7], chose not to call SPASUR a radar, but a radio detection and location, or "radal," system. Their rational for this distinction—and for the basic SPASUR design—is as follows:

The *radal* (radio detection and location) system described differs from a pulse radar in principle and in detail differs greatly. Pulse radar ... uses a single installation from which energy is transmitted and received. The location of reflecting objects is inferred from the angle of arrival and from the time delay between transmission and reception. ... Pulse radars were developed for detecting and tracking ships and aircraft, objects which have great maneuverability but low speeds.

The radal system is designed to detect and locate objects having great speed but very limited maneuverability. For such objects a great range capability but a modest number of sightings serves to determine the path of the object for days to come.

To satisfy the requirement of detection at great ranges the system has been designed to maximize range capability. Since a larger average power can be generated economically at CW than with pulses, a CW system is indicated. To use antennas having large capture areas without unusably small beamwidths, a low frequency is used. The antenna beamwidths are designed to detect over a large angle in one direction and a very narrow angle in the other. This technique permits detection of objects passing through an area of large dimensions but small volume.

Thus Easton and Fleming join Guerlac ([25], p. 163) and Watson-Watt [170] in this early, countervailing viewpoint. However significant this viewpoint may be, their SPASUR system design insight is clearly significant.

A second U.S. multistatic fan beam radar fence, called DOPLOC (Doppler phase lock), was also deployed in an "interim" configuration in the late 1950s, again for satellite tracking [181, 182]. It was configured for both beacon and skin tracking, the latter using a 50-kW CW ground-based transmitter operating on 108

MHz at Ft. Sill, Oaklahoma. Receiving sites were located at White Sands Missile Range (WSMR), New Mexico, and at Forrest City, Arkansas, about 1600 km apart, with the transmitting site roughly midway between the receiving sites. Each receiving site generated three fan beams, one directly overhead and one to either side of vertical. The interim system was constrained to three fan beams "to conserve power" [181].

The DOPLOC receivers used a narrow band (1- to 50-Hz) phase-locked loop that tracked the doppler frequency shift from the satellite [182]. The time-varying doppler shift was recorded as the satellite passed through each beam. From the single-pass data, the satellite ephemeris was estimated using iterative calculations, while imposing satellite elliptic motion as a constraint [181].

The system was exercised against many satellites in the late 1950s. Patton reports [181] that a relatively accurate set of orbital parameters could be generated with as little as 1.5 to 3 minutes of intermittent observations from a single receiving site. (The WSMR site was inoperative during the reported test period.) Apparently, an operational configuration of DOPLOC was not deployed.

2.3 SECOND RESURGENCE

Development of the antiradiation missile [220] is probably the event that triggered the second bistatic resurgence in the 1970s and 1980s. The argument was that the effectiveness of an ARM attacking a radar could be reduced by moving the transmitter from the battle area into a sanctuary, possibly a satellite-based sanctuary, which is less vulnerable to attack [44]. A second threat that began to emerge at this time was the retrodirective jammer, or retrojammer, in which a high-gain jamming antenna is directed toward the monostatic radar by the jammer's receiver [192]. Effectiveness of the retrojammer could be reduced by selecting the bistatic geometry (the bistatic angle) such that the receiving site would lie outside of the retrojammer's mainbeam, as detailed in Section 12.3.

An important change had to be made in the configuration of bistatic radars before these counter-ARM and counter-retrojamming concepts could be realized. Up to this time, bistatic radars were designed for limited spatial coverage (fixed-beam fences) or cued operation (instrumentation systems and semiactive homing missiles), again requiring limited spatial coverage. Because ARMs and retrojammers were of principal concern to monostatic, microwave air defense radars, usually with large spatial coverge capabilities, the microwave air defense bistatic radar had to emulate, or at least approximate, the monostatic coverage. As pointed out by Skolnik in 1961 [15], this task is particularly difficult for a bistatic radar. The British Chain Home HF and VHF air defense radars in World War II did have the capability to operate in a bistatic mode with an adjacent site [28, 157], but their problem was eased considerably by the use of wide beamwidths and relatively short baselines. A number of programs were established during the second resurgence to assess the difficulty of the bistatic microwave air defense task [18, 44-50]. One program, *Sanctuary* [45], carried the assessment through field tests; it is described in Section 2.3.1.

Compounding the spatial coverage problem were beam scan-on-scan problems, which arise when high-gain transmitting and receiving antennas are used to achieve adequate range performance, and clutter suppression, which is a critical requirement for air defense operations and a particularly vexing problem in a bistatic radar. One solution to the former problem is pulse chasing, which was identified in the 1960s [16], but not tested until 1975 [129]. In pulse chasing, the receiving beam rapidly scans across the transmitting beam, "chasing" the pulse as it propagates from the transmitter. Beam scan-on-scan and pulse chasing are detailed in Sections 13.1 and 13.2, respectively.

Clutter suppression requirements are typically established by modeling the anticipated geometry to calculate the clutter doppler shift and spread, along with the expected target doppler shift, as detailed in Chapter 6. Fortunately, digital computers were both available and reasonably friendly at this time, easing the modeling task. When two or three of the sites—target, transmitter, and receiver—are moving, the clutter calculations can become messy. An attack on this problem by Lorti and Balser [51] yielded a new bistatic radar concept, which the author calls *clutter tuning*, that is possible when both the transmitter and receiver are moving. [51-53, 173-175]. When the separate velocity vectors of the transmitter and receiver or their relative geometries are changed, the clutter doppler and doppler spread can be controlled, i.e., tuned. One potential implementation of this concept allows the receiver to generate a SAR map directly on its velocity vector—an impossible task for the monostatic SAR. Clutter tuning combined with the sanctuary concept protects the transmitter while allowing the receiving platform to fly toward the target with no radar emissions. Clutter tuning is detailed in Section 12.4

In addition to these new air defense and clutter tuning concepts, the early bistatic hitchhiking and forward-scatter fence concepts were revisited in new configurations; multistatic instrumentation radar development (and deployment) continued; and nonmilitary bistatic radar applications were developed and tested.

2.3.1 Air Defense

A major—at least in terms of cost—U.S. experimental bistatic radar air defense program was called Sanctuary [45, 174-175]. Work began in 1977 and flight tests were conducted in 1980 at the U.S. Pacific Missile Test Center, Point Mugu, California. The system used a standoff, dedicated transmitter as defined in Section 4.4 and a ground-based receiver as shown in Figure 2.9. The transmitter was carried

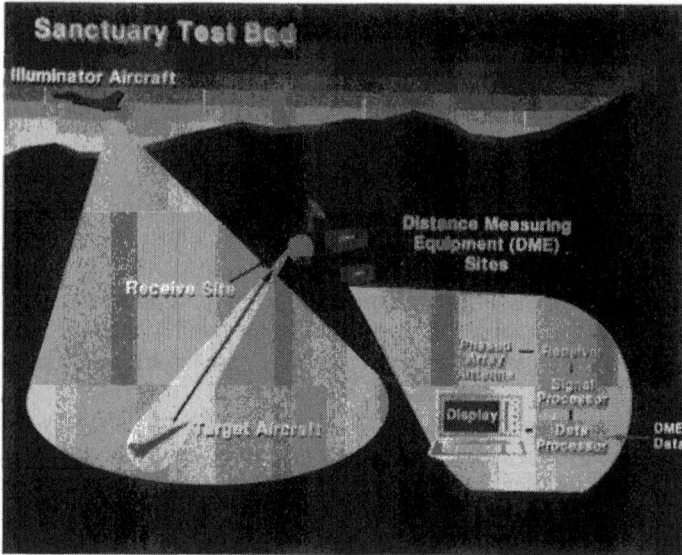

Figure 2.9. Sanctuary test bed configuration. (Courtesy of Technology Service Corp.)

by an A-3 aircraft and used a 1.7-kW CW solid-state device operating on 1385 MHz, and a "floodlight" antenna with a 14.5-dBi gain. The floodlight transmitting antenna finessed the beam scan-on-scan problem, but at a penalty in detection range. Transmitter coordinates were provided by trilateration *distance measurement equipment* (DME). The receiver was synchronized in time and phase to the transmitting signal via temperature-controlled crystal oscillators. The waveform consisted of 128 repetitions of a 1024-element binary phase code during an 87-ms coherent dwell.

The receiving antenna (Figure 2.10) was a four-beam phased-array antenna with 27.2-dBi gain. It used a sidelobe canceller for direct path cancellation and a sidelobe blanker for excising clutter discretes. The canceller and blanker antennas are in the two small radomes on the top of the array [176]. The signal processor was in a dual transform configuration, where pulse compression was performed in the frequency domain following doppler filtering. It consisted of 11,500 integrated circuits and consumed 10 kW of power [177].

The major objective of the experiment was to control doppler-modulated mainbeam and sidelobe clutter (Chapter 6), thus generating a "clutter-free" display for air defense operations. The system was, of course, required to demonstrate coherent operation with a moving transmitter, adequate synchronization, and sufficient direct path cancellation. These supporting objectives were reported [45] to have been successfully met in July 1980. A maximum detection range of >100 km

Figure 2.10 Sanctuary receiving antenna. (Courtesy of Technology Service Corp.)

against tactical aircraft has also been reported over an unspecified wide range of bistatic angles [54].

The bistatic maximum range of Sanctuary can be estimated by using Equation (4.1a) with parameters given in the previous paragraphs and the following assumptions:

$$\sigma_B = 6 \text{ m}^2,$$

$$F_T = F_R = 1,$$

$$kT_s = 6 \times 10^{-21} \text{ W/Hz} (\approx 2 \text{ dB noise figure}),$$

$$(S/N)_{min} = 16 \text{ dB and}$$

$$L_T L_R = 10 \text{ dB}$$

Thus, a bistatic maximum range product, κ, of 1.14×10^4 km^2 is available. The equivalent monostatic range, $\sqrt{\kappa}$, is 107 km, which matches the lower bound of the reported "maximum detection range," assuming that range is the equivalent monostatic range.

While the Sanctuary tests demonstrated a modest air defense surveillance capability—modest in the sense that the equivalent monostatic range was a factor of 2 to 3 less than typical monostatic surveillance radars—the concept apparently was not pursued further. Reasons for ending the program have not been documented. Possible significant factors in the decision were (1) receiving site retrofit complexity, (2) the requirement to deploy and support an airborne or satellite-based transmitter, and (3) competition from other, possibly less expensive, counter-ARM techniques, such as decoys and blinking between adjacent monostatic radar sites.

2.3.2 Clutter Tuning

Two clutter tuning experiments have been reported [52, 53, 173–175] with results, including bistatic SAR imagery, given by Auterman [53]. The geometry for Auterman's X-band ($\lambda = 3.2$ cm) experiment is shown in Figure 2.11. In this experiment the geometry was changed while the velocity vectors were held constant. Two Convair CV-580 aircraft were flown on constant velocity, 180-kt flight paths, offset from the target by 7–10 km. These paths were symmetric about a perpendicular to the flight line, which is also the bisector of the bistatic angle. SAR images were taken at three bistatic angles, β, of 2° (near-monostatic), 40°, and 80°, as shown in the figure. In this geometry the transmitter and receiver contribute equal clutter doppler spread for bistatic isorange resolution, which is analogous to monostatic

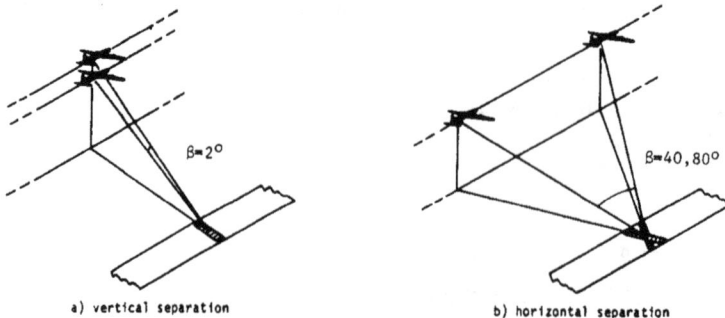

a) vertical separation   b) horizontal separation

Figure 2.11 Geometry for bistatic SAR experiment. (© 1984 IEEE [53].)

SAR cross-range resolution (Section 7.4). The array, or coherent integration, time, T = ⅔ s, was selected to generate good quality SAR imagery, in terms of low integrated sidelobe ratios (Section 13.5).

Inserting the foregoing parameters into the equation for monostatic SAR cross-range resolution, τ_i, Equation (7.10), yields τ_i = 2.3 m at a monostatic range of 9 km. The signal bandwidth for the experiment was 40 MHz, yielding a monostatic range cell, or down-range dimension, of 3.8 m.

For the bistatic SAR case, these dimensions will be degraded (i.e., increased) from the monostatic case due to geometry. Both down-range and cross-range dimension will be increased by $\cos^{-1}(\beta/2)$, and the cross-range dimension will be further increased by a reduction in the transmitter and receiver angular rates about the target, as shown by Equations (4.11b) and (7.17), respectively. Calculations for the three bistatic angles used in the experiment are shown in Table 2.1. Auterman [53] reports that the imagery was processed for ΔR_B = 3.8 m and τ_i = 2.1 m.

Table 2.1
Bistatic SAR Resolution Cell Dimensions for Auterman's [53] Clutter Tuning Experiment

Bistatic Angle, β (°)	Bistatic Range Cell, ΔR_B ($\approx$ Down-Range, m)	Bistatic Isorange Resolution, τ_i ($\approx$ Cross-Range, m)
2	3.8	2.4
40	4.0	2.7
80	4.9	4.1

Time synchronization to within a few microseconds (Section 13.4) was accomplished by observing the transmitter signal over the direct path (baseline) and using transponders to measure the baseline. Phase synchronization (Section 13.5) was achieved by matched quartz oscillators in each aircraft. Antenna alignment to the target area (6° and 5° transmitting and receiving antenna beamwidths, respectively) was by inertial navigation and radio communication.

Results of the bistatic SAR experiments are shown in Figures 2.12(a), (b), and (c). The images are of the Willow Run airport region near Ypsilanti, Michigan. The flight line was from left to right, across the top of the image. Belleville Lake is on the left with the I-94 freeway between the lake and the airport. Auterman reports that because all the images were (optically) processed identically, the β = 40° and 80° images show a reduced range swath (vertical axis) due to decreased bistatic footprint (intersection of the transmitting and receiving beams) and an increase in the bistatic range cell, ΔR_B. A similar effect occurs for the cross-range swath (horizontal axis). For comparison, approximate areas common to each image are shown in Figure 2.12(a), the near-monostatic SAR image. Note that the β = 40° and 80° areas extend below the β = 2° area.

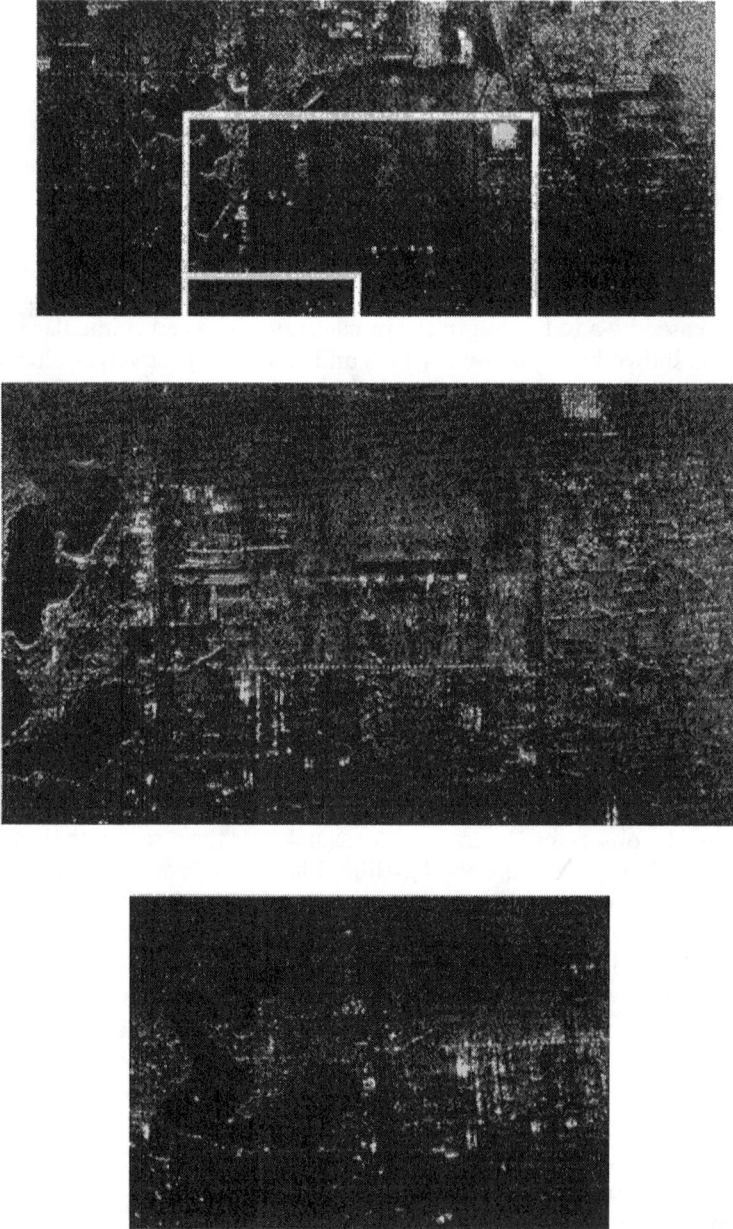

Figure 2.12 Imagery from bistatic SAR experiment: (a) $\beta = 2°$, (b) $\beta = 40°$, and (c) $\beta = 80°$. (© 1984 IEEE [53].)

Auterman [53] observes that when comparing total returns in each image,

There is little or no difference in the appearance of clutter with three bistatic angles and only minor differences in the appearance of cultural targets. The latter may be due to the symmetry of the mapping geometry combined with most structural targets being aligned with the flight line, resulting in equal angles of incidence and reflection for many surfaces.

An additional observation can be made by comparing areas common to each image. Specifically when Figure 2.12(b) ($\beta = 40°$) is compared to its common area in Figure 2.12(a) ($\beta = 2°$) more clutter and urban detail appears in the $\beta = 40°$ image than in the $\beta = 2°$ image. This effect can be caused by a smaller dynamic range of bistatic echoes in the image. In monostatic and small bistatic angle geometries, retroreflectors occur, particularly in urban terrain. They usually occur less frequently at larger bistatic angles (Chapter 8). A typical example is the square, grid-like return near the right-center of Figure 2.12(a) and again near the upper right-hand corner of Figure 2.12(b) ($\beta = 40°$). The return appears to be much stronger in Figure 2.12(a), suggesting that it is comprised of retroreflectors. Thus, if a monostatic SAR image has a 60-dB dynamic range, a bistatic SAR image might have a 40- to 50-dB dynamic image. Because the dynamic range of displays is typically 20 dB or less, only a small window in either the monostatic or bistatic SAR image can be displayed at one time. If the display window is set to capture the strongest returns, the setting will be 10–20 dB lower in a bistatic image than in a monostatic image, thus allowing returns of lower amplitude to be displayed in the bistatic image.

Little difference is evident in detail between Figure 2.12(c) ($\beta = 80°$) and its common area in Figure 2.12(b) ($\beta = 40°$), which suggests that the strongest returns are approximately equal, and thus the display window settings are about the same. Note that in Figure 2.12(c) the imagery appears somewhat more blurred, as would be expected because the bistatic isorange resolution for $\beta = 80°$ is degraded by a factor of 1.5 and the bistatic range cell is increased by a factor of 1.2 when compared to the $\beta = 40°$ imagery.

2.3.3 Hitchhiking

The concept of using a monostatic radar as the bistatic transmitter and a hitchhiking bistatic receiver was first implemented by Germany as the Klein Heidelberg in World War II. The transmitter can be either a cooperative or noncooperative signal source, including a communication transmitter, although a radar is usually preferred. Since World War II the concept has been reinvented for many applications, with reinvention denoting the fact that the current inventors were not aware of the German system. (The author was a co-reinventor of one concept.)

In one U.S. experimental program [54, 174] conducted in the early 1980s, a bistatic radar test bed called Bistatic Alerting and Cueing (BAC) (Figure 2.13) used the E-3A Airborne Warning and Control System (AWACS) and an emulation of the airborne Joint Surveillance and Target Attack Radar System (JSTARS) as cooperative transmitters. The mobile ground-based receiver detected and coherently processed returns from short-range air and moving ground targets and is shown in Figure 2.14. Its purpose, as the name suggests, was to alert and cue mobile, autonomous, short-range air defense and ground surveillance systems to improve their survivability and target acquisition performance. In this configuration, both site registration and data time-delay problems inherent in using remote data sources were eliminated. The receiver is also immune to an ARM attack. The airborne radar usually was located some distance behind the bistatic receiver, in a relatively more secure sanctuary position; thus, the bistatic receiver operated in an over-the-shoulder geometry. This geometry simplifies synchronization, location, and clutter suppression tasks. Field test results of the BAC concept have not been reported.

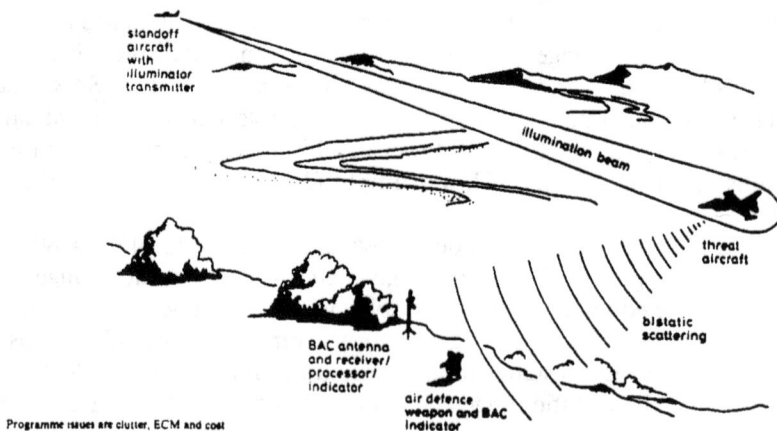

Figure 2.13 The BAC concept. (Courtesy of Cardif Publishing Co. [54].)

When the length of the baseline is greater than the monostatic radar's own detection range, the bistatic radar can operate to extend the monostatic radar's coverage. This concept is detailed in Section 12.1.1.

In the United Kingdom, the hitchhiking concept was extended to a commercial television station that served as a bistatic transmitter. Griffiths et al. [55] reported limited-funding air defense experiments using the Crystal Palace TV transmitters in London and a bistatic receiver located at the University College London, 11.8 km from the transmitter. The transmitter operates on four TV chan-

Figure 2.14 The BAC experimental receiving system. Large shelter contains the receiver and processor; small shelters contain primary power and power conversion equipment. (Courtesy of Technology Service Corporation.)

nels between 487 and 567 MHz with a total *effective radiated power* (ERP) of 1 MW over a 360° azimuth angle. However, the vertical plane radiation pattern is optimized for below-horizon coverage, with the main beam pointed downward by about 1°, and following a cosecant-squared pattern at lower angles. Furthermore, a -15-dB null in the pattern occurs at about 2.5° above the horizon. Thus, for aircraft flying above the transmitter's horizon, the available total ERP is reduced to "at best" 250 kW. Also, because the system was designed to operate on only one channel at a time, the ERP is further reduced by a factor of 4, to 62.5 kW.

The receivers were modified commercial TV tuners and IF circuits. One receiver used a 10-element Yagi antenna for time synchronization over the direct path. A second used four 17-element Yagis with 17-dBi gain for target reception. It was shielded by a building from the direct path and tilted upward to reduce ground clutter. On some tests, an azimuth rotator was used with a video camera on the mast to track targets manually. The predicted bistatic maximum range product, κ, Equation (4.1b), was ≈ 200 km^2. The equivalent monostatic range, $\sqrt{\kappa} \simeq$ 14 km. Receiver dynamic range was 48 dB, which was set by the eight-bit analog-to-digital (A/D) converters.

For the experiments, two TV test patterns that emulated a pulsed waveform were used. The first pattern was the line sync pedestal and pulse, the dominant feature of the TV signal, called "sync-pulse-white." It consists of a sync pulse, which amplitude modulates the carrier at a 100% level, separated in time by a pure white modulation at a 20% level. The autocorrelation function of this waveform has relatively high range sidelobes (≈ 5 dB). ambiguities every 9.6 km, and modest range resolution with a pulsewidth of 12 μs. However, it was considered adequate for short-range detection and two-pulse noncoherent MTI tests because it allowed simple time-domain processing, without the need for a coherent direct path reference. TV transmitter stability, 0.25 parts per million, was found to be more than adequate for the minimum required subclutter visibility of about 50 dB. Ground targets such as buildings were detected on an A-scope and commercial aircraft from Heathrow Airport were detected with the MTI canceller, although not consistently.

The second TV test pattern was multiburst, consisting of a black-and-white video picture containing narrow vertical stripes, which was transmitted over a single sideband of the carrier. The range resolution was improved by a factor of 2 over the sync-pulse-white waveform. Again, ground targets were detected, but air targets were not, probably due to at least a 7 dB lower SNR.

Griffiths *et al.* [55] provided the following analysis of the experiments:

> The experimental results proved negative for the most part, and it is necessary to account for this. Part of the explanation is due to the difficulties in capturing adequately long data records, but mostly it is because the dynamic range of the MTI cancellation system used (48 dB) is inadequate to cope with the high clutter/signal ratios of the quasi-CW radar system, which may be as much as 40 dB higher than this value. The positive results obtained occurred on the few occasions when the target aspect gave a high value of bistatic cross-section, the receive antenna direction coincided with that of the target and the data were satisfactorily captured.

They then concluded that:

> The bistatic radar function is dictated by the nature of the illuminating signal, in terms of frequency, modulation bandwidth, transmit power and antenna directionality. The simplest case is when a pulsed radar transmitter is available as the illuminator. Here the receiver signal processing will be similar to that of the illuminating radar, and the expected system performance can be readily calculated from standard expressions.
>
> The situation when the illuminator is not radar-like is more difficult. Of key importance is the autocorrelation function (and hence ambiguity function) of the transmitted waveform. . . .
>
> In most of these respects the television waveform is not ideal. The autocorrelation function for a TV waveform shows broad peaks at 64 μs intervals corresponding to the line sync pulses; if the sync pulses are gated out, the autocorrelation function peaks will be sharper (depending on the picture content), but will still recur at 64 μs intervals. In the case of any quasi-CW illu-

minator, the range sidelobe level will be high, and the radar best suited for Doppler rather than range measurements. The television transmit power is high and the azimuth coverage omnidirectional, but the elevation plane coverage is deliberately restricted.

The signal processing in the receiver is also dictated by the nature of the illuminating waveform. It is necessary either to know explicitly the form of this signal, or to receive separately an undistorted direct version of it. In the case of an unpredictable modulation scheme, such as ad hoc TV picture material or other broadcast signals, direct signal reception is mandatory.

In the case of quasi-CW waveforms, such as television transmissions, performance is severely limited by the unfavourable target/clutter ratio. . . . This ratio is improved if a higher receive antenna gain is employed, both because the target signal level is higher and because the clutter contribution is reduced by the narrower antenna beamwidth. . . .

In spite of the problems encountered, bistatic radar based on illuminators of opportunity has substantial attractions. While television transmissions are in several ways not ideal for this purpose, and require substantial processing to extract target echoes, a system of adequate dynamic range using real-time crosscorrelation would represent an intriguing prospect.

Bistatic radars using space-based transmitters and receivers that are either space-based, airborne, or ground-based have also been studied [3, 56–59]. In another hitchhiker configuration, limited field tests were conducted in the United States using communication satellites (NATO IIIb and DSCS II) as the CW transmitter and a ground-based coherent receiver to detect and process returns from commercial aircraft [58, 173]. Because the effective radiated power levels of the satellites were modest (40–47 dBW) and the transmitter-to-target ranges were large, target-to-receiver detection ranges were small, <4 km unless a very large receiving aperture was used.

2.3.4 Forward-Scatter Fences

A 5.8-GHz hybrid, monostatic-bistatic radar fence, called the Aircraft Security Radar (ASR), was developed and tested in the United States to protect military aircraft on the ground from intruders [60]. The bistatic mode was configured for near-forward-scatter operation, and used a variation of the beat frequency method to detect slowly moving targets. As reported by Walker and Callahan [60]:

The direct pulse from the transmitter and the echo pulse from the target are mixed in the receiver by a square law detector. The difference [beat] frequency (video) contains the target Doppler in the form of amplitude modulation of the envelope of the direct pulse. The sample video is doppler processed to reject stationary targets.

Five small, portable transmitter-receiver units were located around the aircraft with typical separations of 65 m, as shown in Figure 2.15. Each transmitter serviced an adjacent receiver. By time-gating the receiver to process only returns within 5 ns of the leading edge of the 6-ns direct path (transmitted) pulse, the bistatic mode constrained target detections to the highly eccentric ($e = 0.977$) ellipsoid, as shown in Figure 2.15. No spatial discrimination was needed or used; antennas were omnidirectional in azimuth. A transmitting-receiving unit is shown in Figure 2.16.

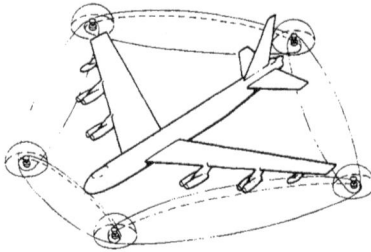

Figure 2.15 Aircraft security radars deployed around a B-52. (© 1985 IEEE [60].)

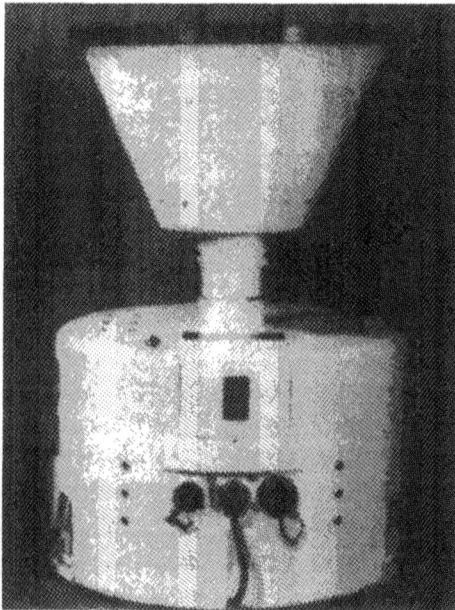

Figure 2.16 Aircraft security radar. (© 1985 IEEE [60].)

Each unit also operated in a monostatic "gap filler" mode to extend coverage around the unit where the bistatic ellipses become small—just the opposite of the configuration used in the DEW line, where the bistatic system (AN/FPS-23 Fluttar) filled in the low-altitude gaps between monostatic radars. A separate monostatic waveform, interleaved with the bistatic waveform, was used to detect moving targets between 2 and 3 m from the unit.

The bistatic mode was designed to reject target dopplers greater than 5 Hz in order to reduce false alarms from extraneous targets, such as birds and wind-blown debris. When these targets are on the extended baseline and near a transmitting or receiving site, they will generate 27 dB higher signal returns than when they are on an ellipse midway between the transmitter and the receiver, assuming an ellipse eccentricity of 0.977 (Section 4.7.2).

In field tests the radar detected moving targets at velocities as low as 2 cm/s. Higher speed targets, such as vehicles, generated bistatic target dopplers greater than the 5-Hz cutoff frequency. However, these targets were usually large enough to block the direct path signal—just as the Dorchester broke the Taylor and Young VHF beam in 1922—and this loss of synchronization was processed as a detection. The ASR system is reported to operate on 1 W of prime power [60]. Deployment plans have not been reported.

2.3.5 Multistatic Radars

The Multistatic Measurement System was installed at the U.S. Kwajalein Missile Range in 1980 [14]. Its purpose is to collect bistatic signature data at bistatic and target aspect angles as large as 130° and to perform high accuracy, range, and doppler tracking of ballistic missile re-entry vehicles [13]. The MMS uses two monostatic radars, the L-band TRADEX and the UHF ALTAIR on Roi-Namur Island, as illuminators and their monostatic receivers as one set of multistatic receivers. A second set of unmanned receivers is installed on Gellinam Island, about 40 km away, and a third unmanned L-band receiver is installed on Illeginni Island about 35 km away. The geometry is shown in Figure 2.17.

The monostatic radars acquire and track the target and point the high-gain (36 dBi at L-band and 24 dBi at UHF) multistatic receiving antennas. The remote stations receive the target returns, digitize their phase and amplitude, and transmit the data back to the master monostatic radar site via a microwave link. The system coherently integrates these returns in 100-ms intervals and then computes TDOA contours, i.e., range difference hyperbolas (Chapter 3) with accuracies of the order of 0.5 ns. These data are combined with monostatic data to compute the target trajectory. Doppler measurements are used to calculate the vector velocity and acceleration of the target.

Phase synchronization is accomplished by establishing a 10-MHz phase-locked loop over a two-way RF link between the master site and each remote site.

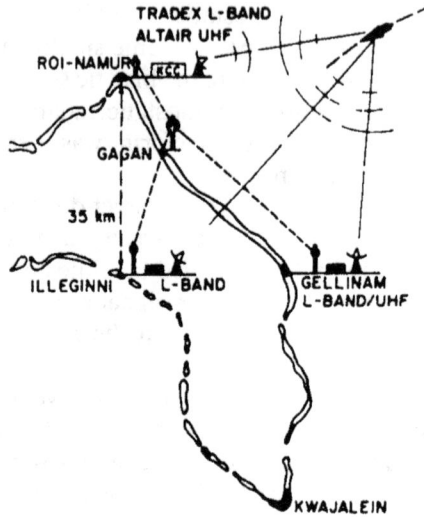

Figure 2.17 Overall layout of the MMS at the Kwajalein Missile Range. (© 1980 IEEE [13].)

Expected variations (≈ 20 ns) in atmospheric delays across the link are compensated to within about 0.2 ns, with phase stability ranging from 0.01 ns over minutes to 0.1 ns over hours. A similar dual-path compensation technique is used for time synchronization. In addition, time-delay variations over the remote-to-master site metric link, which transmits the TOA data, are compensated to within 0.5 ns by a dual-path technique using a calibration pulse from the TRADEX transmitter. Finally, TDOA measurements are calibrated by operating the MMS as an interferometer (Chapter 11) to cross-correlate radio star signals. Integration times to achieve TDOA measurement accuracies of 0.1 ns range from 100 to 1000 seconds.

Salah and Morriello [13] report that, "Adequate signal-to-noise on a low cross-section target is expected at [$R_T \approx R_R =$] 700 km to allow acquisition of bistatic data at the remote sites." They also project that the system will yield three-dimensional position and doppler estimates with 1σ rms accuracies better than 4 m and 0.1 m/s, respectively, throughout re-entry.

Other multistatic radar concepts have been studied. They include the Doppler Acquisition System (DAS), which uses multiple transmitters and receivers [61], and Distributed Array Radar (DAR) concepts, with large [3] and small [5] spatial separation between receiving sites. The DAS combines data from each site noncoherently; the DAR does so coherently.

With the exception of MMS, these second-resurgence military bistatic concepts appear to have not progressed beyond the research and development stage, principally because their added complexity and need for multiple sites impose difficult logistic and coordination requirements on the operational forces. As with other sensors, their future will be decided on the basis of performance and cost, with other systems competing for the same applications.

2.4 NONMILITARY APPLICATIONS

Bistatic radars have been analyzed, proposed, and in some cases developed for other than military applications, including

1. high-resolution imaging at short ranges (in the near field of the antennas) for use by robotics in an industrial environment [62];
2. airport ground vehicle and aircraft collision warning and avoidance using a baseband bistatic radar [63];
3. planetary surface and environment measurements using a satellite-based transmitter and an earth-based receiver [64–67], or a planet-based transmitter and a satellite-based receiver [68];
4. geological probing of horizontally stratified, underground layers from a transmitter and receiver on the surface, usually operating at frequencies from 100 to 1000 MHz [69];
5. Ocean-wave spectral measurements (wavelength, frequency, and direction of travel) using a Loran-A system [70];
6. detection and soundings of tropospheric layers, ionospheric layers, and high-altitude, clear air, atmospheric targets using ground-based sites [16, 71, 72].

An intriguing variant of the bistatic hitchhiking concept is a commercial system called "Spybuster," a small receiver carried in an automobile to warn the driver of police aircraft patrolling the highway for speeding vehicles. The police aircraft must be illuminated by a 1- to 2-GHz air traffic control radar and be within about six miles of the automobile, so that either the bistatic echo or the beacon response from the aircraft can be received by the Spybuster. (What is not clear from the sales brochure is which signal the Spybuster processes, and because industrial designs are often harder to obtain than military designs, the system would have to be reverse-engineered to resolve the question.) This system cost $299 (U.S.) in 1989.

Chapter 3

COORDINATE SYSTEMS, GEOMETRY, AND EQUATIONS

A two-dimensional, North-referenced coordinate system [73] is the principal coordinate system used throughout this book. Figure 3.1 shows the coordinate system and parameters defining bistatic radar operation in the plane containing the transmitter (T_x), receiver (R_x), and target (Tgt). It is called the bistatic plane [74]. The bistatic triangle lies in the bistatic plane. The distance L between the transmitter and receiver is called the baseline range, or simply baseline. The extended baseline is defined as continuing the baseline beyond either the transmitter or the receiver. The angles θ_T and θ_R are, respectively, the transmitter and receiver look angles, which are taken as positive when measured clockwise from North. They are also called angles of arrival (AOA) or lines of sight (LOS). The bistatic angle β is the angle between the transmitter and receiver with the vertex at the target. Note that $\beta = \theta_T - \theta_R$. It is convenient to use β in calculations of target-related parameters, and θ_T or θ_R in calculations of transmitter- or receiver-related parameters.

When the target lies above the baseline (and extended baseline), such that $-90° < \theta_T < +90°$ and $-90° < \theta_R < +90°$, it is said to be in the "northern hemisphere." The "southern hemisphere" is similarly defined for targets lying below the baseline. In general, bistatic radar operation and performance in the northern and southern hemispheres are equivalent for symmetric geometries. Obvious exceptions include limited scan capabilities for the transmitting or receiving antennas, terrain masking, and variations in bistatic RCS of a target as it moves from one hemisphere to the other.

A bistatic radar usually measures target range as the range sum, $R_T + R_R$. A contour of constant range sum, or isorange contour, is described by an ellipse, which is defined as the locus of points in which the sum of the distances from two fixed points is constant. When the two fixed points are the T_x and R_x sites, $R_T + R_R = 2a$, where a is the semimajor axis of the ellipse. In a monostatic radar, $R_T = R_R$ and its isorange contour is a circle of radius a, the limiting case for a bistatic isorange contour.

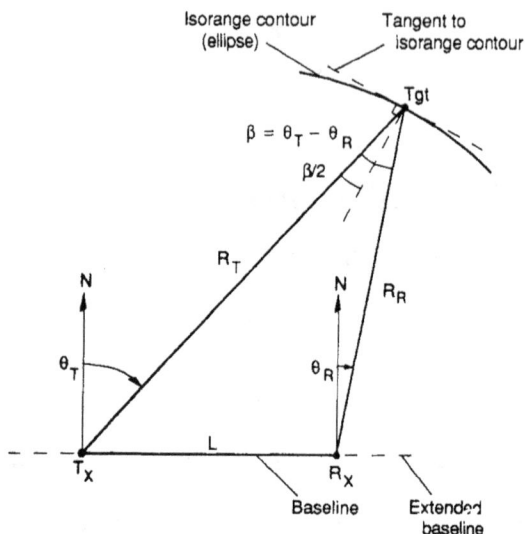

Figure 3.1 Bistatic radar North-referenced coordinate system in two dimensions.

An important relationship is that the bisector of the bistatic angle is orthogonal to the tangent of the isorange contour (ellipse) at any point on the bistatic isorange contour, as shown in Figure 3.1. The tangent is often a reasonable approximation to an isorange contour within the bistatic footprint, the area common to the transmitting and receiving beams. Development of the orthogonal bisector-tangent relationship is given in Appendix F.

The equation for an ellipse, with foci at T_x and R_x, in a rectilinear coordinate system with (x,y) axes located on the bistatic plane and origin located at the midpoint of the baseline (Figure 3.2) is

$$\frac{x^2}{a^2} + \frac{y^2}{b^2} = 1 \qquad (3.1)$$

where a is the semimajor axis of the ellipse and $b = (a^2 - L^2/4)^{1/2}$ is the semiminor axis of the ellipse. All ellipses having common transmitting and receiving site foci, and thus a common baseline, are by definition concentric and will satisfy the relationship:

$$a_i^2 - b_i^2 = L^2/4 \qquad (3.2)$$

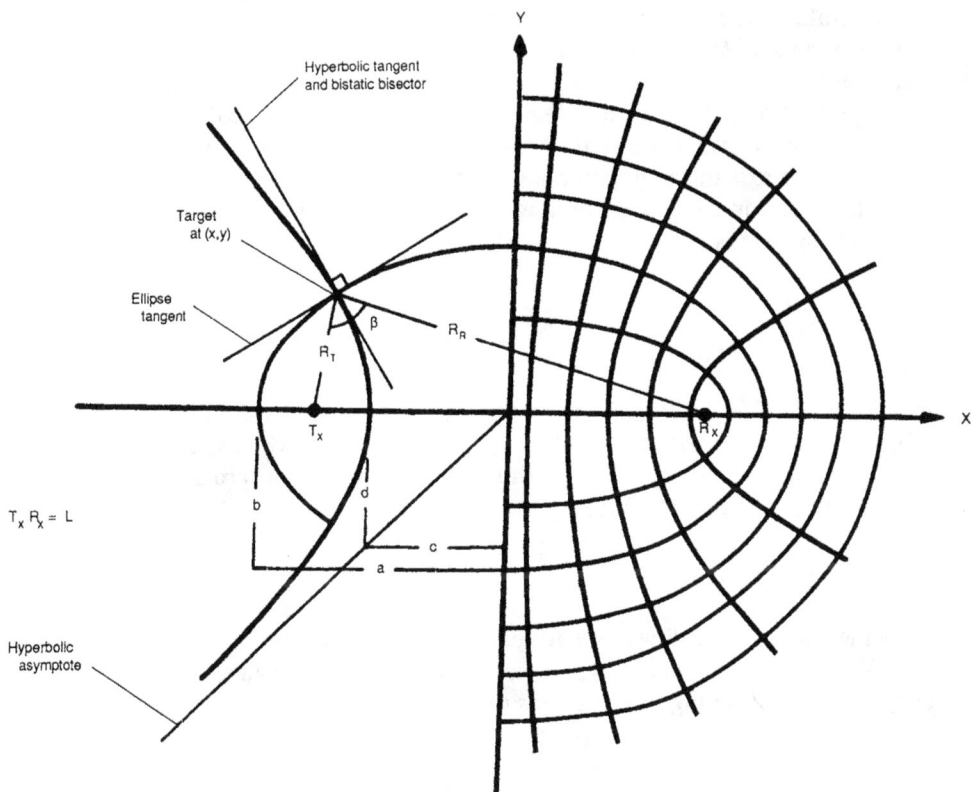

Figure 3.2 Concentric ellipses (isorange contours) and orthogonal, concentric hyperbolas in a rectilinear coordinate system, located on the bistatic plane.

where the subscript i defines axes for each concentric ellipse. Eccentricity, e, of the isorange contour or ellipse is

$$e = L/(R_T + R_R) \tag{3.3a}$$

$$= L/2a \tag{3.3b}$$

$$= \frac{(a^2 - b^2)^{1/2}}{a} \tag{3.3c}$$

Because $(R_T + R_R) \geq L$ for $L \geq 0$, the values of an ellipse's eccentricity lie in the range of $0 \leq e \leq 1$. When $L = 0$, $e = 0$, which is the monostatic case. Equation (3.3) is useful in simplifying expressions involving isorange contours.

A multistatic radar usually measures the target range as a range difference. A contour of constant range difference, or a TDOA contour, is described by a hyperbola, which is defined as the locus of points in which the difference of the distances from two fixed points is constant. When the two fixed points are the T_x and R_x sites, and a second receiver is located at the T_x site, $R_T - R_R = \pm 2c$, where c is the semitransverse axis of the hyperbola (Figure 3.2).

The equation for a hyperbola with foci at T_x and R_x in the same rectilinear coordinate system is

$$\frac{x^2}{c^2} - \frac{y^2}{d^2} = 1 \tag{3.4}$$

where $d = (L^2/4 - c^2)^{1/2}$ is the semiconjugate axis of the hyperbola. The ratio d/c defines the slope of the hyperbolic asymptote. All hyperbolas having common transmitting and receiving sites, and thus a common baseline, are again concentric, and will satisfy the relationship:

$$c_j^2 + d_j^2 = L^2/4 \tag{3.5}$$

where the subscript j defines axes for each concentric hyperbola.

When an ellipse and a hyperbola share common foci (and thus a common baseline), the combining of Equations (3.2) and (3.5) yields

$$a^2 - b^2 = c^2 + d^2 = L^2/4 \tag{3.6}$$

As developed in Appendix F, Equation (3.6) is also the condition for orthogonality. Specifically, the tangents of all concentric hyperbolas are orthogonal to the tangents of all concentric ellipses at their points of intersection when they share common foci (i.e., a common baseline L). A set of these orthogonal ellipses and hyperbolas is shown on the right side of Figure 3.2. Because the bisector of the bistatic angle at any point on an ellipse is also orthogonal to the tangent of the ellipse, it is colinear with the tangent of the orthogonal hyperbola. This relationship is useful in defining maximum bistatic target doppler (Section 6.1).

The bistatic angle is a maximum, β_{max}, when its bisector is also the perpendicular bisector of the baseline. This bisector is also the minor axis of the ellipse. In this case $\theta_T = -\theta_R$ and $R_T = R_R$. Thus,

$$\beta_{max} = 2 \sin^{-1}(L/2R_T) \tag{3.7a}$$

$$= 2 \sin^{-1}(L/2R_R) \tag{3.7b}$$

Combining Equations (3.3a) and (3.7) yields

$$\beta_{max} = 2 \sin^{-1} e \tag{3.8}$$

Because $\theta_T = -\theta_R$ at β_{max}, and $\theta_T - \theta_R = \beta$,

$$\theta_T = \sin^{-1} e \tag{3.9}$$

and

$$\theta_R = \sin^{-1} e \tag{3.10}$$

at β_{max}. In geometries where the target is near the baseline, β_{max} can approach 180°, with $\beta = 180°$ occurring when the target crosses the baseline at any point. This geometry is the forward-scatter region and is discussed in Chapter 8. From Equation (3.8), as $\beta_{max} \rightarrow 180°$, $e \rightarrow 1$. In general, when $\beta_{max} \gtrsim 140°$ and consequently $e \gtrsim 0.95$, many approximations used in estimating bistatic radar performance break down. For example, the approximation to a bistatic range cell, $\Delta R_B \approx c\tau/2 \cos(\beta/2)$, where τ is the compressed pulsewidth, breaks down for $e > 0.95$, with errors exceeding 45% for nearly all pulsewidths and baseline ranges of interest (Section 4.6 and Appendix B). Thus, either large values of β or large values of e should raise an "error flag" whenever they are encountered.

Occasionally bistatic system operation is established, or constrained, by a maximum or minimum permitted bistatic angle. For example, three distinct RCS regions are defined by the bistatic angle, each allowing a unique type of operation (Chapter 8 and Section 12.2). In these cases a plot of a constant β curve on the bistatic plane is required. Such a plot is called a constant β, or iso-β, contour. A family of these contours is shown in Figure 3.3 [73]. Each contour is a circle of radius r_β, where

$$r_\beta = L/(2 \sin\beta) \tag{3.11}$$

The center of the circle is located on the perpendicular bisector of the baseline, displaced from the baseline by a distance d_β, where

$$d_\beta = L/(2 \tan\beta) \tag{3.12}$$

and each circle passes through the transmitting and receiving sites. For large β these constant β contours are called *ogives*. The $\beta = 180°$ contour lies on the baeline; the $\beta = 0°$ contour lies on the extended baseline.

64

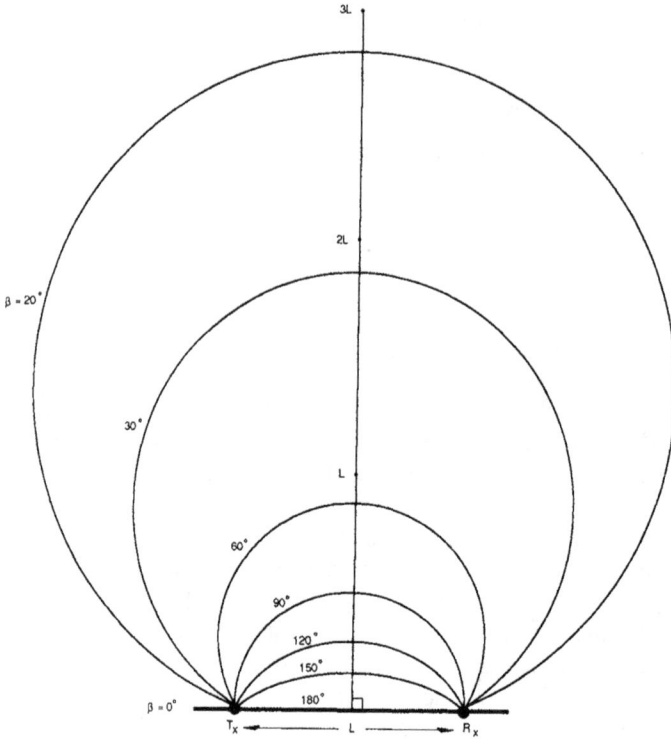

Figure 3.3 Contours of constant bistatic angle, or "constant β contours."

Other useful relationships among parameters on the bistatic plane are developed in Appendix E and summarized in Table 3.1.

As outlined in Chapter 1, geometry often distinguishes bistatic from monostatic radar operation. In these distinguishing cases, equivalent, or corresponding, monostatic operation is defined when the monostatic radar is positioned on the bisector of the bistatic angle [150]. For this geometry bistatic operation is often related to equivalent monostatic operation by the factor $\cos(\beta/2)$. For example, bistatic target doppler is reduced by $\cos(\beta/2)$ compared to equivalent monostatic doppler. The bistatic range cell, along with range, isorange range (synthetic aperture), velocity, and angle resolution, is increased (i.e., degraded) by $\cos(\beta/2)$ compared to equivalent monostatic range cell and resolution. The bistatic RCS of complex targets is equal to the monostatic RCS, but reduced in frequency by $\cos(\beta/2)$ when β is small. These $\cos(\beta/2)$ relationships are developed in subsequent chapters. In all cases bistatic operation collapses to monostatic operation by setting $L = 0$

Table 3.1
Relationships among Parameters on the Bistatic Plane

$$R_R = \frac{(R_T + R_R)^2 - L^2}{2(R_T + R_R + L \sin\theta_R)} \qquad (3.13a)$$

$$= \frac{L(1 - e^2)}{2e(1 + e \sin\theta_R)} \qquad (3.13b)$$

$$R_T = (R_R^2 + L^2 + 2R_R L \sin\theta_R)^{1/2} \qquad (3.14a)$$

$$= \frac{L(e^2 + 1 + 2e \sin\theta_R)}{2e(1 + e \sin\theta_R)} \qquad (3.14b)$$

$$R_T = \frac{(R_T + R_R)^2 - L^2}{2(R_T + R_R - L \sin\theta_T)} \qquad (3.15a)$$

$$= \frac{L(1 - e^2)}{2e(1 - e \sin\theta_T)} \qquad (3.15b)$$

$$R_R = (R_T^2 + L^2 - 2R_T L \sin\theta_T)^{1/2} \qquad (3.16a)$$

$$= \frac{L(e^2 + 1 - 2e \sin\theta_T)}{2e(1 - e \sin\theta_T)} \qquad (3.16b)$$

$$\frac{R_R}{R_T} = \frac{\cos\theta_T}{\cos\theta_R} = \frac{d\theta_T}{d\theta_R} = \frac{1 - e \sin\theta_T}{1 + e \sin\theta_R}$$

$$= \frac{1 - e^2}{1 + e^2 + 2e \sin\theta_R} = \frac{1 + e^2 - 2e \sin\theta_T}{1 - e^2} \qquad (3.17)$$

$$R_T = L \cos\theta_R/\sin\beta \qquad (3.18)$$

$$R_R = L \cos\theta_T/\sin\beta \qquad (3.19)$$

$$L = (R_T^2 + R_R^2 - 2R_T R_R \cos\beta)^{1/2} \qquad (3.20)$$

or $R_T = R_R$ and $\beta = 0°$ in bistatic equations. This rather mundane observation is surprisingly useful as a "sanity check" when working with bistatic radar geometry.

The North-referenced coordinate system is convenient for developing simple equations for bistatic geometry. When the bistatic radar's receiving antenna is a phased array, and the array normal is also normal to the baseline, θ_R is measured directly by the antenna in any bistatic plane. This fortuitous situation is caused by conic distortion, which is inherent in any phased-array antenna. However, when the array normal is not normal to the baseline, or when the receiving antenna is mechanically steered or scanned, θ_R is not measured directly. A mechanically steered antenna typically measures AOA in terms of azimuth and elevation angles, referenced to a three-dimensional (x-y-z) coordinate system, centered on the receiving site. A phased-array antenna may also be required to convert its conically distorted measurements to an (x-y-z) coordinate system for operation with other radars or systems. The conversion between AOA measurements in an (x-y-z) coordinate system and the North-referenced coordinate system, or bistatic plane, is developed in Section 5.3. Other coordinate systems have been developed for specific bistatic radar applications, and are documented elsewhere [16, 46, 48, 75, 76, 199, 209, 217]. A polar coordinate system with the (r,θ) coordinates located on the bistatic plane and origin located at the midpoint of the baseline can be used to plot ovals of Cassini (Section 4.3) and is shown in Figure 4.1.

Chapter 4
RANGE RELATIONSHIPS

The bistatic range equation is similar in form to the monostatic range equation and is derived in a similar process. The principal—and obvious—difference in the equations is that $R_T R_R$ replaces R_M^2, where R_T is the bistatic transmitter-to-target range, R_R is the bistatic receiver-to-target range, and R_M is the monostatic transmitter- and receiver-to-target range. This simple difference causes significant differences in monostatic and bistatic radar operation. One major difference is that bistatic thermal noise-limited detection contours are defined by ovals of Cassini, rather than by circles for the simplest monostatic case. These ovals are particularly useful in defining regions where the bistatic radar can operate.

A second difference is that bistatic constant range sum contours, defined by $R_T + R_R$ as ellipses, are not colinear with bistatic constant detection contours, defined by ovals of Cassini, whereas for the monostatic case they are both colinear circles. This difference sets operating limits for the bistatic radar, and also causes the target's S/N to vary as a function of its position on a constant range sum contour—unlike the monostatic case.

A third difference is that the width of a bistatic range cell changes as a function of the target's position on a constant range sum contour, or ellipse, whereas it is a constant value for the monostatic case. A fourth difference is that the bistatic transmitting and receiving pattern propagation factors, F_T and F_R, respectively, can be significantly different, whereas they are usually identical for the monostatic case, and only deviate when the monostatic radar employs separate transmitting and receiving antenna patterns or operates in a nonreciprocal propagation medium. The purpose of this chapter is to quantify these and associated differences.

4.1 RANGE EQUATION

The range equation [1, 3, 5, 15, 16, 77, 78] for a bistatic radar is derived in a manner completely analogous to that for a monostatic radar. With this analog, the bistatic radar maximum range equation can be written as

---

$$(R_T R_R)_{max} = \left[\frac{P_T G_T G_R \lambda^2 \sigma_B F_T^2 F_R^2}{(4\pi)^3 k T_s B_n (S/N)_{min} L_T L_R} \right]^{1/2} \qquad (4.1a)$$

or

$$(R_T R_R)_{max} = \kappa \qquad (4.1b)$$

where

R_T = transmitter-to-target range,
R_R = receiver-to-target range,
P_T = transmitter power output,
G_T = transmitting antenna power gain,
G_R = receiving antenna power gain,
λ = wavelength
σ_B = bistatic radar target cross section,
F_T = pattern propagation factor for transmitter-to-target path,
F_R = pattern propagation factor for target-to-receiver path,
k = Boltzmann's constant,
T_s = receiving system noise temperature,
B_n = noise bandwidth of receiver's predetection filter, sufficient to pass all spectral components of the transmitted signal,
$(S/N)_{min}$ = signal-to-noise power ratio required for detection,
L_T = transmitting system losses (>1) not included in other parameters,
L_R = receiving system loss (>1) not included in other parameters,
κ = bistatic maximum range product.

Equation (4.1a) is related to the corresponding monostatic maximum range equation by the following: $\sigma_M = \sigma_B$, $L_T L_R = L_M$, and $R_T^2 R_R^2 = R_M^4$. Similarly, (4.1b) is related to the corresponding monostatic maximum range equation by $(R_M)_{max} = \sqrt{\kappa}$. The term $\sqrt{\kappa}$ is sometimes called the *equivalent monostatic range,* which occurs when the transmitting and receiving sites are colocated [44].

Equation (4.1a) applies for all types of waveforms, CW, amplitude-modulated (AM), frequency-modulated (FM), or pulsed. More specific formulations of the maximum range equation, as given in Blake [136], also apply to the bistatic radar case. Equation (4.1a) is used because it more clearly illustrates the utility of constant S/N contours (ovals of Cassini) and other geometric relationships.

4.2 PATTERN PROPAGATION FACTORS

The transmitting and receiving pattern propagation factors, F_T and F_R, of (4.1a) are defined as $F_T = F_T' f_T$ and $F_R = F_R' f_R$, where F_T' and F_R' are the propagation factors and f_T and f_R are the antenna pattern factors.

The antenna pattern factors f_T and f_R are defined by Blake [136] as the relative strength of the free-space field radiated by an antenna as a function of pointing angles in the antenna's coordinate system. Pointing angles are defined in Section 5.3 (coordinate conversion) and typically are given in terms of an azimuth and elevation angle referenced to true north and a local vertical. These factors are applied whenever the target is not at the peak of the transmitting and receiving beams. (An additional case is considered for a target and a jammer in Chapter 10.)

The propagation factors F_T' and F_R' customarily include the effects of multipath, diffraction, and refraction, with atmospheric absorption effects included in the loss terms, L_T and L_R, of (4.1a). As with a monostatic radar, bistatic radar propagation requires a suitable path from the transmitter to the target and the target to the receiver. In contrast to a monostatic radar, however, propagation effects can be significantly different over the two bistatic paths and must be treated separately. Multipath is a typical example, where the target can be in a multipath lobe on one path and a multipath null on the other, depending on antenna and target altitude and terrain conditions.*

In the special case of signal propagation through the ionosphere, such as a space-based radar would experience, amplitude and phase scintillation as well as angle fluctuations will occur. This effect most often occurs at VHF and UHF frequencies, but if the ionosphere is highly disturbed, frequencies as high as 7–8 GHz can be affected [77]. Amplitude scintillation, or signal fading, is of major concern to detection performance. Dana and Knepp [77] estimate that in a highly disturbed ionosphere with a slowly fading signal (a large signal decorrelation time with respect to the coherent integration time), a bistatic radar by virtue of its separate propagation paths suffers a loss in detection performance of 7 dB, compared to a monostatic loss of 11 dB. For fast fading, both the bistatic and monostatic radars suffer additional, and roughly equal, losses depending on the ratio of signal decorrelation time to coherent integration time.

If propagation occurs through rain, the use of circular polarization by a monostatic radar can achieve 15–20 dB of rain cancellation compared to linear polarization [125]. In a bistatic radar, this level of circular polarization cancellation is achieved for bistatic angles up to 60° at frequencies below 10 GHz. The cancellation advantage diminishes as the bistatic angle increases beyond 60°, and at about 160° there is no cancellation advantage at any frequency [78].

---

*In a monostatic radar, separate pattern propagation factors are used principally to account for different antenna patterns, which, of course, also applies to a bistatic radar. These separate factors are also used to account for nonreciprocal propagation effects. Examples of nonreciprocal propagation are gyrotropic media, such as ferrite materials and the ionosphere [103].

4.3 OVALS OF CASSINI

The formal definition of an oval of Cassini is the locus of the vertex of a triangle when the product of the sides adjacent to the vertex is constant and the length of the opposite side is fixed [137]. When applied to the bistatic triangle of Figure 3.1, the vertex is at the target site, R_T and R_R are the sides adjacent to the vertex, and the baseline L is the length of the fixed, opposite side. Thus, for specific transmitting and receiving sites, which establish L, and for a constant bistatic maximum range product, κ, (4.1b) is the expression for a bistatic maximum range oval of Cassini.

It is sometimes instructive to plot ovals of Cassini as a function of available signal-to-noise power ratio, S/N, rather than just the ratio required for detection, $(S/N)_{min}$, used in (4.1a) and (4.1b). In this case (4.1a) and (4.1b) are modified simply by dropping the "max" and "min" designation for $(R_T R_R)$ and S/N, respectively, and solving for S/N. Thus,

$$S/N = \frac{K}{R_T^2 R_R^2} \tag{4.2}$$

where S/N is the signal-to-noise power ratio at range R_T and R_R, and

$$K = \frac{P_T G_T G_R \lambda^2 \sigma_B F_T^2 F_R^2}{(4\pi)^3 k T_s B_n L_T L_R} \tag{4.3}$$

The term K is called the *bistatic radar constant*, and is related to the bistatic maximum range product, κ, as follows:

$$K = \kappa^2 (S/N)_{min} \tag{4.4}$$

A convenient way to plot ovals of Cassini is in a polar coordinate (r, θ) system, as shown in Figure 4.1. Converting R_T and R_R to polar coordinates yields

$$R_T^2 R_R^2 = (r^2 + L^2/4)^2 - r^2 L^2 \cos^2 \theta \tag{4.5}$$

Substituting (4.5) into (4.2) yields

$$(S/N) = \frac{K}{(r^2 + L^2/4)^2 - r^2 L^2 \cos^2 \theta} \tag{4.6}$$

Equation (4.6) is plotted in Figure 4.2 for 10 dB $\leq$ S/N $\leq$ 30 dB and K arbitrarily set to $30L^4$.

$$R_R^2 = (r^2 + L^2/4) - rL\cos\theta$$

$$R_T^2 = (r^2 + L^2/4) + rL\cos\theta$$

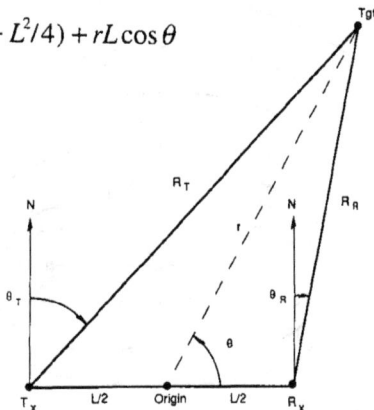

Figure 4.1 Geometry for converting North-referenced coordinates into polar coordinates (r,θ).

The ovals of Figure 4.2 are contours of a constant SNR on any bistatic plane. They assume that an adequate LOS exists on the transmitter-to-target path and receiver-to-target path, and that σ_B, F_T, and F_R are invariant with r and θ, which is usually not the case. However, this simplifying assumption is useful in understanding basic relationships and constraints. As S/N increases, the ovals shrink, finally collapsing around the transmitting and receiving sites. The same effect occurs when the baseline L increases: a constant S/N oval shrinks, again collapsing around the transmitting and receiving sites. The point on the baseline where the oval breaks into two parts is called the *cusp*. The oval is called a *lemniscate* (of two parts) at this S/N. At this point, $r = 0$, and from (4.6),

$$(S/N) = 16K/L^4 \tag{4.7}$$

Substituting (4.4) into (4.7) and assuming that $(S/N) = (S/N)_{min}$ for this lemniscate, yields

$$L = 2\sqrt{\kappa} \tag{4.8}$$

That is, a maximum range oval of Cassini breaks into two parts when $L > 2\sqrt{\kappa}$. It remains a single oval for $L < 2\sqrt{\kappa}$ and is a lemniscate for $L = 2\sqrt{\kappa}$. The maximum range oval of Cassini is plotted for any L by combining (4.1b) and (4.5). Note also that when $L = 0$, $R_T R_R = r^2$, which is the monostatic case where the ovals become circles of radius r.

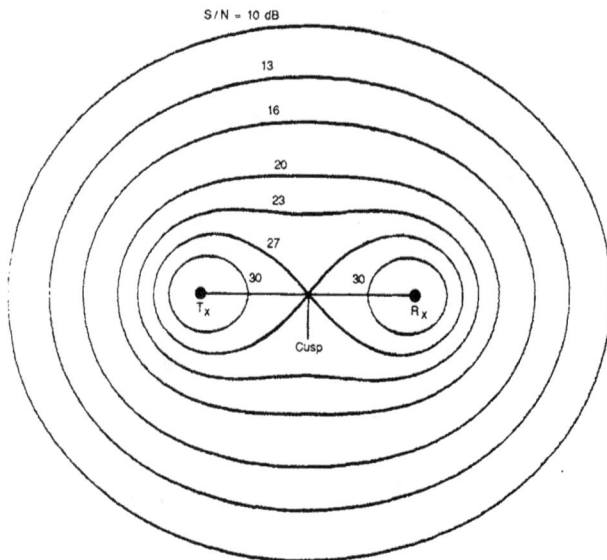

Figure 4.2 Contours of a constant SNR—ovals of Cassini, with $K = 30L^4$.

4.4 OPERATING REGIONS

Ovals of Cassini define three distinct operating regions for a bistatic radar: (1) the receiver-centered region, the small oval around the receiver in Figure 4.2: (2) the transmitter-centered region, the small oval around the transmitter; and (3) the receiver-transmitter-centered region, called the *cosite region,* any of the ovals surrounding both transmitter and receiver.

The receiver-centered region occurs when $L > 2\sqrt{\kappa}$ and $R_T \gg R_R$. It can be used for (1) air-to-ground operations using a standoff transmitter and a penetrating aircraft with a silient receiver; (2) semiactive homing missiles that lock-on-after-launch (LOAL); (3) short-range ground-based air defense or ground surveillance again with a standoff transmitter, sometimes called over-the-shoulder operation; and (4) obtaining an indication of target activity near an airborne receiving platform, which is also called *passive situation awareness.*

The transmitter-centered region occurs when $L > 2\sqrt{\kappa}$ and $R_R \gg R_T$. It can be used for (1) monitoring of activity near a noncooperative transmitter and (2) alerting of a missile launch from a hostile platform that illuminates the missile with its own (noncooperative) transmitter. Passive situation awareness can also use a noncooperative transmitter.

The cosite region occurs when $L < 2\sqrt{\kappa}$. It can be used for (1) medium-range air defense; (2) satellite or missile tracking from ground-based sites; (3) semiactive homing missiles that lock on before launch (LOBL); and (4) "tripwire" forward-scatter fences for detecting intruding aircraft or ground targets.

Critical to the selection of these operating regions is the value of the bistatic radar constant, K, that is available in a given scenario. Many of the terms in (4.3) are transmitter-controlled. Three transmitter configurations often control the operating regions: dedicated, cooperative, and noncooperative. The dedicated transmitter is defined as being under both design and operational control of the bistatic radar system; the cooperative transmitter is designed for other functions but found suitable to support bistatic operations and can be controlled to do so; and the noncooperative transmitter, while suitable for bistatic operations, cannot be controlled, for example, when it is a hostile or neutral transmitter. The bistatic receiver is sometimes said to "hitchhike" off a cooperative or noncooperative transmitter, usually a monostatic radar.

Table 4.1 summarizes useful bistatic radar applications permitted by operating regions and transmitter configurations. The two omitted entries on the row labeled "Transmitter Centered" are operational constraints: A dedicated or cooperative transmitter can usually gather nearby data in a monostatic radar mode easier than can a remote, bistatic receiver. The two omitted entries on the row labeled "Cosite" are technical constraints: To generate a sufficiently large bistatic radar constant (or maximum range product) for cosite operation, the transmitter design and operation must be optimized for bistatic radar use; hence, the dedicated transmitter is often the only viable cosite configuration. Exceptions to this rule include exploiting HF ground-wave propagation and occasional atmospheric ducting.

4.5 TARGET PATH DYNAMIC RANGE

Another method of characterizing constant S/N contours from (4.6) is shown on Figure 4.3, where S/N is plotted as a function of bistatic ranges, R_T and R_R, each normalized to the baseline range, L. Again, the bistatic radar constant, K, is arbitrarily set to $30L^4$, with σ_B, F_T, and F_R held constant. In this plot, the constant S/N contours become hyperbolas. Equivalent monostatic operation (in this case defined as monostatic radar constant equal to bistatic radar constant, and $R_M^4 = R_T^2 R_R^2$) is a line passing through the origin with slope $R_T/R_R = 1$. The monostatic radar is located at either the transmitting or receiving site.

Figure 4.3 can be used to determine the dynamic range required of a bistatic receiver to accommodate a given target path or trajectory. Target path dynamic range is defined as the difference between maximum and minimum S/N along the target path, where thermal noise, N, is assumed to be constant. For example, four

Table 4.1
Bistatic Radar Applications

Bistatic Radar Operating Regions	Range Relationships	Transmitter Configuration		
		Dedicated	Cooperative	Noncooperative
Receiver Centered	$L > 2\sqrt{\kappa}$ $R_T \gg R_R$	• Air-to-ground attack (silent penetration) • Semiactive homing missile (LOAL)[a]	• Over-the-shoulder operation • Passive situation awareness	• Passive situation awareness
Transmitter Centered	$L > 2\sqrt{\kappa}$ $R_R \gg R_T$	—	—	• Monitoring • Launch alert
Cosite	$L < 2\sqrt{\kappa}$	• Medium-range air defense • Satellite tracking • Range instrumentation • Semiactive homing missile (LOBL)[b] • Intrusion detection	—	—

[a]LOAL = Lock-on-after-launch.
[b]LOBL = Lock-on-before-launch.

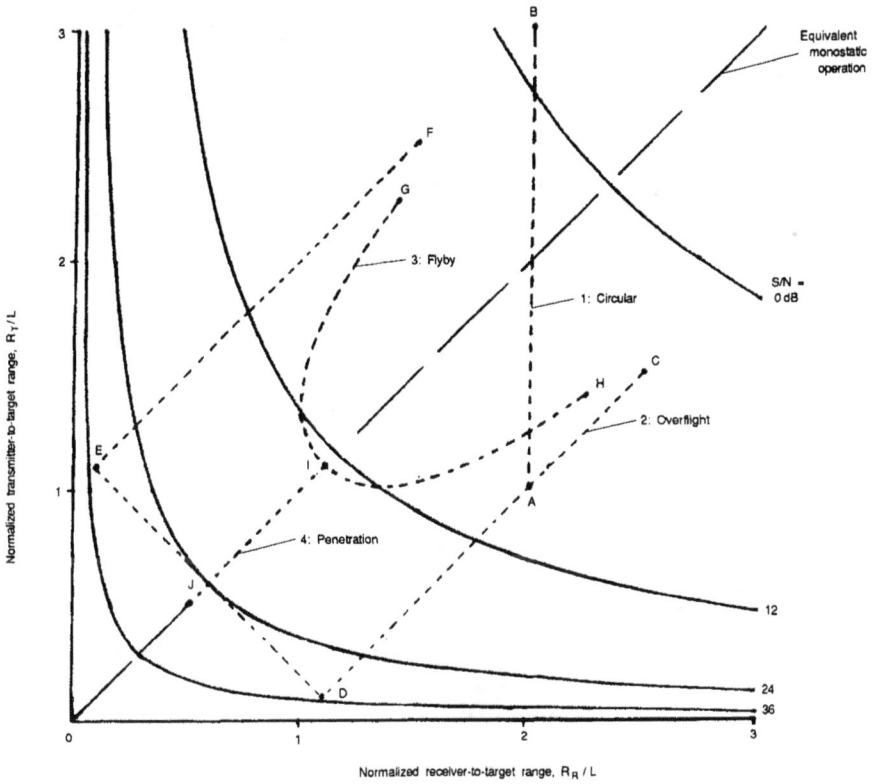

Figure 4.3 Contours of constant SNR—hyperbolas, with $K = 30L^4$.

target paths are shown in Figure 4.4, with their equivalent paths mapped as dashed curves onto Figure 4.3. In case 1 a circular path of $2L$ radius about the receiving site (Figure 4.4) is plotted as the line segment $\overline{AB}$ on Figure 4.3. The target path dynamic range is $8.8-(-0.8) = 9.6$ dB. A monostatic radar located at the receiving site has a constant S/N of 2.7 dB at all points on the path. Results for the remaining cases are shown in Figure 4.4. In general, the equivalent monostatic target path dynamic range is greater than the bistatic target path dynamic range, which in turn requires a greater monostatic receiver dynamic range. This requirement is sometimes given as a bistatic radar advantage. Note, however, that special target paths exist where the bistatic dynamic range is greater than (case 1) or equal to (case 4) the monostatic dynamic range.

	Equivalent Path on Figure 4-3	Bistatic Dynamic Range (dB)	Monostatic Dynamic Range (dB)
CASE 1 : Circular	Line segment $\overline{AB}$	9 6	0
CASE 2 : Overflight (4 L path length at 0.1 L altitude)	Line segments $\overline{CD} + \overline{DE} + \overline{EF}$	31 5	55 5
CASE 3 : Flyby (3 L path length, offset by L)	Parabola with end points G, H	7	14
CASE 4 : Penetration (2 L path length, bisecting baseline)	Line segment $\overline{IJ}$	14	14

(CASE 1 diagram: Target path, 2 L radius, T_x L R_x)

* Located at receiving site

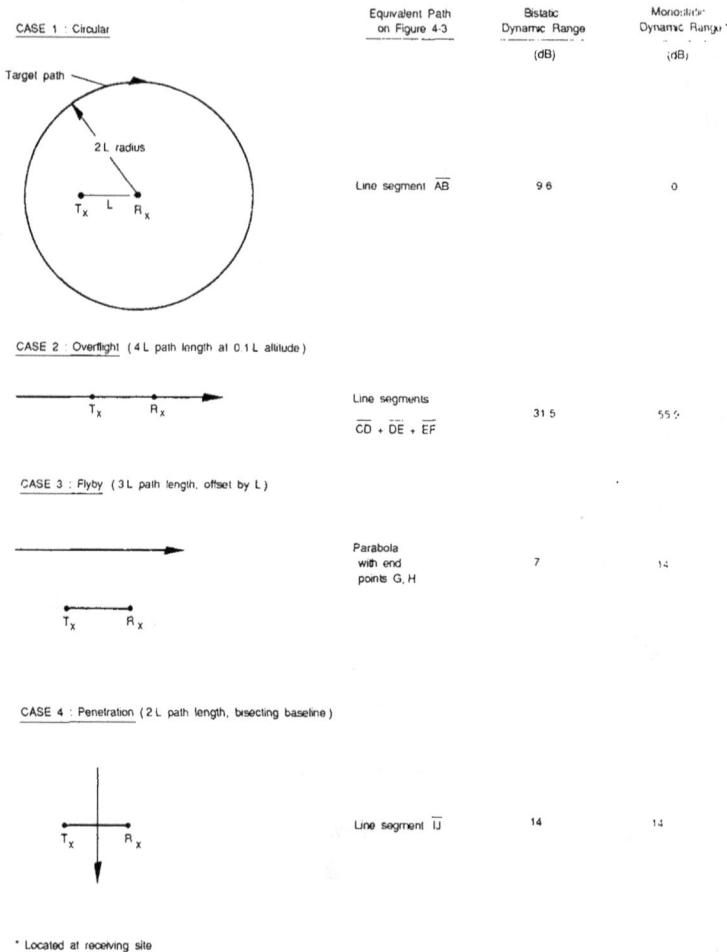

Figure 4.4 Target geometries for dynamic range examples with baseline range $= L$, bistatic $=$ monostatic radar constant $= 30L^4$.

4.6 ISORANGE ELLIPSOIDS, ISORANGE CONTOURS, AND RANGE CELLS

The transmitter-to-target-to-receiver range measured by a bistatic radar is the sum $(R_T + R_R) = 2a$. This sum locates the target somewhere on the surface of an ellipsoid, the foci of which are the transmitting and receiving sites, separated by the baseline L, and with major axis of $2a$. The ellipsoid is completely defined by the two parameters, a and L, and is the isorange ellipsoid of constant range sum $2a$.

A third measurement, usually orthogonal AOAs, for example, azimuth and elevation angles, is required to locate the target on the isorange ellipsoid. With the target location established, the bistatic plane—along with the bistatic triangle—is defined by the baseline and target position. The intersection of the bistatic plane with the isorange ellipsoid produces an ellipse of constant range sum, or *isorange contour*, as shown in Figure 4.5. Because the isorange contour is coplanar but not colinear with the corresponding oval of Cassini passing through the target position, the S/N changes for different target positions on the isorange contour. This variation in S/N is treated in Section 4.7.

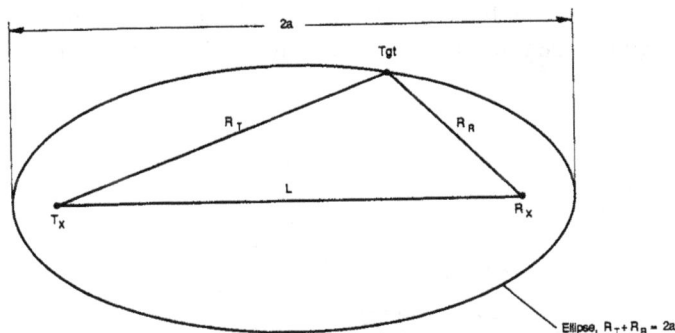

Figure 4.5 Ellipse of constant range sum, or isorange contour, with bistatic plane in the plane of the paper.

The monostatic analog to the bistatic isorange ellipsoid is a sphere of radius a (and $L = 0$). When the target is located on the sphere, any plane containing the target and monostatic radar position intersects the sphere to produce circles of constant range, the monostatic isorange contour. The analogy is not exact because an infinite number of equal monostatic isorange contours exist for each target position, whereas only one bistatic isorange range contour exists for each target position. The analogy is useful, however, to visualize the special bistatic geometry.

The separation, ΔR_M, between two concentric monostatic isorange contours establishes the width of a monostatic range cell,† and is

---

†A monostatic range cell is sometimes called a monostatic range *resolution* cell. The term range cell is used here to distinguish cell width from range resolution of closely spaced targets, which is treated in Chapter 7. This terminology also applies to the bistatic case.

$$\Delta R_M = c\tau/2 \qquad\qquad (4.9)$$

where

c = speed of light
τ = compressed pulsewidth.

For example, the inner monostatic isorange contour has radius a; the outer contour has radius $a' = a + c\tau/2$. Similarly, the width of a bistatic range cell is defined as the separation, ΔR_B, between two concentric bistatic isorange contours, or ellipses. The separation, however, between concentric ellipses varies depending on the target position. The separation is a minimum on the extended baseline and maximum on the perpendicular bisector of the baseline, as shown in Figure 4.6. Apparently a rigorous definition or development of this separation has not been published.

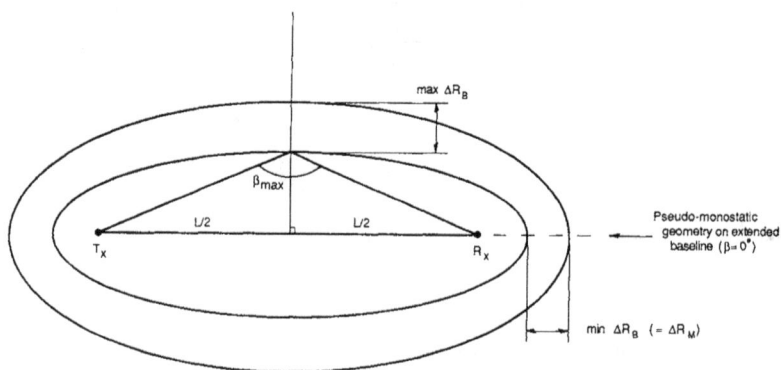

Figure 4.6 Maximum and minimum bistatic range cells.

A bistatic range cell, ΔR_B, is defined as the separation between two confocal ellipses, $(R_T + R_R) = 2a$ and $(R_T' + R_R') = 2a'$, $a' > a$ (both with foci at the transmitting and receiving sites with baseline L), where the separation is measured along the bisector of the bistatic angle β on the inner ellipse; Figure 4.7 shows the geometry. The pseudomonostatic range cell, ΔR_M, is defined as the separation between confocal ellipses on the extended baseline:

$$\Delta R_M = (a' - a) \qquad\qquad (4.10)$$

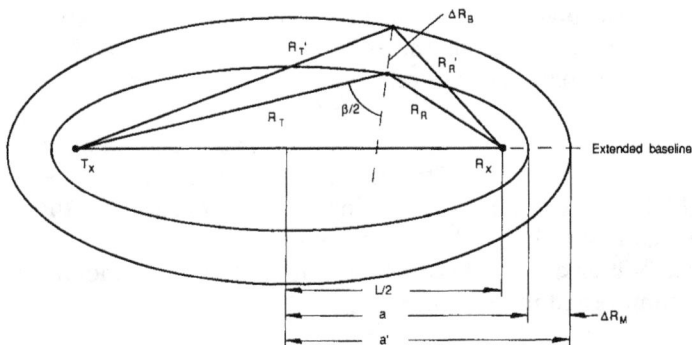

Figure 4.7 Geometry for bistatic range cell, ΔR_B.

which, from (4.9), $= c\tau/2$. Appendix B develops an exact, but implicit, expression for ΔR_B. An approximation to ΔR_B is

$$\Delta R_B = \Delta R_M/\cos(\beta/2) \tag{4.11a}$$

$$= \frac{c\tau}{2\cos(\beta/2)} \tag{4.11b}$$

The error between the exact and approximate expressions is also developed in Appendix B. The maximum error, ϵ_{max}, occurs on the perpendicular bisector of the baseline, where β is a maximum. It is given as

$$\epsilon_{max} = \frac{a(a' - a)}{b(b' - b)} - 1 \tag{4.12}$$

where

a = semimajor axis of inner ellipse,
a' = semimajor axis of outer ellipse,
b = semiminor axis of inner ellipse = $(a^2 - L^2/4)^{1/2}$,
b' = semiminor axis of outer ellipse = $(a'^2 - L^2/4)^{1/2}$.

Thus, when the baseline, L, the target range sum, $(R_T + R_R) = 2a$, and the compressed pulsewidth $\tau = 2(a' - a)/c$ are defined, ϵ_{max} can be calculated by (4.12). Appendix B shows that ϵ_{max} is positive; that is, the approximation to ΔR_B, (4.11), yields a separation that is greater than or equal to the exact expression for all β. (Equality occurs when $\beta = 0$ on the extended baseline.) Thus, (4.11) represents an upper bound to the bistatic range cell, ΔR_B.

This bound is reasonably "tight" for (1) an inner ellipse that is not too "flat" (i.e., not too eccentric), where $e = L/2a$, and (2) a pseudomonostatic range cell, ΔR_M, that is a small fraction of the baseline, L. For example, when $e < 0.5$ and $\Delta R_M < 0.1L$ the maximum error in using the approximation of (4.11) is less than 1.5%. Note that as $e \to 1$, $\beta \to 180°$, and from (4.11) $\Delta R_B \to \infty$. That is, the approximation to ΔR_B breaks down because the true separation cannot approach infinity for a finite compressed pulsewidth.

4.7 OPERATING LIMITS AND INSTANTANEOUS S/N DYNAMIC RANGE

For a given baseline, L, and bistatic maximum range product, κ, a maximum range oval of Cassini, or contour of constant $(S/N)_{min}$, is defined by (4.1b). Associated with this oval is a family of permissible isorange contours, depending on the target position on the oval, as shown in Figure 4.8.

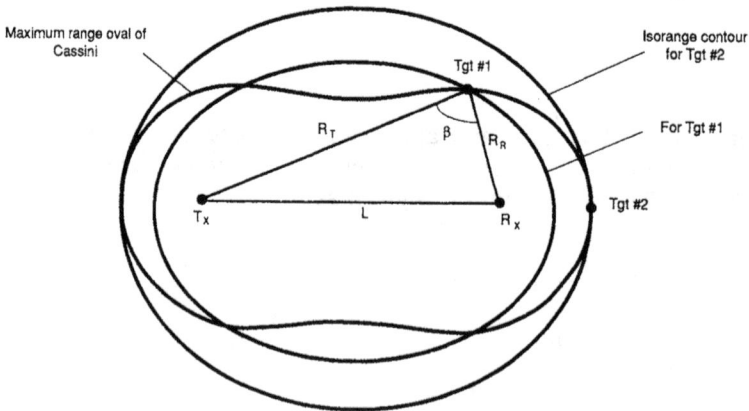

Figure 4.8 Isorange contours associated with an oval of Cassini.

An expression for the isorange contour permitted at an arbitrary target position on the oval, such as target 1 in Figure 4.8, is developed as follows. From the law of cosines:

$$L^2 = R_T^2 + R_R^2 - 2R_T R_R \cos \beta \qquad (4.13)$$

Solving for $(R_T + R_R)$, by completing the squares, yields

$$(R_T + R_R) = [L^2 + 2R_TR_R(1 + \cos \beta)]^{1/2} \tag{4.14}$$

Because the oval, defined as $(R_TR_R)_{\max} = \kappa$, and the isorange contour intersect at the target position, established by the bistatic angle, β, (4.14) becomes

$$(R_T + R_R)_{\max} = [L^2 + 2\kappa(1 + \cos \beta)]^{1/2} \tag{4.15}$$

Thus, a family of maximum range, confocal ellipses is defined by (4.15) for each target position defined by β on the maximum range oval of Cassini, (4.1b).

4.7.1 Operating Limits

Equation (4.15) establishes many operating limits for a bistatic radar, given κ and specific combinations of L and β; they are shown on Table 4.2. The pseudomonostatic point defines bistatic operation on either end of the extended baseline ($\beta = 0°$), or in practice near the extended baseline ($0° < \beta \lesssim 5°$). It is of interest because the bistatic radar operates much like a monostatic radar, except that $R_T \neq R_R$ for these geometries. The forward-scatter point defines bistatic operation on the baseline ($\beta = 180°$), or in practice near the baseline ($180° < \beta \lesssim 160°$), and is discussed in Section 12.2. The cusp defines the point at which the oval of Cassini breaks into two parts. For $L > 2\sqrt{\kappa}$, all bistatic angles are possible on the oval (and associated ellipses). For $L < 2\sqrt{\kappa}$, a maximum bistatic angle is permitted, which is shown as the last entry in Table 4.2. The equation $(R_T + R_R)_{\max} = 2\sqrt{\kappa}$ is common to many of these operating points. It is called the *characteristic ellipse* for a bistatic radar. The characteristic ellipse is a circle radius $2\sqrt{\kappa}$ for the monostatic case; a line of length $2\sqrt{\kappa}$ for the cusp; and an ellipse of major axis length $2\sqrt{\kappa}$ for the maximum β case.

4.7.2 Instantaneous S/N Dynamic Range

Because the constant range sum isorange contours and the constant S/N ovals of Cassini are not colinear, the target's S/N will vary for each target position on the isorange contour. This variation can be important when a bistatic radar uses range-gating; that is, when all target returns from a given range cell are processed. A range cell is defined by two confocal isorange contours (Section 4.6). The target returns are weighted by the transmitting-receiving beam patterns and doppler filters, if appropriate.

The variation in S/N on an isorange contour is developed as follows. Substituting (4.2) into (4.14) and solving for S/N yields

Table 4.2
Bistatic Radar Operating Limits

$\beta =$	$L =$	$(R_T + R_R) =$	Operating Limits
$0°$	0	R_T / R_R — T_X, R_X ... Tgt — $\boxed{2\sqrt{\kappa}}$	Monostatic ($R_M = \sqrt{\kappa}$)
$0°$	> 0	R_T^* / R_R^* — T_X ... R_X Tgt — $\boxed{\sqrt{L^2 + 4\kappa}}$	Pseudo-monostatic (max range sum on extended baseline)
$180°$	> 0	R_T R_R — T_X ... Tgt R_X — $\boxed{L}$	Forward scatter (min range sum on baseline)
$180°$	$2\sqrt{\kappa}$	R_T R_R — T_X ... Tgt R_X — $\boxed{2\sqrt{\kappa}}$	Cusp on baseline (also min range sum, where $R_T = R_R = \sqrt{\kappa}$)
$\cos^{-1}(1 - L^2/2\kappa)$ **	$< 2\sqrt{\kappa}$	Tgt R_T / R_R, β, T_X ... R_X, $\boxed{2\sqrt{\kappa}}$, Baseline bisector	Max β on oval where $R_T = R_R = \sqrt{\kappa}$

$\left.\begin{array}{l} * \; R_T = \sqrt{\kappa + L^2/4} \;\; \pm L/2 \\ * \; R_R = \sqrt{\kappa + L^2/4} \;\; + L/2 \end{array}\right\}$ paired maximum and minimum values for (R_T, R_R)

** or $2\sin^{-1}(L/2\sqrt{\kappa})$

$$S/N = \frac{4K \ (1 + \cos \beta)^2}{[(R_T + R_R)^2 - L^2]^2} \qquad (4.16)$$

where K is the bistatic radar constant. The denominator defines the isorange contour and β defines the target's position on the isorange contour.

This equation can be simplified by recalling from (3.2) that the square of the ellipse semiminor axis $b^2 = a^2 - L^2/4 = (R_T + R_R)^2/4 - L^2/4$. Substituting this expression into (4.16) yields

$$S/N = \frac{K}{4b^4} \ (1 + \cos \beta)^2 \qquad (4.17)$$

The $(S/N)_{max}$ occurs at $\beta = 0°$; thus,

$$(S/N)_{max} = K/b^4 \qquad (4.18)$$

The $(S/N)_{min}$ occurs at β_{max}, at the intersection of the perpendicular bisector of the baseline and the ellipse. From (3.8), $\beta_{max} = 2 \sin^{-1} e$. Thus,

$$(S/N)_{min} = (K/4b^4) \ [1 + \cos (2 \sin^{-1} e)]^2 \qquad (4.19)$$

The instantaneous S/N dynamic range on an ellipse, $\Delta(S/N)$ is defined as

$$\Delta(S/N) = (S/N)_{max}/(S/N)_{min} \qquad (4.20)$$

Substituting (4.18) and (4.19) into (4.20) yields

$$\Delta(S/N) = 4[1 + \cos (2 \sin^{-1} e)]^{-2} \qquad (4.21)$$

Equation (4.21) is plotted on Figure 4.9, and can be clarified with an example. Assume that a bistatic radar is configured for forward-scatter fence operation, Figure 4.10 and Section 12.2. Large bistatic angles are required for enhanced detection of target #1, which is located, for example, halfway between the transmitter and receiver, at $\beta_{max} = 145°$. At this β_{max}, the eccentricity of the isorange contour, or ellipse, is 0.954, from (3.8). The instantaneous S/N dynamic range, $\Delta(S/N)$ is 20.8 dB, from (4.21) or Figure 4.9.

Now the simultaneous return from target #2 of equal RCS, possibly discrete clutter, located at the intersection of the isorange contour and extended baseline will be 20.8 dB stronger than that of the intended target #1. The reason is simply that target #2 is on an oval of Cassini with a 20.8 dB higher S/N than that of target #1. Note that target #2 can be at either end of the extended baseline—near the receiving site or the transmitting site.

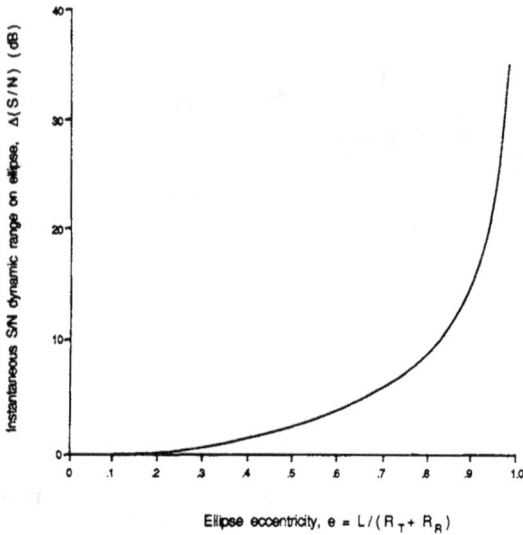

Figure 4.9 Ratio of maximum to minimum target S/N on an isorange contour, defined as an ellipse of eccentricity, e.

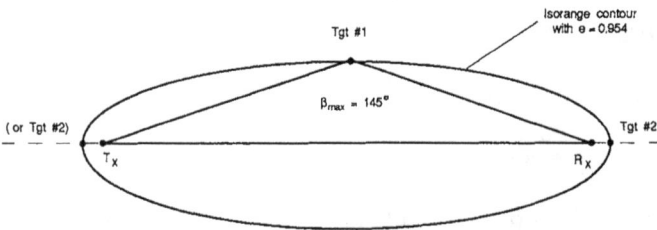

Figure 4.10 Forward-scatter fence configuration illustrating large target S/N dynamic range.

In contrast, range-gated monostatic radars process target returns on circular isorange contours, which are also contours of constant S/N. Thus, $e = 0$ and, from (4.21), $\Delta(S/N) = 0$ dB, as expected. Of course, the RCS of the second target can be higher than that of the first, which proportionally increases the instantaneous S/N dynamic range for both monostatic and bistatic radars. The return from the second target can be reduced by the receiving (or transmitting) antenna patterns and doppler filtering, but care must be taken to ensure that the attenuation is sufficient to prevent masking of the intended target, especially in the bistatic case.

Chapter 5
LOCATION AND AREA RELATIONSHIPS

With one exception the topics covered in this chapter are unique to bistatic radars. The exception is noise-limited errors associated with measuring the target range sum, doppler, and angle; these errors are the same as for monostatic measurements. For a bistatic radar, these measurements, along with a measure or estimate of transmitter position, are combined in various ways to solve the bistatic triangle and thus estimate the target position, or location, and the error associated with this location estimate. These topics are covered in Sections 5.1 and 5.2. Often, the measurements are taken in a local coordinate system centered at the receiving (or transmitting) site. These measurements usually must be converted to a transmitter-receiver-referenced coordinate system (the North-referenced coordinate system is used in this book) in order to establish target location, as discussed in Section 5.3. When bistatic range sum and an azimuth angle are displayed directly on a plan position indicator, which normally displays monostatic range versus azimuth angle, the display will be distorted, but can be corrected for most geometries, Section 5.4. Both sensitivity- and LOS-constrained coverage of a bistatic radar can deviate significantly from that of a monostatic radar, in which coverage is defined as the region or area on the bistatic plane where the target is "visible," i.e., detectable by the receiver and within LOS of both the transmitter and receiver. These altered coverage patterns often control bistatic radar operation, Section 5.5. Bistatic radar performance in clutter depends in part on the clutter cell area, which again can deviate significantly from that of a monostatic radar, depending on the bistatic geometry, Section 5.6. Geometry also controls the maximum unambiguous range and PRF. Further, in some bistatic geometries, the area, or volume, common to the bistatic transmitting and receiving beams will be small, which can cause beam scan-on-scan problems (Section 13.1), but also can allow the bistatic radar to operate with a higher pulse repetition frequency before encountering range ambiguities (Section 5.7).

5.1 TARGET LOCATION

Target position from the receiving site (θ_R, R_R) is usually required in a bistatic radar. The receiver-to-target range, R_R, cannot be measured directly, but can be calculated by the bistatic receiver by solving the bistatic triangle of Fig. 3.1. A typical solution [1] in the North-referenced coordinate system (on the bistatic plane) is developed in Appendix E as:

$$R_R = \frac{(R_T + R_R)^2 - L^2}{2(R_T + R_R + L \sin\theta_R)} \tag{5.1}$$

This solution first requires establishing transmitter location with respect to the receiver. Transmitter coordinates can be sent to the receiver by a dedicated or cooperative transmitter. Alternatively, transmitter location can be estimated by an emitter location system, usually colocated with the receiver. The receiver then establishes the transmitter-receiver baseline, which in turn provides an estimate of the baseline range, L, and establishes the North-referenced coordinate system. The receiver look angle, θ_R, is either measured directly or target azimuth and elevation measurements are converted to θ_R, as in Section 5.3. Beam-splitting techniques can be used to increase the measurement accuracy.

The range sum $(R_T + R_R)$ can be estimated by two methods, using the timing sequence in Figure 5.1(a). In the direct method of Figure 5.1(b), the receiver measures the time interval, ΔT_n, between reception of the transmitted pulse and reception of the target echo. It then calculates the range sum as

$$(R_T + R_R) = c\Delta T_n + L \tag{5.2}$$

This method can be used with any transmitter configuration, given an adequate LOS between transmitter and receiver.

In the indirect method of Figure 5.1(c), synchronized stable clocks are used by the receiver and a dedicated transmitter. The receiver measures the time interval ΔT_u between transmission of the pulse and reception of the target echo. It then calculates the range sum as

$$(R_T + R_R) = c\Delta T_u \tag{5.3}$$

A transmitter-to-receiver LOS is not required for this measurement. However, a LOS will be required if periodic clock synchronization is implemented over the direct path (Section 13.4), or if the receiver is required to measure the baseline directly.

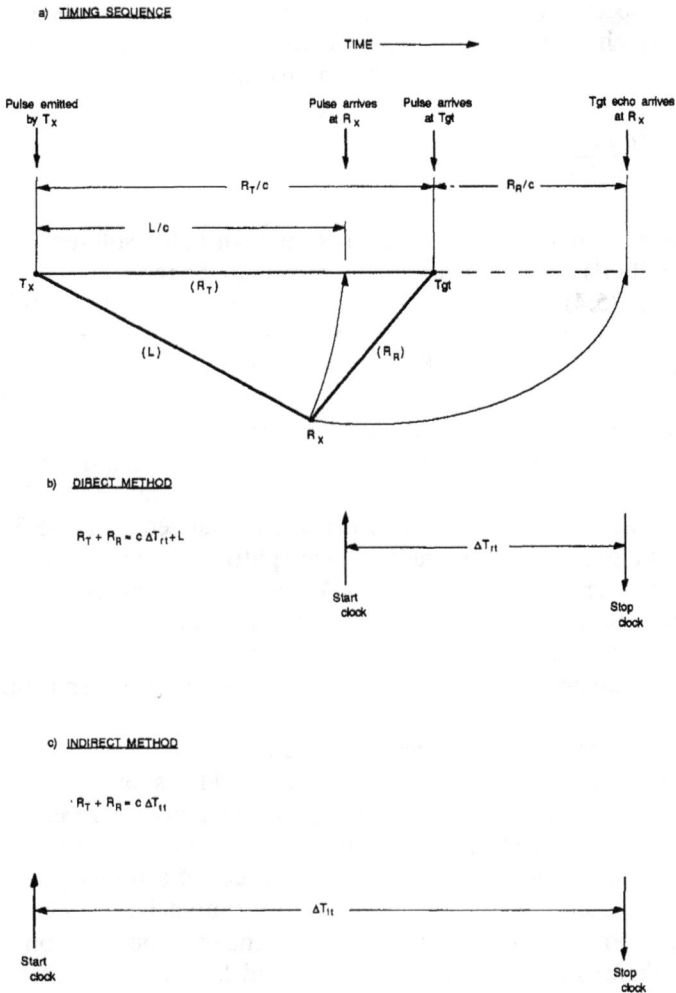

Figure 5.1 Timing diagram for two methods of calculating range sum, $R_T + R_R$.

For both the direct and indirect methods, (5.1) is valid for all target locations, except when the target lies on the baseline, between the transmitter and receiver. In this case $(R_T + R_R) = L$, $\theta_R = -90°$, and R_R is indeterminate. That is, the transmitted pulse and target echo arrive simultaneously at the receiver, and R_R cannot be measured.

For the special case of a bistatic radar using the direct range sum estimation method, in which the range sum $(R_T + R_R)$ is only slightly greater than the baseline L, Appendix C shows that (5.1) can be approximated as

$$R_R \simeq \frac{c\Delta T_n}{1 + \sin\theta_R} \tag{5.4}$$

This approximation does not require an estimate of L, but still requires an estimate of the transmitter's angular position with respect to the receiver in order to measure θ_R. The error in (5.4) is less than 10% for $0° < \theta_R < 180°$, and $L \geq 0.82(R_T + R_R)$, or from (5.2), $L \geq 4.56c\Delta T_n$. The error grows rapidly for $\theta_R < 0°$. This approximation is useful when the bistatic receiver is operating in a short-range over-the-shoulder geometry.

In a simple variant of (5.1), the bistatic transmitter look angle, θ_T, can be used in place of θ_R [48]. The transmitter-to-target range, R_T, can also be calculated by a method similar to that for R_R, using either θ_T or θ_R measurements. These location equations are developed in Appendix E and summarized in Table 3.1. When the transmitter is also a monostatic radar, beam-splitting techniques can again be used to increase the accuracy of θ_T estimates. When the transmitter simply illuminates the target, θ_T accuracy is reduced to the order of the transmitting beamwidth, which in turn degrades target position estimates, in terms of θ_T, R_R, or R_T. As a consequence, a θ_R measurement is usually preferred, where beam splitting can nearly always be invoked.

Many other target location techniques are possible [15, 16, 18, 81, 181, 217]. Doppler-only measurements can be used. Skolnik [15] shows that a minimum of five sequential doppler frequency measurements by a bistatic receiver will locate a target's range, azimuth, and elevation coordinates. (The doppler location equation given in [15] has five unknown variables.) Because of symmetry, however, using only doppler measurements, a target trajectory cannot be distinguished from its mirror image about a vertical plane that bisects the baseline. Dawson [61] provides similar calculations for a single transmitter and five receiving sites. The system operates coherently between each transmitting-receiving pair, but noncoherently between receiving sites. Dawson's location equations are summarized by Caspers [16].

In the special case of a target on a ballistic trajectory or in orbit, Patton [181] provides a multistatic solution using doppler-only measurements. His computations consist of "improving approximations for initial position and velocity components by successive differential corrections which are obtained from a least squares treatment of an overdetermined system of condition equations while imposing elliptic [or ballistic] motion as a constraint." Patton reports that in field tests a relatively accurate set of orbital parameters can be generated by the

DOPLOC system, Section 2.3, with as little as 1.5 to 3 minutes of intermittent observations from a single (bistatic) receiving site.

In a method analogous to estimating AOA of a target by measuring the phase difference between widely spaced antennas, as in an interferometer antenna, Skolnik [15] shows that two closely spaced CW frequencies can be used to estimate the range sum, $R_T + R_R$. Long pulsewidths can also be used in place of CW waveforms. This development is an extension of the monostatic two-frequency technique described by Hansen in Section 5.7 of Ridenour [186]. The range measurement accuracy is improved compared to a pulsed system with a bandwidth that is equal to the separation between the two frequencies. (In the monostatic case, the improvement is a factor of $\pi \sqrt{2}$.) Like the interferometer system, however, the range measurements are ambiguous, unless three or more frequencies are used. Ultimately, performance of the pulsed- and multiple-frequency techniques converges, as summarized by Skolnik [15]:

> Both the interferometer and the multi-frequency distance-measuring technique fail if more than one target is present, unless the targets can be resolved by some other means, such as by differences in their Doppler frequencies or differences in angle. The accuracy of an angle measurement with an interferometer depends upon the distance between the extremes of the antenna. The space in between may be empty as it is in a two-element interferometer. However, if two or more targets are to be resolved from one another, the entire aperture must be filled in, as with a closely spaced array antenna or a reflector antenna. Similarly, in the multiple-frequency technique, the accuracy is determined by the difference between the largest and the smallest frequency. Additional frequencies must be added in between if the measurement is to be unambiguous. The entire spectrum must be continuous if targets are to be both resolved and unambiguous.

Range sum location equations based on two-frequency measurements are given in the appendix to [15].

A "theta-theta" location technique uses the angles θ_T and θ_R and an estimate of L, where θ_T is typically provided by a monostatic radar, which acts as a cooperative bistatic transmitter.* One variant of this technique uses angle measurements from two receiving sites in a multistatic configuration; SPASUR, discussed in Section 2.3, operates this way. A second variant of this technique uses the angles θ_T and θ_R and a measure of ΔT_{rt}, (5.2). Combining (5.2) with (3.18) and (3.19) and solving for R_R yields

$$R_R = \frac{c\Delta T_{rt} \cos\theta_T}{\cos\theta_T + \cos\theta_T - \sin(\theta_T - \theta_R)} \tag{5.5}$$

---

* $R_R = L\cos\theta_T / \sin(\theta_T - \theta_R)$

where $\theta_T - \theta_R = \beta$. The baseline, L, can be calculated in a similar manner [217]. This technique might be used by a bistatic hitchhiker. When the transmitter is a cooperative monostatic radar, θ_T can be sent to the bistatic receiver directly. When θ_T is not made available, but the monostatic radar operates with a predictable antenna scan pattern, the bistatic receiver can estimate θ_T with some loss in target location accuracy. In general, (5.1), which uses a measure of L in place of θ_T for the calculation of R_R, is a preferred approach.

A hyperbolic measurement system can also be used for target location. In this system a receiver measures the difference in propagation times from two separate transmitters. The loci of target positions now lie on a hyperbola, and the intersection of the receiver's AOA estimate with the hyperbola establishes the target position. Use of a third transmitter provides a full hyperbolic fix on the target. Coarse target velocity vector estimates can be made using a grid of forward-scatter trip wire fences, such as the French *"maille en Z"* system. Equations are developed in Section 12.2.2 and require that the target velocity vector does not change while the target passes through the grid. As with a monostatic radar, when successive data measurements (or redundant data) are available to a bistatic or multistatic radar using any of these location techniques, target state estimates can be made with Kalman or other types of filters [80, 81, 209, 210].

5.2   MEASUREMENT AND LOCATION ERRORS

Errors in measuring the target range sum, $R_T + R_R$ (or time delay), target doppler, f_{Tgt}, and receiver look angle, θ_R, are developed in a process identical to that of a monostatic radar. In general, the root-mean-square (rms) error, dM, of a monostatic or bistatic radar measurement can be expressed as

$$dM = \frac{\eta M}{\sqrt{2S/N}} \tag{5.6}$$

where S/N is the signal-to-noise ratio available for the measurement and η is a constant whose value is of the order of unity. For a time-delay measurement, η depends on the shape of the frequency spectrum and M is the rise time of the pulse; for a doppler frequency measurement, η depends on the shape of the time waveform and M is the reciprocal of the observation time; and for an angle measurement, η depends on the shape of the aperture illumination and M is the beamwidth [1]. These errors are treated extensively for the monostatic case, for example, by Skolnik [1], and Barton and Ward [202] and also apply to the bistatic case.

Combining measurements into a target location estimate, and estimating errors associated with it, can be significantly more complicated for a bistatic radar because the bistatic measurements often are not orthogonal. A typical example is range sum and receiver look angle, which are orthogonal in a monostatic radar, but only occasionally so in a bistatic radar, for example, when the target is located on the extended baseline. Target location errors associated with nonorthogonal measurements have been developed for specific applications [203–206], including bistatic [209] and multistatic [210] radars.

As outlined in Section 5.1, receiver-to-target range R_R is often required from a bistatic radar, but this parameter is not measured directly. Only the range sum $R_T + R_R$ is measured, with a typical solution for R_R given by (5.1). This section develops the geometry-dependent errors associated with (5.1) and provides examples for typical geometries.

Equation (5.1) is a function of the receiver look angle, θ_R, range sum, $R_T + R_R$, and baseline, L, measurements, which are assumed to be independent. That is, the errors associated with these measurements are uncorrelated, which is usually the case. Further, assume that (1) the S/N is large in each measurement with receiver noise setting the value for N and (2) the magnitude of the measurement error is the rms value of the difference between the measured value and the true value, with bias errors removed. Restated another way, the measurement errors are assumed to be uncorrelated, zero-mean random processes, having a Gaussian distribution with standard deviation equal to the rms measurement error.

In estimating the receiver-to-target range, R_R, the root-sum-squared (rss) error, dR_R, is

$$dR_R = \left\{ \left[\frac{\partial R_R}{\partial (R_T + R_R)} \, d(R_T + R_R) \right]^2 \right. $$
$$\left. + \left(\frac{\partial R_R}{\partial L} \, dL \right)^2 + \left(\frac{\partial R_R}{\partial \theta_R} \, d\theta_R \right)^2 \right\}^{1/2} \quad (5.7)$$

where $d(R_T + R_R)$, dL, and $d\theta_R$ are the rms errors in measuring $(R_T + R_R)$, L, and θ_R, respectively, and are established by (5.6). The unique bistatic characteristic in this analysis is the dependence of the R_R estimate on $R_T + R_R$, L, and θ_R measurements, and thus the contribution of their partial derivatives to the rss error estimate of R_R. Now

$$\frac{\partial R_R}{\partial (R_T + R_R)} = \frac{(R_T + R_R)^2 + L^2 + 2L(R_T + R_R)\sin\theta_R}{2(R_T + R_R + L\sin\theta_R)^2} \quad (5.8a)$$

$$= \frac{1 + e^2 + 2e\sin\theta_R}{2(1 + e\sin\theta_R)^2} \quad (5.8b)$$

$$\frac{\partial R_R}{\partial L} = -\frac{[L^2 + (R_T + R_R)^2]\sin\theta_R + 2L(R_T + R_R)}{2(R_T + R_R + L\sin\theta_R)^2} \tag{5.9a}$$

$$= -\frac{(e^2 + 1)\sin\theta_R + 2e}{2(1 + e\sin\theta_R)^2} \tag{5.9b}$$

$$\frac{\partial R_R}{\partial \theta_R} = -\frac{L[(R_T + R_R)^2 - L^2]\cos\theta_R}{2(R_T + R_R + L\sin\theta_R)^2} \tag{5.10a}$$

$$= -\frac{L(1 - e^2)\cos\theta_R}{2(1 + e\sin\theta_R)^2} \tag{5.10b}$$

where the isorange contour (ellipse) eccentricity $e = L/(R_T + R_R)$ and the partial derivatives represent the slope of the error surface in the plane of the respective variable.

Before proceeding with an evaluation of dR_R, (5.7), analyzing each of these error slopes is useful. Figures 5.2 and 5.3 plot error slopes for the range sum, $R_T + R_R$, and baseline L error contributions to the dR_R estimate, as a function of θ_R, and three values of e. Figure 5.4 is a similar plot of the receiver look angle θ_R error contribution, but normalized to the baseline, L.

Two trends are evident in these figures. First, when $\theta_R = +90°$, the error slopes in each figure converge to a constant value for all e. A similar convergence occurs for $\theta_R = -90°$ and $e < 1$. (The $\theta_R = -90°$, $e = 1$ point is the forward-

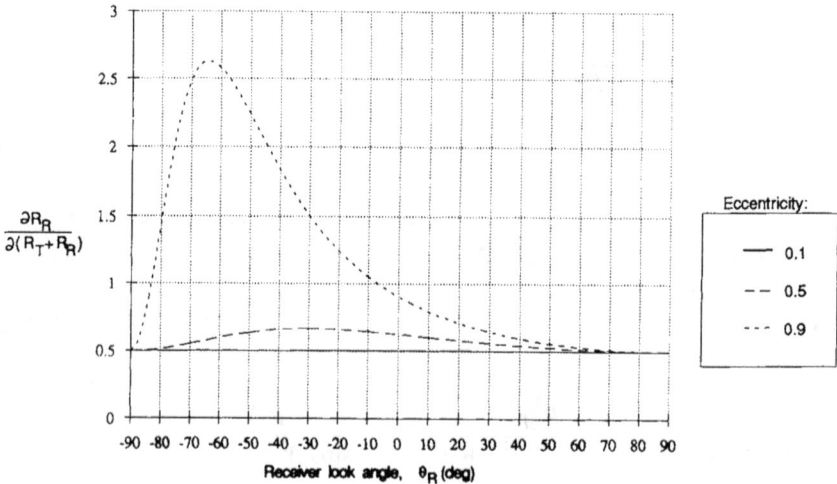

Figure 5.2 Range sum error slope or differential error contribution to receiver-to-target range, R_R, estimate, (5.8).

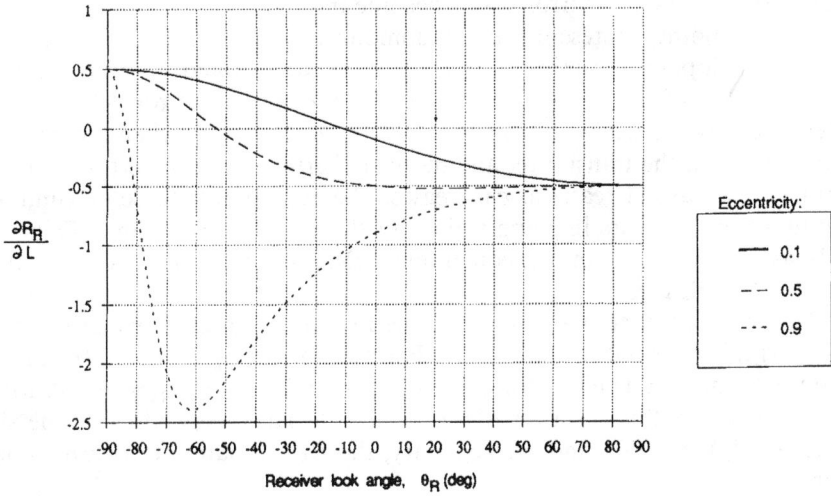

Figure 5.3 Baseline error slope or differential error contribution to receiver-to-target range, R_R, estimate, (5.9).

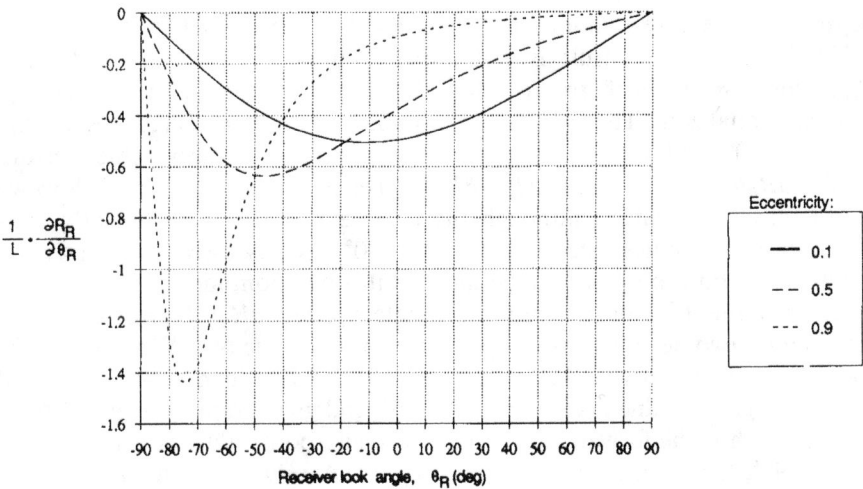

Figure 5.4 Normalized receiving look angle error slope or differential error contribution to receiver-to-target range, R_R, estimate, (5.10).

scatter case, in which the error slopes become infinite or indeterminate.) These θ_R = $\pm 90°$ end points represent the pseudomonostatic case, in which the $R_T + R_R$ and θ_R error slopes are a minimum. Second, maximum and minimum values of the error slopes occur for $-90° < \theta_R < +90°$. In general, the error slope maximum (either positive or negative) occurs for $-90° < \theta_R < 0°$, i.e., when the target is in a region between the transmitter and receiver. Furthermore, the error slope maximum nearly always increases as e increases. The exception to these maximum and minimum trends occurs in some regions of the baseline error slope (Figure 5.3), which, as will be shown in subsequent examples, is not usually a significant contributor to the R_R error.

Note that at some fixed θ_R values the magnitude of a particular error slope can be larger for a smaller eccentricity. For example, at $\theta_R = 60°$, $(1/L)(\partial R_R/\partial \theta_R)$ (Figure 5.4) shows this effect for $e = 0.9$, 0.5, and 0.1; At $\theta_R = -50°$, $\partial R_R/dL$ (Figure 5.3) shows this effect for $e = 0.5$ and 0.1. Thus, depending on the values of $d(R_T + R_R)$, dL, $d\theta_R$, and the geometry, the R_R estimation error can be larger for smaller eccentricities.

The maximum and minimum values for these error slopes can be calculated by taking second partial derivatives of (5.8b), (5.9b), and (5.10b). Results are summarized in Table 5.1, which includes the special cases for $e = 1$, the forward-scatter case; and for $\theta_R = \pm 90°$, the pseudomonostatic case as previously discussed; and for $e = 0$, the monostatic case.

In the monostatic case, the error slope for $\partial R_R/\partial(R_T + R_R)$ of 0.5 is expected because monostatic range (and range error) is half of the time delay (and time-delay error) times the speed of light [1]. However, the error slopes for $\partial R_R/\partial L$ and $(1/L)(\partial R_R/\partial \theta_R)$ require elaboration. When $e = 0$, either $L = 0$ or $(R_T + R_R) = \infty$. Clearly, the latter result is not possible; thus, $L = 0$, and $\partial R_R/\partial \theta_R = -(L/2)\cos\theta_R = 0$. Furthermore, when $L = 0$, the error, dL, in measuring L should also equal zero. Thus $(\partial R_R/\partial L) dL = 0$. As a result, (5.7) reduces to $dR_R = 0.5d(R_T + R_R) = dR_M$, where dR_M is the monostatic range measurement error.

The $\partial R_R/\partial(R_T + R_R)$ and $(1/L)(\partial R_R/\partial \theta_R)$ error slopes in Table 5.1 show well-behaved maximum values, although the latter expression is particularly messy. Their minimum values occur at $\theta_R = \pm 90°$, as discussed previously. The $(1/L)(\partial R_R/\partial \theta_R)$ minimum is zero, because, in this pseudomonostatic geometry, an angle measurement is not required to estimate R_R. The $\partial R_R/\partial L$ error slope maximum assumes two negative values depending on the value of e. When $e \leq \sqrt{2} - 1$, the maximum negative value occurs at $\theta_R = +90°$. When $e > \sqrt{2} - 1$, the maximum negative value occurs at $\theta_R < +90°$, and grows rapidly as $e \to 1$ and $\theta_R \to -90°$, which is the forward-scatter case. Because both values converge to -0.5 at $\theta_R = +90°$, the value is called a *local negative maximum* for the $e > \sqrt{2} - 1$ case. For all $e < 1$, the $\partial R_R/\partial L$ error slope assumes a minimum (zero) value, depending on e. In this specific geometry, the baseline measurement error does not contribute to the R_R error. Except for the special case of $\theta_R = \pm 90°$, all of these. error slope maxima and minima occur at different values of θ_R. Consequently, their

Table 5.1
Summary of Conditions for Maximum and Minimum Error Slopes in the Receiver-to-Target Range Equation (5.1)

θ_R	e	Error Slope	Comments
$\dfrac{\partial R_R}{\partial (R_T + R_R)}$ SUMMARY			
All	0	0.5	Monostatic case
$+90°$	All	0.5	Minimum error[a]
$-90°$	<1	0.5	Minimum error[a]
$-90°$	1	$+\infty$	Forward-scatter case
$-\sin^{-1}e$	<1	$\dfrac{1}{2(1-e^2)}$	Maximum (positive) error at β_{max}
$\dfrac{\partial R_R}{\partial L}$ SUMMARY			
All	0	$-0.5\sin\theta_R$	Monostatic case (see text)
$+90°$	$\leq\sqrt{2}-1$	-0.5	Maximum (negative) error[a]
$+90°$	$>\sqrt{2}-1$	-0.5	Local maximum (negative) error[a]
$-90°$	<1	0.5	Maximum (positive) error[a]
$-90°$	1	$-\infty$	Forward-scatter case
$-\sin^{-1}\left(\dfrac{2e}{e^2+1}\right)$	<1	0	Minimum error
$\sin^{-1}\left(\dfrac{1-3e^2}{e+e^3}\right)$	$>\sqrt{2}-1$	$-\dfrac{(e^2+1)^2}{8e(1-e^2)}$	Maximum (negative) error
$\left(\dfrac{1}{L}\right)\left(\dfrac{\partial R_R}{\partial\theta_R}\right)$ SUMMARY			
All	0	$-(\tfrac{1}{2})\cos\theta_R$	Monostatic case ($=$ 0 because $L=0$)
$+90°$	All	0	Minimum error[a]
$-90°$	<1	0	Minimum error[a]
$-90°$	1	Indeterminate	Forward-scatter case
$\sin^{-1}\left[\dfrac{1}{2e}(1-\sqrt{8e^2+1})\right]$	<1	$\dfrac{(1-e^2)(\sqrt{8e^2+1}-2e^2-1)^{1/2}}{\sqrt{2}e(3\sqrt{8e^2+1}-4e^2-5)}$	Maximum (negative) error

[a]On extended baseline; pseudomonostatic case.

contribution to the ∂R estimate changes as the geometry changes, which requires evaluation for specific bistatic radar configurations and geometries.

Consider the following examples of a ground-based, air surveillance bistatic radar. Assume (1) a receiving antenna with beam splitting, yielding $d\theta_R \approx 0.005$ rad rms; (2) a range sum measurement accuracy of $d(R_T + R_R) \approx 25$ m rms; and

(3) a map-reading accuracy, $dL = 10$ m rms. Assume also that the measurements are noise limited and unbiased.

Figure 5.5 is a plot of (5.7) for these assumptions, a baseline $L = 10$ km and a target located on the $R_T + R_R = 100$ km isorange contour. The eccentricity is 0.1. The top curve is the rss error in the R_R estimate, with a peak error of about 28 m at $\theta_R \approx -10°$. In the region $\theta_R \approx \pm60°$, the R_R error is dominated by the angle term, and near the extended baseline, by the range sum term. The baseline term has little effect on the R_R error because dL is small.

As the target approaches the baseline, isorange contour eccentricity increases, which increases the error in estimating R_R. Figure 5.6 shows this effect now with the target moved to the $R_T + R_R = 25$ km isorange contour, with an eccentricity of 0.4. The peak R_R error has increased to about 33 m at $\theta_R \approx -40°$. Note that this result is different from that of a monostatic radar, where S/N-independent range errors remain constant. Like a monostatic radar, however, the R_R error remains constant on the extended baseline, the pseudomonostatic operating point.

A similar effect occurs when the baseline is increased, which again increases eccentricity and the R_R error. Figure 5.7 is a variation of Figure 5.5—the baseline is increased from 10 to 50 km and the eccentricity is 0.5. The R_R error is almost totally dominated by the angle term, with the peak error increased to about 160 m at $\theta_R \approx -50°$.

These examples show that when the baseline is a significant fraction of the range sum, such that the isorange contour eccentricity is $\gtrsim 0.2$, the R_R errors are large for targets in a region between the transmitter and receiver, and significantly smaller for targets elsewhere. The former condition represents typical cosite operation; the latter represents typical over-the-shoulder operation, as defined in Section 4.4. Thus, depending on the operating region, a design trade-off exists between angle and range sum errors to achieve an acceptable R_R error. Specifically, a larger receiving aperture or increased beam-splitting capability might be more useful for cosite operation than for over-the-shoulder operation. This trade-off should also consider opposing system requirements that benefit from a larger receiving aperture—such as generating an acceptable bistatic maximum range constant, (4.1b), operating in a clutter environment (Section 5.6), and operating in an ECM environment (Chapter 10)—versus those that benefit from a smaller receiving aperture, such as beam scan-on-scan remedies (Section 13.1) and pulse chasing (Section 13.2).

In general, the baseline error contribution can be considered insignificant—given that, for example, maps are read correctly and navigation systems are calibrated properly, such that large bias errors are not generated. An exception to this rule is the special case of a coherent multistatic radar, for which each site is treated as one element in a thinned array. Now errors of the order of a wavelength in location of the sites (elements) generate phase errors in the array. If the array is not rigid (e.g., airborne) or if the elements are moving, the errors will be time varying.

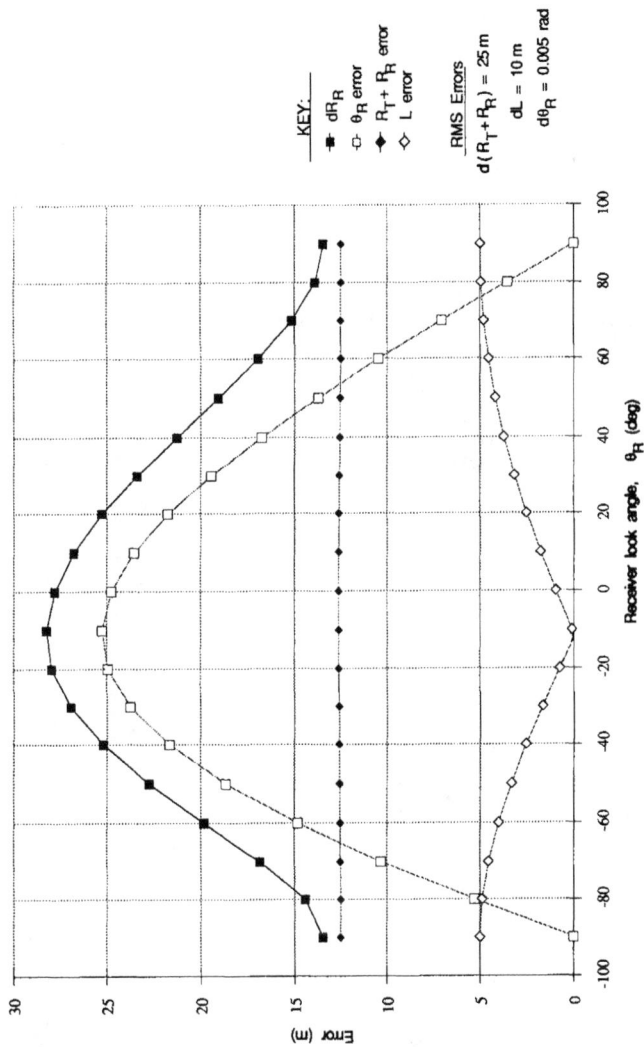

Figure 5.5 The rss error, dR_R, in estimating receiver-to-target range, R_R, (5.7), for the following conditions: $R_T + R_R = 100$ km, $L = 10$ km, and $e = 0.1$.

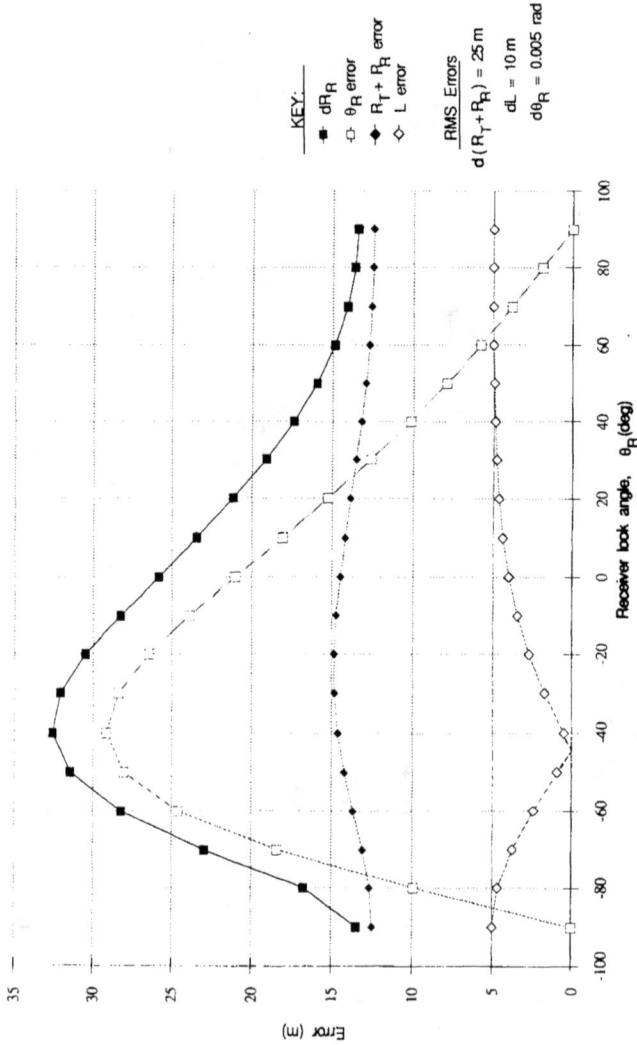

Figure 5.6 The rss error, dR_R, in estimating receiver-to-target range, R_R, (5.7), for the following conditions: $R_T + R_R = 25$ km, $L = 10$ km, and $e = 0.4$.

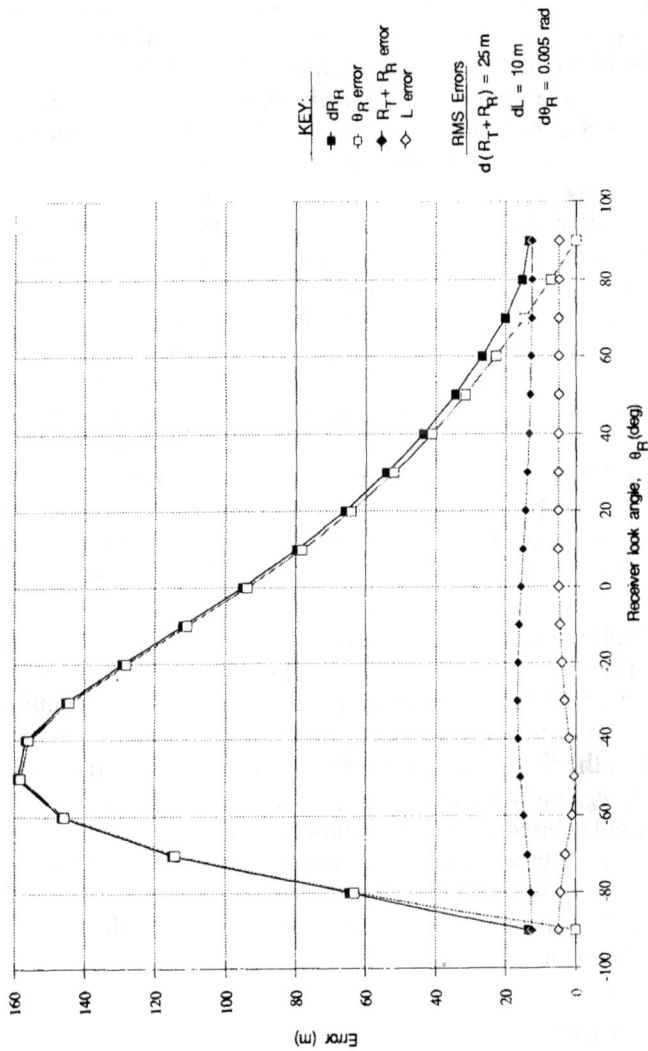

Figure 5.7 The rss error, dR_R, in estimating receiver-to-target range, R_R, (5.7), for the following conditions: $R_T + R_R = 100$ km, $L = 50$ km, and $e = 0.5$.

Although phase errors can be generated from many sources, such as phase shifters, variable transmission line lengths, mutual coupling, and propagation turbulence, site location is the dominant source, particularly for open-loop multistatic beam forming. The most sensitive antenna parameters to phase errors are mainbeam gain, sidelobe levels, and pointing angle. Beamwidth and beamshape are not significantly affected [4]. (Amplitude errors, the most severe of which is the loss of an element or site, will also affect array performance, most notably in increased sidelobe levels.) A quantitative analysis of these errors for coherent multistatic radar operation is given by Steinberg [4]. If the array is self-cohering (Chapter 11), the tolerance in element placing can be relaxed by two or more orders of magnitude because the self-cohering process removes most of the location error effects [4].

We must emphasize that the foregoing error analysis is for geometry-dependent errors, unique to a bistatic radar. Additional θ_R and $R_T + R_R$ error sources that contribute to the R_R error must be factored into the estimate. Typical sources include propagation effects, clutter, and ECM. When the direct range sum estimation method, Section 5.1, is used to estimate $R_T + R_R$, error sources include transmitted pulse instabilities, multipath over the direct path, and interference from the direct path signal, particularly at large isorange contour eccentricities. This last source is analogous to eclipsing and will also affect θ_R measurements. Compounding the eclipsing problem is direct path interference from range sidelobes when pulse compression is used by the transmitter. If linear FM (LFM) pulse compression is used, Hamming or cosine-squared time-domain weighting by the receiver improves near-in sidelobe suppression by about 5 dB, when compared to the same type of frequency-domain weighting [79]. These types of errors can be reduced when using the indirect range sum estimation method—synchronized stable clocks—and when the direct path signal can be masked, for example, by terrain.

Target location errors have been developed for other location techniques. For a hyperbolic location system, target location errors are reported [18] to decrease as the target approaches the line joining the two transmitters. For a theta-theta location system using θ_T, θ_R, and L estimates, the error is a minimum when the target lies on the perpendicular bisector of the baseline with $\beta = 45°$, and increases elsewhere [18]. For a theta-theta location system using θ_T, θ_R, and ΔT_n estimates, (5.5), the error is a minimum when the target lies near the extended baseline and increases rapidly as the target approaches the baseline [217]. The trend in these errors is similar to the errors for (5.1).

5.3 COORDINATE CONVERSION

When a bistatic receiver uses a mechanically steered or scanned antenna, the target AOA is typically measured in terms of azimuth and elevation angles referenced to a local coordinate system centered on the receiving site. This local coordinate system is often specified as a local vertical (z-axis) and an (x-y) plane orthogonal to the z-axis, passing through the receiving site, as shown in Figure 5.8. Initially, the x-axis will be taken as colinear with the bistatic baseline. This alignment is called the *baseline-referenced local coordinate system*. For a given target position, the azimuth and elevation angles, A_R' and E_R' respectively, are measured as shown in Figure 5.8, where, using the same convention as the bistatic North-referenced coordinate system, A_R' is positive when measured clockwise from the baseline in the (x-y) plane.

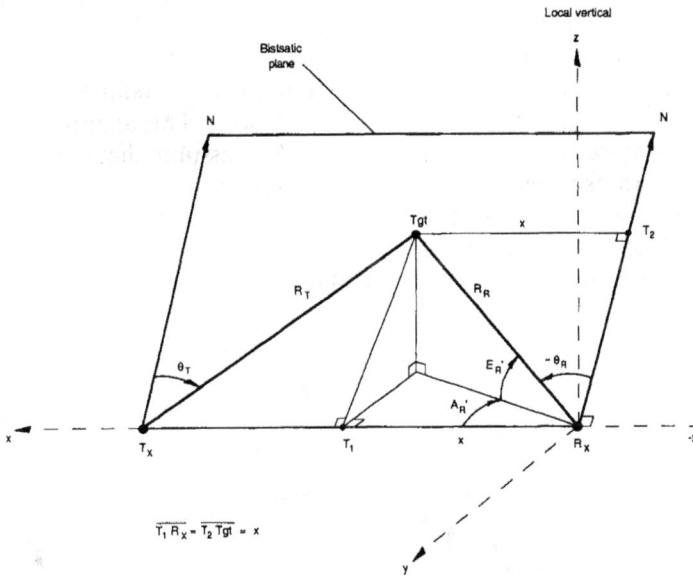

Figure 5.8 Geometry to convert receiver's baseline-referenced AOA measurements (A_R', E_R') into the receiving look angle, θ_R, on the bistatic plane, for the special case of the transmitter located on the x-axis of the receiver's local coordinate system.

The baseline-referenced AOA measurements, A_R' and E_R', are related to the receiver look angle in the bistatic plane, θ_R, as follows:

$$\sin(-\theta_R) = x/R_R \qquad (5.11)$$

where, from direction cosines,

$$x = R_R \cos E_R' \cos A_R' \qquad (5.12)$$

Combining (5.11) and (5.12) yields

$$\theta_R = -\sin^{-1}(\cos E_R' \cos A_R')$$ (5.13)

Note that for this coordinate alignment the North-axis of the bistatic plane is in the $(y\text{-}z)$ plane of the baseline-referenced local coordinate system. Also for $90° < A_R' < 270°$, $\theta_R > 0°$, as is defined for the bistatic plane.

Equation (5.13) is the special case of the baseline colinear with the x-axis of the receiver's local coordinate system, i.e., the transmitter located on the x-axis. Usually, the transmitter will be off-axis and above or below the $(x\text{-}y)$ plane. Thus, (5.13) must be modified as follows to account for this rotation in coordinates.

Assume that the receiver has independently established its local coordinate system with the z-axis on the local vertical and the x-axis pointed in a convenient direction, such as toward the North Pole (true North). This alignment is called the *true North-referenced local coordinate system*. Also assume that the receiver has an estimate of the transmitter's angular position, A_{RT} and E_{RT}, as shown in Figure 5.9. This estimate might be provided by tracking the transmitter if airborne, a survey if ground-based, or GPS via a data link. [In Figure 5.9, the transmitter is positioned above the $(x\text{-}y)$ plane and out of the $(x\text{-}z)$ plane, toward the viewer. The target is

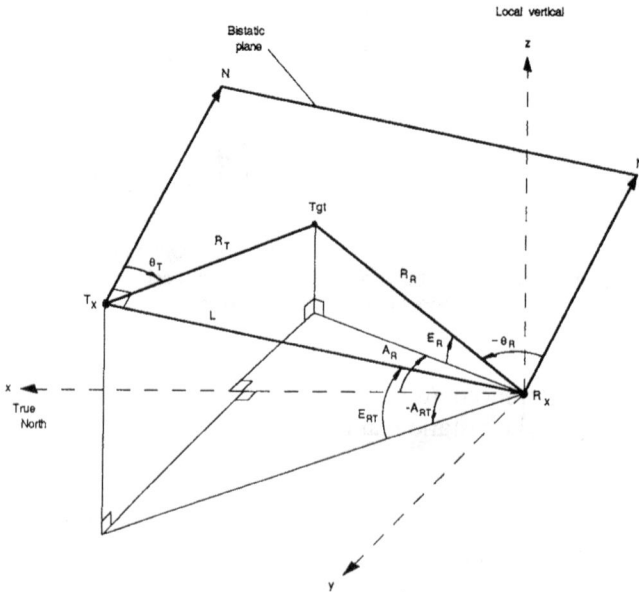

Figure 5.9 Geometry to convert receiver's true North-referenced AOA measurements of target (A_R, E_R) and transmitter (A_{RT}, E_{RT}) position into receiver look angle (θ_R) on the bistatic plane.

also above the $(x\text{-}y)$ plane and out of the $(x\text{-}z)$ plane, but away from the viewer.] The receiver look angle, θ_R, in the bistatic plane becomes [151]:

$$\theta_R = -\sin^{-1}[\cos E_R \cos E_{RT} \cos(A_R - A_{RT}) + \sin E_R \sin E_{RT}] \qquad (5.14)$$

where A_{RT} and E_{RT} are the rotation angles to map the receiver's true North-referenced measurements into the baseline-referenced local coordinate system, and A_R and E_R are the receiver's AOA measurements of the target position in the true North-referenced local coordinate system. When $A_{RT} = E_{RT} = 0°$, $A_R = A'_R$, $E_R = E'_R$, and (5.14) reduces to (5.13). Note that for this coordinate alignment the North axis of the bistatic plane is not in the ($y\text{-}z$) plane of the true North-referenced local coordinate system.

When a bistatic receiver uses a phased-array antenna that electronically scans in two dimensions, for example, a planar array, the AOA is typically measured in terms of θ and ϕ, as shown in Figure 5.10, where θ is positive when measured clockwise from the y' axis. The reference vectors, x' and y', for θ and ϕ measurements, lie in the $(x\text{-}y)$ and ($y\text{-}z$) plane, respectively. The x' vector is thus normal to the z-axis, and the y' vector is normal to the x-axis. The measurement of θ is sometimes called the *conically distorted azimuth measurement*.

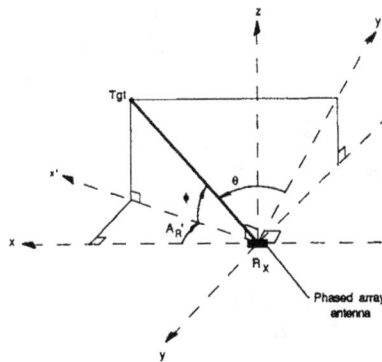

Figure 5.10 Geometry for phased-array antenna AOA measurements.

When the bistatic transmitter is located on the x-axis, as in the baseline-referenced local coordinate system, the plane defined by $\overline{R_x\text{Tgt}}$ and y' is the bistatic plane. Hence, $\theta = \theta_R$ and $\phi = E'_R$. From (5.13) A'_R is calculated as

$$A'_R = \cos^{-1}(\sin - \theta_R / \cos E'_R) \qquad (5.15a)$$

$$= \cos^{-1}(\sin - \theta / \cos \phi) \qquad (5.15b)$$

The development is similar for a receiver using the true North-referenced local coordinate system, i.e., for the general case of the transmitter not on the x-axis. In this case $\phi = E_R$ and $A_R = \cos^{-1}(\sin-\theta/\cos\phi)$. When the transmitter's angular position, A_{RT} and E_{RT}, is known, the receiver look angle, θ_R, is calculated by (5.14).

5.4 DISPLAY CORRECTION

The typical bistatic radar location equation, (5.1), is an expression for R_R in elliptic coordinates centered at the receiver site, where the coordinates are in terms of a polar angle θ_R and a range sum $R_T + R_R$. A *plan position indicator* (PPI) can display these data in polar coordinates, θ_R and R_R, by calculating R_R in terms of θ_R and a propagation time-dely factor C. For the direct method of estimating $R_T + R_R$, Section 5.1, the expression is obtained by combining (5.1) and (5.2) and normalizing R_R to the baseline length L [73]:

$$\frac{R_R}{L} = \frac{C(C + 2)}{2(C + 1 + \sin\theta_R)} \tag{5.16}$$

where $C = c\Delta T_{rt}/L$, the range corresponding to the time interval between reception of the transmitted pulse and reception of the target echo, normalized to the baseline.

Equation (5.16) is plotted in Figure 5.11 for specific values of θ_R. The receiver site is at the origin. The range sweep starts at the origin and is triggered by the transmitting signal when it arrives at the receiver site. Equivalent monostatic operation occurs at $\theta_R = 90°$, where the receiving beam is pointed directly away from the transmitter. At this beam position,

$$(R_R/L)_{90°} = C/2 \tag{5.17}$$

Subtracting (5.17) from (5.16) gives the normalized bistatic range correction factor, $\delta R_R/L$, to convert a PPI display from monostatic to bistatic operation:

$$\frac{\delta R_R}{L} = \frac{C(1 - \sin\theta_R)}{2(C + 1 + \sin\theta_R)} \tag{5.18}$$

Equation (5.18) is just the y-axis difference between the curve for a given θ_R and the equivalent monostatic curve ($\theta_R = 90°$) in Figure 5.11.

For a fixed θ_R, the radial sweep rate, $\partial R_R/\partial t$, for a PPI displaying bistatic data is obtained by differentiating (5.16) with respect to time:

$$\frac{\partial R_R}{\partial t} = \frac{c}{2}\left[1 + \frac{\cos^2\theta_R}{(C + 1 + \sin\theta_R)^2}\right] \tag{5.19}$$

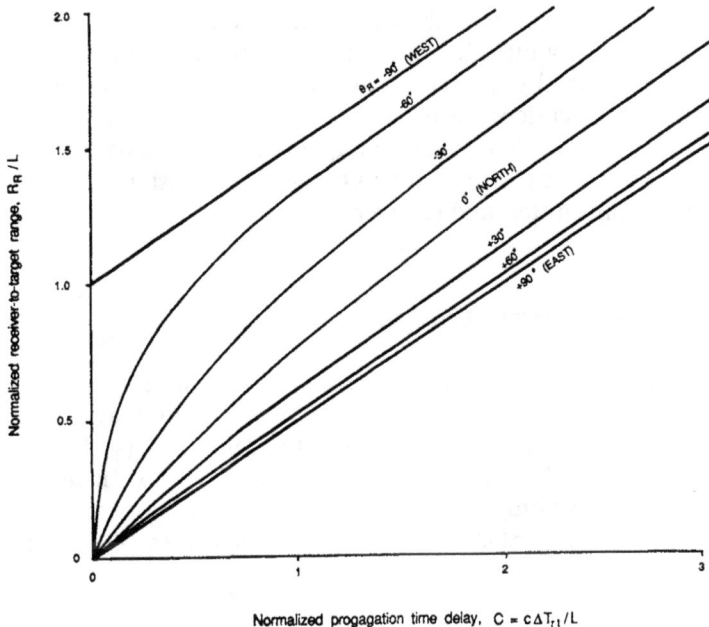

Figure 5.11 PPI polar range sweep required to display elliptic coordinates with the receiver at the origin and the baseline oriented East-West. (Courtesy of IEE [73].)

For large C, $\partial R_R/\partial t$ approaches $c/2$, the monostatic sweep rate. Also for $C \neq 0$ and $\theta_R = \pm 90°$, $\partial R_R/\partial t = c/2$. At the start of the sweep, $C = 0$ and the radial sweep rate becomes $c/(1 + \sin\theta_R)$. For $0° > \theta_R > -90°$, this initial sweep rate becomes large, as the slope of the curves in Figure 5.11 shows. At $C = 0$ and $\theta_R = -90°$, the rate is infinite.

The processing needed to implement this type of display correction has been developed in analog [83] and digital [46] form. A method of storing video for one pulse repetition interval (PRI) and replaying a correctly delayed version on the next PRI to eliminate the need for large sweep rates has also been developed [47]. Display correction equations for three dimensions are also available [46, 48].

5.5 COVERAGE

Bistatic radar coverage is defined as the region or area on the bistatic plane where the target is "visible," i.e., detectable by the receiver and within LOS of both transmitter and receiver. Bistatic detection coverage is established by the maximum range oval of Cassini, (4.1a) and (4.1b). Bistatic LOS coverage is affected by multipath, refraction, diffraction, and shadowing, including earth curvature. The first three effects are usually included in the pattern propagation factor and loss terms of (4.1a), and modified by special bistatic propagation conditions outlined in Section 4.2. The last effect depends on the geometry between target, transmitter, and receiver. This section treats two cases unique to bistatic radars: detection coverage constrained by ovals of Cassini, and LOS coverage constrained by the geometry between target, transmitter, and receiver.

5.5.1 Detection-Constrained Coverage

Appendix D develops expressions for the area, A_B, within a maximum range oval of Cassini. This oval is the maximum range detection contour available to a bistatic radar with a bistatic maximum range product κ, defined by (4.1b) as $(R_T R_R)_{max} = \kappa$. Two expressions apply, the first when baseline $L < 2\sqrt{\kappa}$. In this case a single, cosite, oval surrounds both the transmitter and receiver, for example, any of the ovals for 10 dB $\leq (S/N) \leq 23$ dB in Figure 4.2. The area, A_{B1}, within the single oval is

$$A_{B1} = \pi\kappa \left[1 - \left(\frac{1}{2}\right)^2 \left(\frac{L^4}{16\kappa^2}\right)\left(\frac{1}{1}\right) - \left(\frac{1\cdot 3}{2\cdot 4}\right)^2 \left(\frac{L^4}{16\kappa^2}\right)^2 \left(\frac{1}{3}\right) - \cdots \right] \quad (5.20a)$$

$$\approx \pi\kappa \left[1 - (1/64)(L^4/\kappa^2) - (3/16384)(L^8/\kappa^4) \right] \quad (5.20b)$$

The second expression applies when $L \geq 2\sqrt{\kappa}$. In this case two identical ovals exist, one surrounding the transmitter and one surrounding the receiver, for example, the ovals for $(S/N) = 27$, and 30 dB in Figure 4.2. The area A_{B2}, within both ovals is

$$A_{B2} = \frac{2\pi\kappa^2}{L^2} \left[1 + \left(\frac{1}{2^2\cdot 2!}\right)\left(\frac{16\kappa^2}{L^4}\right) + \left(\frac{3^2}{2^4\cdot 3!\cdot 2!}\right)\left(\frac{16\kappa^2}{L^4}\right)^2 \right.$$
$$\left. + \left(\frac{3^2\cdot 5^2}{2^6\cdot 4!\cdot 3!}\right)\left(\frac{16\kappa^2}{L^4}\right)^3 + \cdots \right] \quad (5.21a)$$

$$\approx \frac{2\pi\kappa^2}{L^2} (1 + 2\,\kappa^2/L^4 + 12\,\kappa^4/L^8 + 100\,\kappa^6/L^{12}) \quad (5.21b)$$

The area for the lemniscate or "cusp oval" where $L = 2\sqrt{\kappa}$, for example, the oval with $(S/N) = 27$ dB in Figure 4.2, is 2κ exactly.

For the monostatic case, $L = 0$ and (5.20a) becomes

$$A_M = \pi \kappa \tag{5.22a}$$

$$= \pi (R_M)^2_{max} \tag{5.22b}$$

as expected because the oval becomes a circle of radius $(R_M)_{max}$ and area A_M.

Figure 5.12 plots (5.20a) and (5.21a) normalized with respect to the monostatic area A_M, as a function of the baseline L normalized with respect to $\sqrt{\kappa}$, or the equivalent monostatic range R_M. For values nears $L/\sqrt{\kappa} = 2$, which is the ovals' cusp, more terms than those shown in (5.20a) and (5.21a) are usually needed for a good estimate of the area. The ratio A_{B2}/A_M at $L/\sqrt{\kappa} = 2$ is $2\kappa/\pi\kappa = 2/\pi = 0.637$, as shown in the figure.

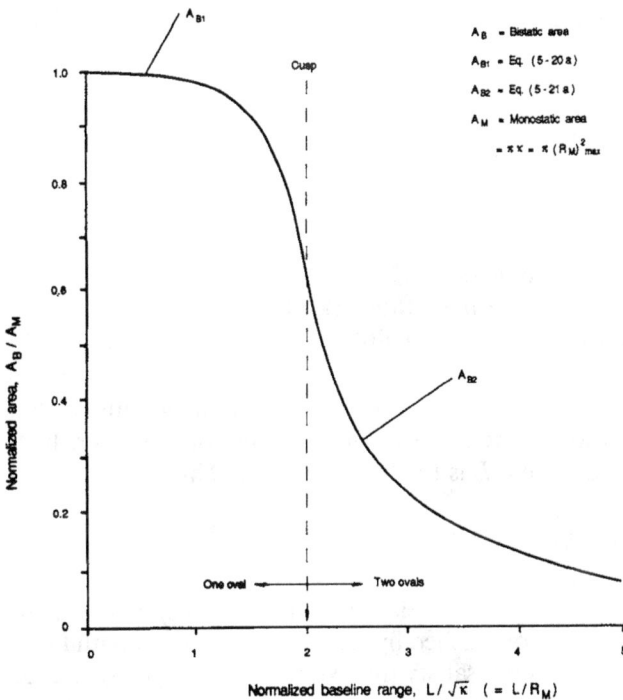

Figure 5.12 Ratio of bistatic area (oval of Cassini) to monostatic area (circle) for a bistatic radar with maximum range product κ, and a monostatic radar with equivalent monostatic maximum range $\sqrt{\kappa}$.

Figure 5.12 demonstrates that $A_M > A_B$ for the case of a bistatic radar with maximum range product, κ, and a monostatic radar with equivalent monostatic maximum range $\sqrt{\kappa}$. In all cases the expressions for A_M and A_B assume that a suitable LOS exists between target, transmitter, and receiver. This topic is treated in Section 5.5.2.

5.5.2 Line-of-Sight-Constrained Coverage

For given target, transmitter, and receiver altitudes, the target must be simultaneously within LOS to both the transmitter and receiver sites. For a smooth earth, these LOS requirements are established by coverage circles centered at each site. Targets in the area common to both circles, A_C, have a LOS to both sites as shown in Figure 5.13. For a 4/3 earth model and ignoring multipath lobing, the radius of these coverage circles, in kilometers is approximated by [16]:

$$r_R = 130(\sqrt{h_t} + \sqrt{h_R}) \tag{5.23}$$

and

$$r_T = 130(\sqrt{h_t} + \sqrt{h_T}) \tag{5.24}$$

where

h_t = target altitude (km),
h_R = receiving antenna altitude (km),
h_T = transmitting antenna altitude (km).

If the receiver establishes synchronization via a direct path RF link, then an adequate LOS is also required between transmitter and receiver. In this case $h_t = 0$ and $r_R + r_T \geq L$, where L is the baseline range. Thus,

$$L \leq 130(\sqrt{h_R} + \sqrt{h_T}) \tag{5.25}$$

This LOS requirement can be relaxed if an RF tropospheric scatter path is used. If synchronization is accomplished by stable clocks, a direct path LOS is not required, and the system must only satisfy the requirements of (5.23) and (5.24).

The common coverage area, A_C, is the intersection of the two coverage circles:

$$A_C = \tfrac{1}{2}[r_R^2(\phi_R - \sin\phi_R) + r_T^2(\phi_T - \sin\phi_T)] \tag{5.26}$$

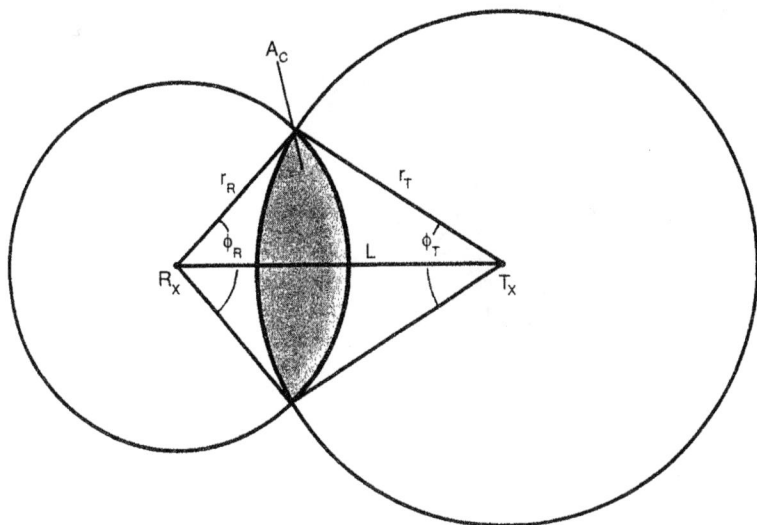

Figure 5.13 Geometry for common coverage area, A_C.

where ϕ_R and ϕ_T are (as shown in Figure 5.13):

$$\phi_R = 2 \cos^{-1} \left(\frac{r_R^2 - r_T^2 + L^2}{2r_R L} \right) \tag{5.27}$$

$$\phi_T = 2 \cos^{-1} \left(\frac{r_T^2 - r_R^2 + L^2}{2r_T L} \right) \tag{5.28}$$

Equation (5.26) is valid for $L + r_R > r_T > L - r_R$ or $L + r_T > r_R > L - r_T$. Whenever the right-hand side of either inequality is not satisfied, such that $r_T + r_R \leq L$, then $A_C = 0$; that is, the coverage circles do not intersect. When the left-hand side of the first inequality is not satisfied, such that $r_T \geq L + r_R$, then $A_C = \pi r_R^2$, because the transmitter's coverage circle includes all of the receiver's coverage circle. Similarly, when the left-hand side of the second inequality is not satisfied, such that $r_R \geq L + r_T$, then $A_C = \pi r_T^2$.

In some bistatic radar configurations, maximum transmitter-to-target or receiver-to-target operating ranges and a baseline L are specified. Examples are given in Chapter 12. When the operating ranges are specified as the maximum range allowed by a thermal-noise-limited or a jammer-noise-limited oval of Cassini, such as that given in Table 4.2,

$$(R_T)_{max}, (R_R)_{max} = (\kappa + L^2/4)^{1/2} + L/2 \tag{5.29}$$

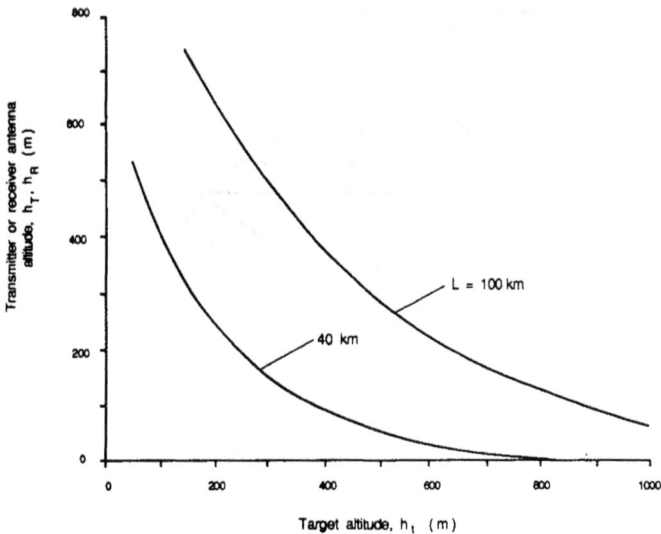

Figure 5.14 Required transmitter or receiver antenna altitudes to establish a noise-limited LOS to a target for a bistatic maximum range constant, $\kappa = 10^4$ km.

where κ is the bistatic maximum range product under thermal noise conditions, (4.1b). Under noise-jamming conditions κ_J, (10.5), is used.

To establish an adequate LOS to a target at altitude h_t, $r_T \geq (R_T)_{max}$ or $r_R \geq (R_R)_{max}$, which, in turn, sets a minimum requirement for h_T or h_R. Combining (5.23), (5.24), and (5.29) and solving for h_T and h_R yields

$$h_T, h_R \geq \left[\frac{(\kappa + L^2/4)^{1/2} + L/2}{130} - \sqrt{h_t} \right]^2 \tag{5.30}$$

Equation (5.30) is plotted in Figure (5.14) for $\kappa = 10^4$ km² (equivalent monostatic range, $\sqrt{\kappa} = 100$ km) and $L = 40$ and 100 km. For $L = 40$ km, the figure shows that an adequate LOS is established for $h_t > 900$ m, when the antenna is at near zero altitude. For $h_t = 500$ m, the antenna must be elevated to 50 m; for $h_t = 100$ m, the antenna must be elevated to 390 m. When the antenna is ground-based, this latter case requires careful siting, for example, by taking advantage of hilly terrain, as is the case with any ground-based radar. In bistatic configurations that specify $(R_T)_{max}$ and $(R_R)_{max}$, both the transmitter and receiver sites must be selected in this manner.

For $L = 100$ km, the antenna altitude requirement becomes more severe: 290 m of elevation for a target altitude of 500 m, and 860 m for a target altitude of 100 m. These requirements obviously constrain antenna placement to fewer sites, if they are available. Absent an appropriate site, the antenna must be sent aloft in an aircraft or aerostat—a more complicated and expensive remedy.

LOS coverage requirements are equally important in a monostatic or bistatic radar net. A typical case is shown in Figure 1.2, which consists of four sites, configured as monostatic radars and combinations of bistatic transmitters and receivers. The generalized coverage equations for monostatic and bistatic radar nets can be developed as follows.

Assume a monostatic radar net consisting of n sites, and a corresponding bistatic radar net consisting of the same n sites used as transmitters. Each bistatic transmitter has a paired receiver, located at n separate sites. For a monostatic radar net, the total coverage, A_{TM}, is the union of all monostatic radar coverage circles: $A_{TM} = \pi(r_{T1}^2 U \ r_{T2}^2 U \ldots r_{Tn}^2)$. For a bistatic radar net, the total coverage, A_{TB}, is the union of all the bistatic radar common coverage areas, i.e., the union of the intersection of each pair of bistatic site coverage circles: $A_{TB} = A_{C1} U A_{C2} U \ldots A_{Cn}$. A_{TB} is maximum when $r_{Ti} \geq r_{Ri} + L$, $i = 1, 2, \ldots, n$. In this case, $A_{Ci} = \pi r_{Ri}^2$ and $A_{TB} = \pi(r_{R1}^2 U \ r_{R2}^2 U \ldots r_{Rn}^2)$. Because $r_{Ti} > r_{Ri}$, $A_{TM} > A_{TB}$; that is, monostatic net coverage is greater than bistatic net coverage.

Terrain and other types of masking or shadowing further degrade both monostatic and bistatic coverage. For ground-based bistatic transmitters and receivers, the degradation can be severe unless both sites are carefully selected [84]. Hypsographic charts or terrain models can be used for this purpose. In the event suitable sites are not available, some air defense bistatic radar concepts resort to the use of an elevated or airborne transmitter [44, 45, 48, 54]. As a general rule, bistatic coverage is less than monostatic coverage in both single and netted configurations.

5.6   CLUTTER CELL AREA

The mainlobe bistatic clutter cell area, A_c is defined, in the broadest sense, as the intersection of the range cell, the doppler cell, and the mainbeam footprint. (A_c is not to be confused with the common coverage area, A_C.) The range and doppler cells are defined by isorange and isodoppler contours, respectively. The mainbeam footprint is the area on the ground, or clutter surface, common to the one-way transmitting and receiving beams, where the beamwidths are conventionally taken at the 3-dB point. Three clutter cell cases are usually of interest: beamwidth-limited, range-limited, and doppler-limited.

5.6.1 Beamwidth-Limited Clutter Cell Area

The beamwidth-limited clutter cell area $(A_c)_b$ is the bistatic mainbeam footprint. It has been evaluated for specific antenna pattern functions and specific geometries by numerical integration techniques [42, 85, 86]. At small grazing angles, a two-dimensional approximation to $(A_c)_b$ is a parallelogram shown as the single-hatched area in Figure 5.15 with area:

$$(A_c)_b \approx \frac{(R_R\,\Delta\theta_R)(R_T\,\Delta\theta_T)}{\sin\beta} \tag{5.31}$$

where $R_R\,\Delta\theta_R$ is the cross-range dimension of the receiving beam at the clutter cell, $R_T\,\Delta\theta_T$ is the corresponding dimension for the transmitting beam, and $\Delta\theta_R$ and $\Delta\theta_T$ are, respectively, the 3-dB beamwidths of the receiving and transmitting beams. Respective transmitting and receiving beam rays are assumed to be parallel, which is a reasonable approximation when the range sum is much greater than the base-line range, $(R_T + R_R) \gg L$. The cell area is a minimum at $\beta = 90°$.

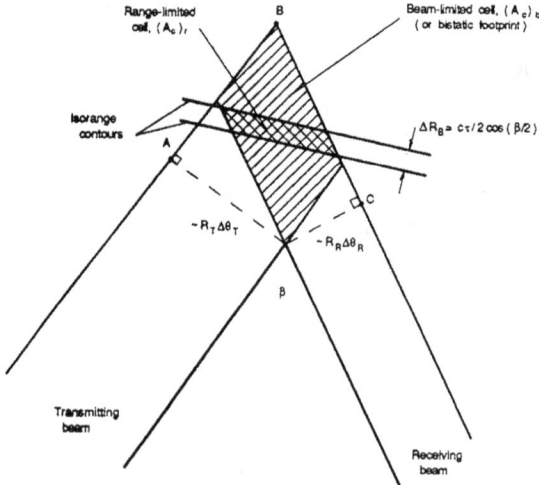

Figure 5.15 Geometry for clutter cell areas.

5.6.2 Range-Limited Clutter Cell Area

The range-limited clutter cell area $(A_c)_r$ has been evaluated at small grazing angles for all geometries of interest [87]. At small grazing angles and at large range sums

$(R_T + R_R \gg L)$, a two-dimensional approximation to $(A_c)_r$ is a parallelogram shown as the double-hatched area in Figure 5.15 with area:

$$(A_c)_r \approx \Delta R_B \left[\frac{R_R \Delta\theta_R}{\cos(\beta/2)} \right] \approx \frac{c\tau R_R \Delta\theta_R}{2\cos^2(\beta/2)} \tag{5.32}$$

where τ is the radar's compressed pulsewidth and ΔR_B is an approximation to the bistatic range cell (Section 4.6). The isorange contours are assumed to be straight lines within the bistatic footprint. For this example, the cross-range dimension of the transmitting beam, $R_T \Delta\theta_T$, is greater than that of the receiving beam, $R_R \Delta\theta_R$, so that the clutter cell is determined by the intersection of the receiving beam and the range cell. For a given geometry, one or the other beam will usually establish the minimum cross-range dimension, and thus the clutter cell area. In either case, the cell area increases as β increases. For small range sums, the cell shape is trapezoidal or triangular at small β, and is rhomboidal or hexagonal at large β [87].

An exact expression for $(A_c)_r$ has been developed [88], again for two dimensions, with the receiving beam and range cell establishing the clutter cell area. With a transformation into the North-referenced coordinate system, results are as follows:

$$(A_c)_r = \frac{L^2\sqrt{1 - e_o^2}}{4e_o^2} \left[f_{e_o}(\theta_{R2}) - f_{e_o}(\theta_{R1}) \right]$$
$$- \frac{L^2\sqrt{1 - e_i^2}}{4e_i^2} \left[f_{e_i}(\theta_{R2}) - f_{e_i}(\theta_{R1}) \right] \tag{5.33}$$

and

$$f_e(\theta) = \tan^{-1} \left\{ \frac{\sqrt{1 + e}\,[1 + \tan(\theta/2)]}{\sqrt{1 - e}\,[1 - \tan(\theta/2)]} \right\}$$
$$+ \frac{e\sqrt{1 - e^2}\,[1 - \tan^2(\theta/2)]}{(1 - e)^2[1 - \tan(\theta/2)]^2 + (1 + e)[1 + \tan(\theta/2)]^2} \tag{5.34}$$

where the appropriate subscripts are added to e and θ in (5.34) to solve (5.33), and

$L =$ baseline range,
$e_i =$ eccentricity of inner isorange contour,
$e_o =$ eccentricity of outer isorange contour $= e_i L/(e_i c\tau + L)$,
$\tau =$ compressed pulsewidth,

and θ_{R1} and θ_{R2} are the pointing angles corresponding to the edges of the receiver's 3-dB beamwidth, $\Delta\theta_R$, where $\Delta\theta_R = \theta_{R2} - \theta_{R1}$.

Equation (5.33) applies directly for all angles contained in the Northern (or Southern) hemisphere, i.e., $\theta_{R1} > -90°$ and $\theta_{R2} < +90°$ for the Northern hemisphere. In regions where θ_{R1} is in one hemisphere and θ_{R2} is in the other, for example, $\theta_{R1} = 85°$ and $\theta_{R2} = 105°$ and $\theta_{R2} = 105°$, (5.33) can be used by summing two calculations, one for ($\theta_{R1} = 85°$, $\theta_{R2} = 90°$) and the second for ($\theta_{R1} = 90°$, $\theta_{R2} = 105°$). Equation (5.33) also applies to the case when the transmitting beam establishes the clutter cell area: $\Delta\theta_T = \theta_{T2} - \theta_{T1}$ is used in (5.33) and (5.34).

The relative errors between (5.32) and (5.33) were calculated for $\Delta\theta_R = 7.5°$, $L = 25$ km, and ellipse separation ΔR_B on the extended baseline ($\beta = 0°$) of 50 m ($\tau = 0.33$ μs), such that $L \gg \Delta R_B$, a typical case. Relative errors were found to be less than 1% for inner loop eccentricity $e_i < 0.9$ and $\theta_R > -70°$, i.e., θ_R pointed away from the transmitter site. For $e_i > 0.9$ and $\theta_R < -70°$, the relative error increased significantly, exceeding 20% for $e_i > 0.95$ and $\theta_R < -87°$. Thus, (5.33) should be used when the target is near the site whose beamwidth is *not* used in establishing the clutter cell area. It should also be used when the target is near the baseline, because $e \to 1$, $\beta \to 180°$, and the approximation to ΔR_B in (5.32) breaks down.

5.6.3 Doppler-Limited Clutter Cell Area

The doppler-limited clutter cell area $(A_c)_d$ has been determined by numerical integration techniques when it is bounded by a range cell [51, 89]. No convenient algebraic expression has been developed for the cell area because the doppler cell size and orientation with respect to the baseline change as the transmitter and receiver velocity vectors and look angles change. In the special case where the transmitter and receiver velocity vectors are equal and the bistatic angle is large, the isorange and isodoppler contours are essentially parallel, creating very large clutter cell areas [59].

5.7 MAXIMUM UNAMBIGUOUS RANGE AND PRF

In a monostatic radar, the range beyond which targets appear as second-time-around echoes is called the *maximum unambiguous range,* $(R_M)_u$, which is [1]

$$(R_M)_u = \frac{c}{2\text{PRF}} \tag{5.35}$$

where PRF is the *pulse repetition frequency* in Hz. The corresponding bistatic radar maximum unambiguous range is $(R_T + R_R)_u$, where

$$(R_T + R_R)_u = \frac{c}{\text{PRF}} \tag{5.36}$$

which is an ellipse, or isorange contour, of major axis length c/PRF.

Equation (5.36) applies for cosite operation and when targets generating second-time-around echoes lie in both the transmitting and the receiving beams. This geometry occurs (a) on or near the extended baseline, (b) where a broad or floodlight beam is used by the transmitter or receiver, or both. When high-gain, narrow beams are used by both the transmitter and receiver and operation is away from the extended baseline, second-time-around echoes fall in one or both antenna sidelobes and thus are attenuated. Furthermore, the PRF can be significantly increased when operation is constrained to the small common beam volume, as detailed later in this section. However, when these high-gain beams scan, beam scan-on-scan problems can arise (Section 13.1). One solution is *pulse chasing* (Section 13.2), which, in turn, establishes its own requirements on PRF, as also detailed in this section.

When a monostatic radar and a bistatic radar operate at maximum range, a maximum range-unambiguous pulse repetition frequency, PRF_u, is established. Consider the special case where the radars operate with equivalent parameters so that $(R_M)_{\max} = \sqrt{\kappa}$ and $(R_T R_R)_{\max} = \kappa$, where κ is the bistatic maximum range product and $\sqrt{\kappa}$ is the equivalent monostatic range. For the monostatic radar, $(R_M)_u = (R_M)_{\max} = \sqrt{\kappa}$ so that

$$(\text{PRF}_M)_u = \frac{c}{2\sqrt{\kappa}} \tag{5.37}$$

where $(\text{PRF}_M)_u$ is the maximum range-unambiguous PRF for a monostatic radar operating at maximum range $\sqrt{\kappa}$. For a bistatic radar again operating in the cosite region, PRF_u varies as a function of the geometry. Specifically, PRF_u is established by the intersection of the isorange contour (5.36), where $(R_T + R_R)_u = (R_T + R_R)_{\max}$, with the maximum range oval of Cassini $(R_T R_R)_{\max} = \kappa$, as defined in Section 4.7, and specifically by (4.15). Combination of (4.15) and (5.36) yields

$$(\text{PRF}_B)_u = c/(R_T + R_R)_{\max} \tag{5.38a}$$

$$= c[L^2 + 2\kappa(1 + \cos\beta)]^{-1/2} \tag{5.38b}$$

where $(\text{PRF}_B)_u$ is the maximum range-unambiguous PRF for the bistatic radar operating at a maximum range product κ. Note that $(\text{PRF}_B)_u$ is a minimum when $\beta = 0°$, or when the target is located on the extended baseline, the pseudomonostatic case. This PRF is a maximum at $\beta_{\max}$, or when the target is located on the perpendicular bisector of the baseline; see (3.8) and Table 4.2. Also note that (5.38b) collapses to (5.37) when $L = 0$ and $\beta = 0°$, the sanity check.

The ratio of (5.37) and (5.38b) yields

$$\frac{(PRF_M)_u}{(PRF_B)_u} = [L^2/4\kappa + \cos^2(\beta/2)]^{1/2} \tag{5.39}$$

From Table 4.2, $\cos\beta_{max} = 1 - L^2/2\kappa$; thus, $\cos^2(\beta_{max}/2) = 1 - L^2/4\kappa$ and $(PRF_M)_u = (PRF_B)_u$ at β_{max}. Consequently, $(PRF_B)_u \leqslant (PRF_M)_u$ for a bistatic radar with parameters equivalent to those of a monostatic radar and operating in the cosite region (a) with at least one broad or floodlight beam, or (b) near the extended baseline for all beam configurations.

For a bistatic radar operating in either the transmitter- or receiver-centered region, the small ovals of Cassini surrounding the transmitter or receiver, $PRF_u \approx c/2R_0$, where R_0 is the average oval radius. That is, echoes from all targets inside the receiver-centered oval will arrive at the receiver within a time $2R_0/c$ after the transmitted pulse reaches the receiver. An identical geometry occurs for the transmitter-centered oval. From (12.3), $R_0 \cong \kappa/L$; thus,

$$PRF_u \approx cL/2\kappa \tag{5.40}$$

Because R_0 is by definition small, the bistatic system can operate with very high range-unambiguous PRFs in these regions.

Two special bistatic radar configurations also establish a maximum PRF: high-gain transmitting and receiving beam antennas (Section 13.1), and single-beam pulse chasing (Section 13.2). When the transmitting and receiving beams share a small common volume, or area in two dimensions, such as shown in Figure 5.15 and 13.1, the bistatic radar can operate with a higher PRF before encountering range ambiguities. In this case, $PRF_u = c/\Delta R$, where ΔR is the path length in $\overline{AB} + \overline{BC}$ in Figure 5.15 [73]. For parallel beam rays, $\Delta R \cong (R_T\Delta\theta_T + R_R\Delta\theta_R)/\tan(\beta/2)$ and

$$PRF_u \approx \frac{c\tan(\beta/2)}{(R_T\Delta\theta_T + R_R\Delta\theta_R)} \tag{5.41}$$

For nonparallel beam rays, (5.41) is multiplied by an approximate correction factor $1 - [(\Delta\theta_T + \Delta\theta_R)/2\sin\beta]^2$ [73]. As β becomes large, the common area compresses and PRF_u increases. As β becomes small, the common area elongates and PRF_u is reduced. The expressions break down as β approaches either 0° or 180°.

In single-beam pulse chasing (Section 13.2), the maximum PRF must be selected such that only one pulse traverses the receiver's pulse-chasing coverage area, or scan sector, at one time. This PRF is defined as PRF_1 to distinguish it from the range-unambiguous PRF_u. Two bistatic coverage areas define PRF_1. First, when the bistatic radar operates in the cosite region (Section 4.4):

$$PRF_1 = \frac{c}{(R_{T1} + R_{R1}) - (R_{T2} + R_{R2})} \tag{5.42}$$

where the subscripts 1 and 2 define the receiver's pulse-chasing scan sector, as shown on Figure 5.16. Thus, a pulse traveling over the $(R_{T1} + R_{R1})$ path arrives at the receiver at the same time as the next pulse traveling over the $(R_{T2} + R_{R2})$ path. In this way, only one pulse is chased at a time within the scan sector. In practice, PRF_1 is reduced to account for a finite pulsewidth and any beam-settling times; but this reduction is usually small [231]. Substitution of (3.18) and (3.19) into (5.42) and manipulation of terms yields [231]

$$PRF_1 = \frac{c}{2L} \left[\frac{\cos\frac{1}{2}(\theta_{R1} - \theta_{R2}) - \cos(\theta_T - \theta_{Rc})}{\cos\theta_T \sin\frac{1}{2}(\theta_{R1} - \theta_{R2})} \right] \tag{5.43}$$

where $\theta_{R1} - \theta_{R2}$ defines the pulse-chasing scan sector and $\theta_{Rc} = (\theta_{R1} + \theta_{R2})/2$ is the center of the scan sector, as shown on Figure 5.16.

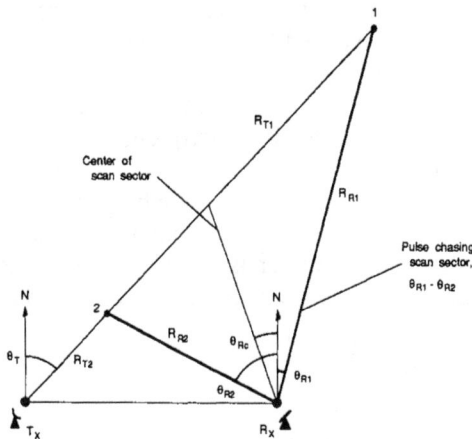

Figure 5.16 Geometry to establish PRF_1 for single-beam pulse chasing in the cosite region.

Now when point 1 on Figure 5.16 lies on the maximum range oval of Cassini—as would be the case when the pulse-chasing system is designed for maximum coverage—$(R_{T1} + R_{R1}) = (R_T + R_R)_{max}$. Thus, from (5.38a) and (5.42), $PRF_1 > (PRF_B)_u$; that is, the maximum PRF allowed by single-beam pulse chasing in the cosite region is greater than the maximum range-unambiguous PRF. Because second-time-around echoes are only a problem near the extended baseline when high-gain, narrow beams are used, the pulse-chasing bistatic system can operate at $(PRF_B)_u$ near the extended baseline and at PRF_1 elsewhere if needed.

Also note that when the transmitter is a monostatic radar operating with a maximum range-unambiguous pulse repetition frequency, $(PRF_M)_u$ and point 1 again lies on the maximum range oval of Cassini, $PRF_1 = (PRF_M)_u$ when $(R_{T2} + R_{R2}) = (R_T + R_R)_{max} - 2(R_M)_{max}$, from Eq (5.42). That is, the left edge of the scan sector, point 2 in Figure 5.16, can be selected so that the single-beam pulse-chasing PRF matches the monostatic radar's maximum range-unambiguous PRF.

When the bistatic radar operates in the transmitter- or receiver-centered region PRF_1 is again calculated with 5.42 and $\theta_T = \pm 90°$ in Figure 5.16. That is, point 2 lies on the baseline and point 1 lies on the extended baseline. In this geometry, $R_{T1} = L + R_0$, $R_{T2} = L - R_0$ and $R_{R1} = R_{R2} = R_0$ for the receiver-centered case, and $R_{R1} = L + R_0$, $R_{R2} = L - R_0$ and $R_{T1} = R_{T2} = R_0$ for the transmitter-centered case; thus, (5.42) becomes

$$PRF_1 = \frac{c}{2R_0} \approx \frac{cL}{2\kappa} \tag{5.44}$$

which is identical to (5.40). Thus, in transmitter- and receiver-centered operating regions, the maximum PRF required for single-beam pulse chasing and the maximum range-unambiguous PRF are equal.

Note that when the transmitter is a monostatic radar operating with a maximum range-unambiguous pulse repetition frequency, $(PRF_M)_u$, $PRF_1 > (PRF_M)_u$. Specifically, the condition for an oval of Cassini to break into two parts is $L \geqslant 2\sqrt{\kappa}$, from Table 4.2; thus, $PRF_1 \geqslant c/\sqrt{\kappa}$, which is greater than $(PRF_M)_u$ (5.37). This condition holds for values of the bistatic maximum range product κ such that $\sqrt{\kappa} < 2(R_M)_{max}$. Consequently, pulse chasing in the transmitter- and receiver-centered operating regions is not usually constrained by a monostatic radar's transmitted PRF.

Chapter 6

DOPPLER RELATIONSHIPS

A canonical definition of bistatic doppler, or doppler shift, f_B, ignoring relativistic effects, is the time rate of change of the total path length of the scattered signal, normalized by the wavelength λ [1]. Because the total path length is the range sum, $R_T + R_R$,

$$f_B = \frac{1}{\lambda}\left[\frac{d}{dt}(R_T + R_R)\right] \tag{6.1a}$$

$$= \frac{1}{\lambda}\left[\frac{dR_T}{dt} + \frac{dR_R}{dt}\right] \tag{6.1b}$$

Figure 6.1 defines the geometry and kinematics for bistatic doppler when the target, transmitter, and receiver are moving. The target's velocity vector projected onto the bistatic plane has magnitude V and aspect angle δ referenced to the bistatic bisector. The aspect angle is positive when measured clockwise from the bistatic bisector. The transmitter and receiver have projected velocity vectors of magnitude V_T and V_R and aspect angles δ_T and δ_R, respectively, per the North-referenced coordinate system of Figure 3.1.

6.1 TARGET DOPPLER

When the transmitter and receiver are stationary ($V_T = V_R = 0$) and the target is moving ($V \neq 0$), the target's bistatic doppler at the receiver site is developed as follows. The term dR_T/dt is the projection of the target velocity vector onto the transmitter-to-target LOS:

$$\frac{dR_T}{dt} = V\cos(\delta - \beta/2) \tag{6.2}$$

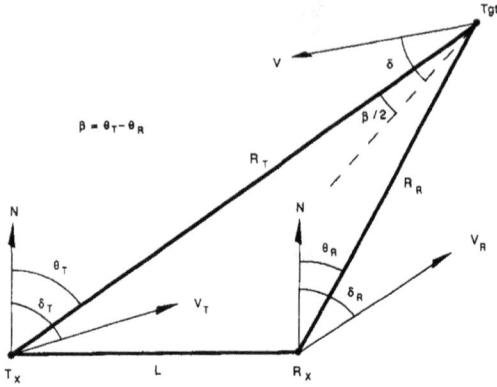

Figure 6.1 Geometry for bistatic doppler.

where δ and β are shown in Figure 6.1. Similarly, dR_R/dt is the projection of the target velocity vector onto the receiver-to-target LOS:

$$\frac{dR_R}{dt} = V \cos(\delta + \beta/2) \tag{6.3}$$

Combining (6.1b), (6.2), and (6.3) yields

$$f_B = f_{Tgt} = (V/\lambda)[\cos(\delta - \beta/2) + \cos(\delta + \beta/2)] \tag{6.4a}$$

$$= (2V/\lambda)\cos\delta \cos(\beta/2) \tag{6.4b}$$

where f_{Tgt} is the bistatic doppler shift caused only by target motion.

Equation (6.4b) is plotted in Figure 6.2 as a function of δ, where the target's bistatic doppler has been normalized by $(2V/\lambda)$. When $\beta = 0°$, (6.4b) reduces to the monostatic case for a monostatic radar located on the bistatic bisector, where δ is now the angle between the target velocity vector and the monostatic radar-to-target LOS. When $\beta = 180°$, the forward-scatter case, $f_{Tgt} = 0$ for any δ. Table 6.1 summarizes the form that (6.4b) can take for specific target aspect angles.

Equation (6.4b) shows that:

- For a given δ, the magnitude of the bistatic target doppler is never greater than that of the monostatic target doppler, when the monostatic radar is located on the bistatic bisector.
- For all β, when $-90° < \delta < +90°$, the bistatic target doppler is positive. Under this definition a closing target referenced to the bistatic bisector generates a positive or "up" doppler.

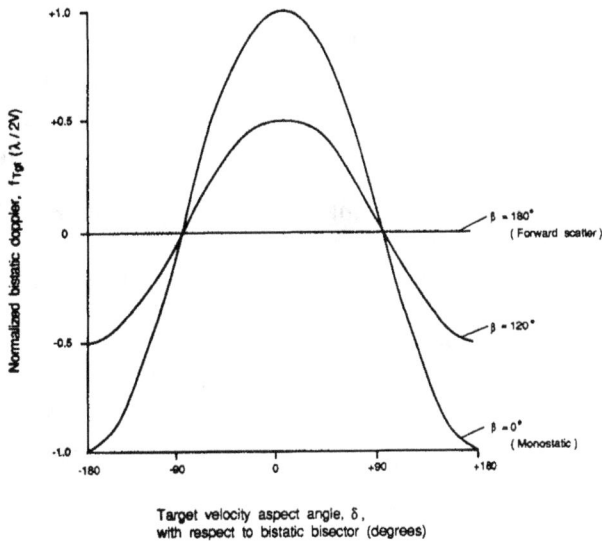

Figure 6.2 Bistatic target doppler for a stationary transmitter, stationary receiver, and a moving target.

- For all β, when the target's velocity vector is normal to the bistatic bisector (δ = $\pm 90°$), the bistatic target doppler is zero. This vector is also colinear with a tangent to an isorange contour drawn through the target position; thus, all isorange contours, including the baseline, are contours of zero bistatic target doppler.
- For all $\beta < 180°$, when the target's velocity vector is colinear with the bistatic bisector, the magnitude of the bistatic target doppler is maximum: maximum positive for $\delta = 0°$, maximum negative for $\delta = 180°$. Because the tangent of an orthogonal hyperbola passing through the target position is also colinear

Table 6.1
Geometry-Dependent Forms of Equation (6.4b)

δ	f_{Tgt}	Geometry
$\beta/2$	$(2V/\lambda)\cos^2(\beta/2)$	V pointed at transmitter
$-\beta/2$	$(2V/\lambda)\cos^2(\beta/2)$	V pointed at receiver
$0°$	$(2V/\lambda)\cos(\beta/2)$	V pointed down the bistatic bisector
$\pm 90°$	0	V perpendicular to the bistatic bisector
$180°$	$-(2V/\lambda)\cos(\beta/2)$	V pointed out the bistatic bisector
$90° \pm \beta/2$	$\mp(V/\lambda)\sin\beta$	V perpendicular to transmitter or receiver LOS

with the bistatic bisector, Figure 3.2, these hyperbolas are contours of maximum bistatic target doppler [20].

For the special case of a monostatic radar located at either the bistatic transmitting or receiving site and $\delta = 90° \pm \beta/2$, such that the target velocity vector is perpendicular to the monostatic radar's LOS, the monostatic target doppler, f_M, is zero. However, the bistatic target doppler, $f_{Tgt} = \mp(V/\lambda)\sin\beta$, which is a potentially useful bistatic doppler tracking enhancement in clutter [160].

Finally, when the monostatic radar is again located at either the bistatic transmitting or receiving site, and a target is traveling on the perpendicular bisector of the baseline such that its velocity vector is colinear with the perpendicular bisector, the monostatic target doppler, f_M, equals the bistatic target doppler, f_{Tgt}, for all target positions on the perpendicular bisector. In the bistatic case, $\delta = 0°$ and $f_{Tgt} = (2V/\lambda)\cos(\beta/2)$. In the monostatic case, $\delta = \beta/2$; thus, $f_M = (2V/\lambda)\cos\delta = (2V/\lambda)\cos(\beta/2)$.

6.2 ISODOPPLER CONTOURS

When the target is stationary (e.g., on the earth) and the transmitter and receiver are moving (e.g., airborne), the bistatic doppler at the receiving site due to the combined transmitter and receiver motion is developed as follows. The terms dR_T/dt and dR_R/dt in (6.1b) are now the projections of the transmitting and receiving velocity vectors onto their respective target LOSs. Thus,

$$f_B = f_{TR} = (V_T/\lambda)\cos(\delta_T - \theta_T) + (V_R/\lambda)\cos(\delta_R - \theta_R) \tag{6.5}$$

where the terms are defined in Figure 6.1, and f_{TR} is the bistatic doppler shift caused by the combined transmitter and receiver motion. In the monostatic case, $V_T = V_R = V_M$, $\delta_T = \delta_R = \delta_M$, $\theta_T = \theta_R = \theta_M$, and (6.5) reduces to the monostatic doppler shift, f_M:

$$f_M = (2V_M/\lambda)\cos(\delta_M - \theta_M) \tag{6.6}$$

where $(\delta_M - \theta_M)$ is the angle between the monostatic radar's velocity vector and the radar-to-stationary-target LOS.

The locus of points for constant doppler shift on the earth's surface is called an *isodoppler contour,* or *"isodop."* In the monostatic case with a flat earth, these isodops are conic sections in three dimensions, in which the cone is defined by the radar's beamwidth. They are radial lines emanating from the radar in two dimensions. In the bistatic case, the isodops are skewed, depending on transmitting and receiving site kinematics, defined by V_T, V_R, δ_T, and δ_R. They can be developed analytically for the simple case of a coplanar flat earth and bistatic plane by setting f_{TR} = constant in (6.5) and solving for θ_R (or θ_T if appropriate):

$$\theta_R = \delta_R \pm \cos^{-1} \left[\frac{f_{TR} - (V_T/\lambda) \cos(\delta_T - \theta_T)}{(V_R/\lambda)} \right] \tag{6.7}$$

The sign of the $\cos^{-1}(\cdot)$ term is selected such that θ_R and θ_T intersect at some point, or points, on the bistatic plane.

Figure 6.3 is a plot of (6.7) in which the transmitter and receiver are at zero, or near zero, altitude and the following conditions exist [90]:

$V_T = V_R = 250$ m/s

$\delta_T = 0°$

$\delta_R = 45°$

$\lambda = 0.03$m

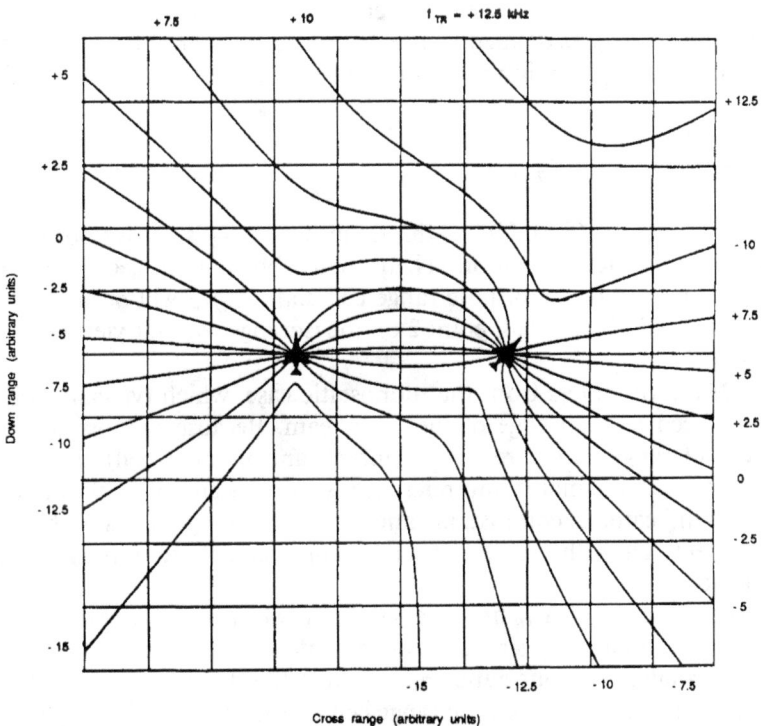

Figure 6.3 Bistatic isodoppler contours for a coplanar flat earth and bistatic plane [90].

The dimension of the grid on the bistatic plane is arbitrary; that is, the isodops are invariant with scale. On the left and right sides of Figure 6.3, the isodops approximate radial lines, which are pseudomonostatic operating points. Note that for some values of f_{TR} and θ_T two solutions for θ_R exist. For example, when $f_{TR} = -12.5$ kHz and $\theta_T = 135°$, $\theta_R = 187.5°$ and $-97.5°$. For other conditions, a single solution, or no solution, for θ_R occurs. The solutions change as the site kinematics change. Isodop calculations in three dimensions over a curved earth are usually done via a computer.

6.3 CLUTTER DOPPLER SPREAD

Bistatic mainbeam clutter doppler spread is defined as the maximum change in clutter doppler (or clutter doppler shift) over a bistatic range cell. The width, or down-range dimension, of a bistatic cell is defined by the separation between concentric isorange contours (ellipses), Section 4.6. The length, or cross-range dimension, of a bistatic range cell is established by the intersection of either the transmitting or receiving beam with the isorange contours. The beam having the smaller cross-range dimension at the target area sets the cell length. The maximum change in clutter doppler, Δf_{TR}, is defined as the difference between the maximum clutter doppler, $(f_{TR})_{max}$, and the minimum clutter doppler, $(f_{TR})_{min}$, over a bistatic range cell:

$$\Delta f_{TR} = (f_{TR})_{max} - (f_{TR})_{min} \tag{6.8}$$

where (f_{TR}) is given by (6.5). Under this definition, Δf_{TR} is always positive, which follows the monostatic convention [1]. In many, but not all, geometries, $(f_{TR})_{max}$ will occur at one end of the bistatic range cell and $(f_{TR})_{min}$ will occur at the other end. For example, in Figure 6.4, point 2 on the isorange contour yields $(f_{TR})_{max}$ and point 1 yields $(f_{TR})_{min}$.

This definition differs from the monostatic case, which typically defines the spread referenced to each edge of the mainbeam. Because monostatic spread is independent of range, either reference (mainbeam or range cell) gives the same results. The monostatic mainbeam reference is used because it applies to all range cells in the mainbeam. In contrast, bistatic spread in a range cell is both range and angle dependent. Thus, the more restrictive range cell definition must be used for bistatic spread.

The development of Δf_{TR} first treats the general case of clutter doppler spread established by each end of a single isorange contour. This case is useful for demonstrating an analog to monostatic spread. Next the development is extended to clutter doppler spread over a bistatic range cell bounded by two isorange contours, and finally to sidelobe clutter doppler spread.

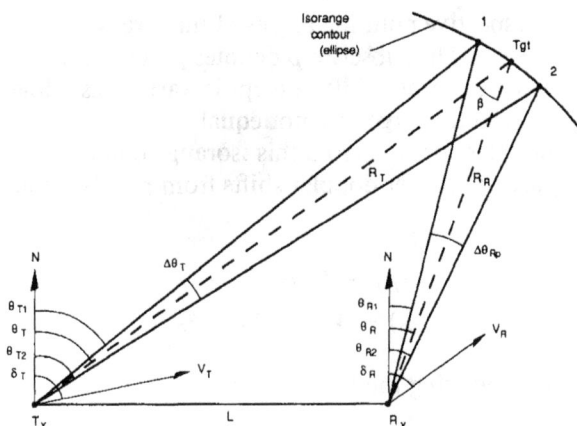

Figure 6.4 Geometry for clutter doppler spread over an isorange contour, with the transmitting beam establishing the clutter doppler spread.

A special case of clutter doppler spread occurs when the transmitter and receiver circle the target in counter-rotation, such that $(\delta_T - \theta_T) = \pm 90°$, and $(\delta_R - \theta_R) = \mp 90°$. For this case, the bistatic doppler from a (stationary) point target, defined by the coordinates (θ_T, θ_R), is zero as shown by (6.5). The bistatic doppler from both ends of the isorange contour can be either positive or negative, depending on the direction of transmitter and receiver rotation. When the bistatic doppler is positive, $(f_{TR})_{min}$ occurs at (θ_T, θ_R), near the center of the isorange contour; $(f_{TR})_{max}$ occurs at the end having the larger value of f_{TR}. When the bistatic doppler is negative, the max and min subscripts are reversed, but from (6.8), Δf_{TR} remains positive. This one-sided clutter doppler spread case is treated in Section 12.4.2.

6.3.1 Clutter Doppler Spread over an Isorange Contour

Figure 6.4 shows the geometry for determining bistatic doppler spread across a single isorange contour when the cross-range dimension of the transmitting beam, $R_T \Delta \theta_T$, is smaller than that of the receiving beam, $R_R \Delta \theta_R$, at the target area. Thus, the transmitting beam establishes the length of the isorange contour for clutter doppler spread, line (1,2) on Figure 6.4. Both transmitter and receiver are moving with respective velocity vectors (V_T, δ_T) and (V_R, δ_R), as shown in Figure 6.4. The target is stationary. The isorange contour is an ellipse of eccentricity $e = L/(R_T + R_R)$, intersecting the transmitting beam at points 1 and 2. The transmitting beam is defined by the conventional 3-dB beamwidth, $\Delta \theta_T$ and pointing angle θ_T. From Figure 6.4, $\Delta \theta_T = \theta_{T2} - \theta_{T1}$. Three rays are drawn from the receiver to points 1, 2, and the target. The respective pointing angles of these rays are θ_{R1}, θ_{R2}, and θ_R, with $\Delta\theta_{Rp}$ defined as $\theta_{R2} - \theta_{R1}$. The subscript p denotes pseudobeamwidth, because $\Delta\theta_{Rp} \neq \Delta\theta_R$, the receiver's 3-dB beamwidth, except in rare cases. Note that the bistatic angles at points 1, 2, and the target are not equal.

The bistatic doppler spread across this isorange contour, Δf_{TR}, is $f_{TR2} - f_{TR1}$, where f_{TR1} and f_{TR2} are the clutter doppler shifts from points 1 and 2 on Figure 6.4. Thus, from (6.5),

$$\Delta f_{TR} = \lambda^{-1}[V_T \cos(\delta_T - \theta_{T2}) + V_R \cos(\delta_R - \theta_{R2}) \\ - V_T \cos(\delta_T - \theta_{T1}) - V_R \cos(\delta_R - \theta_{R1})] \tag{6.9}$$

Rearranging and manipulating the terms yields

$$\Delta f_{TR} = (2/\lambda)\left[V_T \sin\left(\delta_T - \frac{\theta_{T1} + \theta_{T2}}{2}\right) \sin\left(\frac{\theta_{T2} - \theta_{T1}}{2}\right) \\ + V_R \sin\left(\delta_R - \frac{\theta_{R1} + \theta_{R2}}{2}\right) \sin\left(\frac{\theta_{R2} - \theta_{R1}}{2}\right) \right] \tag{6.10}$$

Now $(\theta_{T1} + \theta_{T2})/2 = \theta_T$, $(\theta_{T2} - \theta_{T1})/2 = \Delta\theta_T/2$ and $(\theta_{R2} - \theta_{R1})/2 = \Delta\theta_{Rp}/2$, as defined previously. The term $(\theta_{R1} + \theta_{R2})/2$ is the bisector of $\Delta\theta_{Rp}$, i.e., the center of the receiver's pseudobeamwidth. It is called θ_{Rp}, the receiver's pseudoreceiving beam pointing angle and, except in special geometries, $\theta_{Rp} \neq \theta_R$. Thus,

$$\Delta f_{TR} = (2/\lambda)[V_T \sin(\delta_T - \theta_T) \sin(\Delta\theta_T/2) \\ + V_R \sin(\delta_R - \theta_{Rp}) \sin(\Delta\theta_{Rp}/2)] \tag{6.11}$$

By invoking the small-angle approximation to the sine of an angle, (6.11) becomes

$$\Delta f_{TR} \approx \lambda^{-1}[V_T \sin(\delta_T - \theta_T)\Delta\theta_T + V_R \sin(\delta_R - \theta_{Rp})\Delta\theta_{Rp}] \tag{6.12}$$

In an analogue to the monostatic case (6.12) can be obtained directly by taking the differential of (6.5), where θ_R replaces θ_{Rp} and $\Delta\theta_R$ replaces $\Delta\theta_{Rp}$.* The foregoing brute-force development was given because, except in rare cases, $\theta_R \neq \theta_{Rp}$ and $\Delta\theta_R \neq \Delta\theta_{Rp}$, and depending on the geometry, the errors can be large.

Expressions for θ_{Rp} and $\Delta\theta_{Rp}$, in terms of θ_{R1} and θ_{R2}, which in turn are functions of θ_T, $\Delta\theta_T$, and ellipse eccentricity e, are developed in Appendix E. Results are

$$\theta_{R1} = \sin^{-1}\left[\frac{\sin(\theta_T - \Delta\theta_T/2)(1 + e^2) - 2e}{(1 + e^2) - 2e \sin(\theta_T - \Delta\theta_T/2)}\right] \tag{6.13}$$

$$\theta_{R2} = \sin^{-1}\left[\frac{\sin(\theta_T + \Delta\theta_T/2)(1 + e^2) - 2e}{(1 + e^2) - 2e \sin(\theta_T + \Delta\theta_T/2)}\right] \tag{6.14}$$

*The sign is ignored in this differentiation [1].

where

$$\theta_{Rp} = (\theta_{R1} + \theta_{R2})/2 \qquad (6.15)$$

and

$$\Delta\theta_{Rp} = (\theta_{R2} - \theta_{R1}) \qquad (6.16)$$

These expressions are exact for all θ_T, $\Delta\theta_T$, and $e < 1$. In the monostatic case, $e = 0$, $\theta_{R1} = \theta_T - \Delta\theta_T/2$ and $\theta_{R2} = \theta_T + \Delta\theta_T/2$. Thus, $\theta_{Rp} = \theta_T$ and $\Delta\theta_{Rp} = \Delta\theta_T$, as expected.

For special geometries, θ_{Rp} can be approximated by θ_R, where

$$\theta_{Rp} \approx \theta_R = \cos^{-1}[(R_T/R_R)\cos\theta_T] \qquad (6.17a)$$

$$= \cos^{-1}\left[\frac{(1 - e^2)\cos\theta_T}{(1 + e^2) - 2e\sin\theta_T}\right] \qquad (6.17b)$$

Similarly, $\Delta\theta_{RP}$ can be approximated as

$$\Delta\theta_{Rp} \approx (R_T/R_R)\Delta\theta_T \qquad (6.18a)$$

$$= \frac{(1 - e^2)\Delta\theta_T}{(1 + e^2) - 2e\sin\theta_T} \qquad (6.18b)$$

These approximations are developed in Appendix E and generate differences between the exact and approximate values of θ_{Rp} and $\Delta\theta_{Rp}$ of $<0.1°$ for $\Delta\theta_T \leq 10°$ and $e \leq 0.2$ when $\theta_T \geq 0°$. When $\theta_T < 0°$, values of e can be increased to <1.0 for $\Delta\theta_T \leq 10°$, which yield differences $<0.1°$. For differences other than $0.1°$, the curves given in Appendix E can be used.

The foregoing development assumes that the cross-range dimension of the transmitting beam, $R_T\Delta\theta_T$, is smaller than that of the receiving beam, $R_R\Delta\theta_R$, at the target area, and thus establishes the length of the isorange contour for clutter doppler spread. When the opposite case occurs, the subscripts R and T are interchanged in (6.11) through (6.18), and the $-2e$ terms in (6.13), (6.14), (6.17b), and (6.18b) become $+2e$ terms. Full expressions are given in Appendix E.

6.3.2 Clutter Doppler Spread over a Range Cell

For a given θ_T, $\Delta\theta_T$, V_T, V_R, δ_T, and δ_R, the bistatic doppler spread across any other isorange contour will be different from that calculated by (6.12) because θ_{Rp} and $\Delta\theta_{Rp}$ will change depending on the eccentricity of the new isorange contour. In Section 4.6, the bistatic range cell is defined by two concentric isorange contours separated by $c\tau/2$ where they intersect the extended baseline, and τ is the compressed pulsewidth.

If the eccentricity of the inner and outer isorange contours defining the bistatic range cell are e_i and e_o, respectively, e_o is related to e_i as follows:

$$e_i = L/2a \qquad\qquad (6.19)$$

where $2a$ is major axis of the inner isorange contour. Also,

$$e_o = L/2a' \qquad\qquad (6.20)$$

where $2a'$ is the major axis of the outer isorange contour. From (4.9) and (4.10), $a' - a = c\tau/2$. Thus,

$$e_o = L/(2a + c\tau) \qquad\qquad (6.21a)$$
$$= (e_i^{-1} + c\tau/L)^{-1} \qquad\qquad (6.21b)$$

Therefore, when a bistatic range cell is specified by e_i and $c\tau$ as some fraction of L, e_o can be calculated. Then, values for e_i and e_o are used to calculate respective values of θ_{Rp} and $\Delta\theta_{Rp}$ via (6.13) through (6.16) exactly, or (6.17) and (6.18) approximately. When $c\tau$ is a small fraction of L, the difference in bistatic doppler spread between the inner and outer isorange contours is small and usually can be ignored.

6.3.3 Clutter Doppler Spread through Sidelobes

The foregoing development for bistatic mainbeam clutter doppler spread is also used to calculate clutter doppler spread from a particular bistatic range cell in the receiving antenna sidelobes. For the case of a transmitting main beam illuminating a range cell that is in the receiving antenna's sidelobes, (6.12) applies directly. For the case of a transmitting sidelobe illuminating a range cell that is in the receiving antenna mainlobe or sidelobe, (6.12) is used, with the term θ_T defined as the pointing angle of the transmitting sidelobe and $\Delta\theta_T$ defined as the width of the transmitting sidelobe. This development assumes that $R_T\Delta\theta_T < R_R\Delta\theta_R$. If the opposite case holds, the subscripts and signs are interchanged as outlined previously. The power spectral density of clutter doppler returns through the receiving antenna's mainlobe or sidelobe is calculated by convolving appropriate transmitting and receiving antenna patterns with (6.12). Clutter doppler spread calculations for three dimensions over a curved earth are best done via a computer.

6.4 DOPPLER BEAT FREQUENCY

When the doppler-shifted target signal is mixed with the direct path signal in a bistatic receiver, a doppler beat frequency is produced. For the general case of a moving transmitter, receiver, and target, the beat frequency, or beat note, f_N, is the difference between the doppler shift over the transmitter-to-target-to-receiver path, f_{TTR}, and the doppler shift over the transmitter-to-receiver, or direct, path f_{DP}:

$$f_N = f_{TTR} - f_{DP} \tag{6.22}$$

Because all three sites are moving, f_{TTR} is the sum of (6.4b) and (6.5):

$$f_{TTR} = \frac{1}{\lambda}[2V\cos\delta\cos(\beta/2) + V_T\cos(\delta_T - \theta_T) \\ + V_R\cos(\delta_R - \theta_R)] \tag{6.23}$$

where the terms are defined in Figure 6.1. Equation (6.5) also defines f_{DP}, with $\theta_T = 90°$ and $\theta_R = -90°$. Thus,

$$f_{DP} = \frac{1}{\lambda}[V_T\cos(\delta_T - 90°) + V_R\cos(\delta_R + 90°)] \tag{6.24a}$$

$$= \frac{1}{\lambda}(V_T\sin\delta_T - V_R\sin\delta_R) \tag{6.24b}$$

Combining (6.22), (6.23), and (6.24b) yields

$$f_N = \frac{1}{\lambda}\{2V\cos\delta\cos(\beta/2) + V_T[\cos(\delta_T - \theta_T) - \sin\delta_T] \\ + V_R[\cos(\delta_R - \theta_R) + \sin\delta_R]\} \tag{6.25}$$

When the transmitting and receiving sites are stationary, $V_T = V_R = 0$ and (6.25) reduces to the bistatic doppler shift caused only by target motion, (6.4b).

Chapter 7
TARGET RESOLUTION

The definition of bistatic target resolution is identical to that of monostatic target resolution: the degree to which two or more targets (of approximately equal amplitude and arbitrary constant phase) may be separated in one or more dimensions, such as angle, range, velocity (or doppler), and acceleration [116, 154]. In the monostatic case, target separation is referenced to the radar-to-target LOS. In the bistatic case, target separation can conveniently be referenced to the bistatic bisector.

7.1 RANGE RESOLUTION

For monostatic and bistatic range resolution, an adequate degree of separation between two target echoes at the receiver is conventionally taken to be $c\tau/2$, where τ is the radar's (compressed) pulsewidth [155]. To generate $c\tau/2$ separation at a bistatic receiver, two point-scattering targets, such as targets 1 and 2 in Figure 7.1, must lie on bistatic isorange contours having a separation, ΔR_B, that is approximately $c\tau/2 \cos(\beta/2)$ (Section 4.6). When a line joining the two targets is not colinear with the bistatic bisector, but at an aspect angle ψ with respect to the bistatic bisector, such as for targets 1 and 3 in Figure 7.1, their physical separation ΔR_ψ must be approximately $\Delta R_B/\cos\psi$. Hence,

$$\Delta R_\psi \approx c\tau/[2 \cos(\beta/2) \cos\psi] \tag{7.1}$$

When (7.1) is satisfied, the projection of ΔR_ψ onto the bistatic bisector satisfied the $c\tau/2 \cos(\beta/2)$ requirement, which, in turn, generates a $c\tau/2$ target separation at the bistatic receiver. Obviously, targets with separation $>\Delta R_\psi$ will also be resolved; thus, (7.1) represents a minimum requirement for target separation.

When $\psi = 0°$, $\Delta R_\psi = \Delta R_B \approx c\tau/[2 \cos(\beta/2)]$, the case for targets 1 and 2 in Figure 7.1. When $\psi = \pm\beta/2$, $\Delta R_\psi \approx c\tau/[2 \cos^2(\beta/2)]$, the case for target 3 lying on the transmitter-to-target 1 LOS or the receiver-to-target 1 LOS. Both expressions

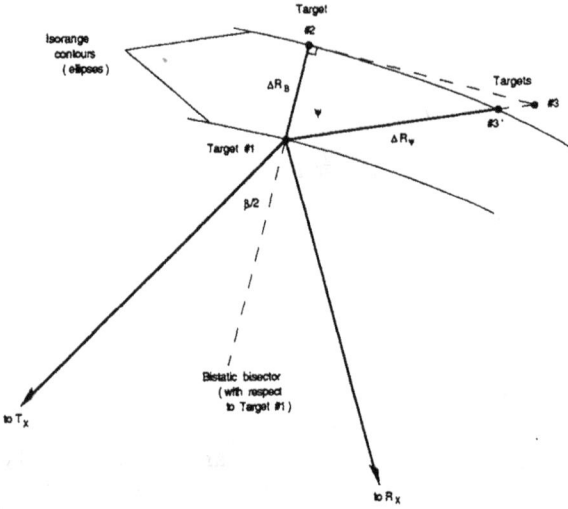

Figure 7.1 Geometry for bistatic range resolution.

have been used for bistatic range resolution, and both are special cases of the general expression (7.1), the required target *spatial* separation to achieve a resolution of $c\tau/2$ at the bistatic receiver. When $\beta = 0$, $\Delta R_\psi = c\tau/(2\cos\psi)$, the monostatic (or pseudomonostatic) case for a monostatic radar located on the bistatic bisector.

Two approximations were used to develop (7.1), one for ΔR_B and one for ΔR_ψ. The ΔR_B approximation is developed in Appendix B and discussed in Section 4.6. It shows that for all β the approximation is always greater than or equal to the exact expression for bistatic isorange contour separation. Also, for most geometries, the approximation represents a reasonably "tight" upper bound to the exact expression. Thus, when the approximate expression is used to estimate bistatic resolution two targets slightly closer together will also be resolved.

In this sense the approximation is a conservative estimate of bistatic resolution capability. Only when the isorange contours (ellipses) become highly eccentric ($e > 0.5$) and the pseudomonostatic range cell, $\Delta R_M = c\tau/2$, becomes a significant fraction of the baseline ($\Delta R_M > 0.1L$) should the exact expression given in Appendix B be used.

The ΔR_ψ approximation assumes that (1) the outer isorange contour (ellipse) of Figure 7.1 is a straight line, specifically a tangent to the outer ellipse at target 2, and (2) the bistatic bisector at target 1 on the inner ellipse is also the bistatic bisector at target 2 on the outer ellipse, and thus perpendicular to the straight line. The errors associated with these approximations involve calculations even more tedious than those for the ΔR_B approximation (Appendix B) and have not been developed.

The straight line assumption sets an upper limit to the length of ΔR_ψ for all points on an outer ellipse of any eccentricity. For example, in Figure 7.1, the distance between targets 1 and 3, established by the approximation, is always greater than the distance between targets 1 and 3′, where target 3′ lies on the outer ellipse. In this sense, the ΔR_ψ approximation is again a conservative estimate of the bistatic resolution capability because two targets slightly closer together will also be resolved, in this case targets 1 and 3′.

The straight line approximation appears to be a reasonably tight upper bound on the exact value of ΔR_ψ when (1) ΔR_ψ is on the order of ΔR_B; or (2) the outer ellipse is large, $(R_T + R_R) \gg L$ (not too eccentric), where $e = L/R_T + R_R) \lesssim 0.1$, and the ellipse approximates a circle; or (3) the outer ellipse is eccentric, $e \gtrsim 0.1$, but the targets lie near the intersection of the perpendicular bisector of the baseline (minor axis) and the ellipse, where the ellipse is relatively flat. Note that for case (3), when the targets lie near the ellipse endpoints at the intersection of the extended baseline (major axis) and the ellipse, the ellipse curvature is large and the approximation can give a value much greater than the exact value.

The bistatic bisector assumption is exact for four points on any pair of ellipses: the intersection of the minor axis and the ellipses, and the intersection of the major axis and the ellipses. At other points the approximation appears reasonable when the inner ellipse is not highly eccentric and the separation between ellipses is small. These conditions are roughly the same as for the ΔR_B approximation, where that approximation was found to be reasonable for $e < 0.5$ and $0.1L > \Delta R_M$.

An equivalent approach to the bistatic bisector assumption is to observe that it is exact when respective rays from the transmitting and receiving beams at targets 1 and 2 are parallel. This situation is approximated when the target is a long distance from both the transmitter and receiver. At long bistatic range $(R_T + R_R \gg L$, e is small, and the ellipses approximate circles, with β approaching zero at all points on both circles.

Finally, when either $\beta/2$ or ψ approach 90°, (7.1) shows ΔR_ψ approaching infinity, which is clearly incorrect because two finite, concentric ellipses will always bound the value of ΔR_ψ. (Note that when $\beta/2$ approaches 90°, ΔR_R also approaches infinity, which is where the approximation to ΔR_R breaks down.) A practical limit to the physical separation between targets, ΔR_ψ, is when ΔR_ψ increases to the point where angle resolution is achieved. This limit is developed in Section 7.3.

7.2 DOPPLER RESOLUTION

For monostatic and bistatic doppler resolution, an adequate degree of doppler separation between two target echoes at the receiver, f_{Tgt1} and f_{Tgt2}, respectively, is conventionally taken to be $1/T$, where T is the receiver's coherent processing interval [155]. Thus, the requirement for doppler resolution is

$$|f_{Tgt1} - f_{Tgt2}| = \frac{1}{T} \tag{7.2}$$

where, again, the equality represents a minimum requirement for doppler separation. In the bistatic case, (6.4b) defines bistatic target doppler as

$$f_{Tgt1} = (2V_1/\lambda) \cos\delta_1 \cos(\beta/2) \tag{7.3}$$

$$f_{Tgt2} = (2V_2/\lambda) \cos\delta_2 \cos(\beta/2) \tag{7.4}$$

The geometry for V_1, V_2, δ_1, and δ_2 is shown in Figure 7.2. The two targets are assumed to be colocated so that they share a common bistatic bisector. Combining (7.2), (7.3), and (7.4) yields

$$\Delta V = (V_1 \cos\delta_1 - V_2 \cos\delta_2) \tag{7.5a}$$

$$= \lambda/[2T \cos(\beta/2)] \tag{7.5b}$$

where ΔV is the required difference between the two target velocity vectors, projected onto the bistatic bisector, for adequate bistatic doppler resolution. When $\beta = 0°$, $\Delta V = \lambda/2T$, the monostatic or pseudomonostatic case.

In the development of (7.5b), the two targets were assumed to be collocated, so that they share a common bistatic bisector. This collocation restriction can be relaxed, as long as (1) their separation is not sufficient to allow resolution in another dimension, such as range or angle, (7.1) and (7.6), respectively; and (2) the angle between their bistatic bisectors is small. The second condition is discussed in Section 7.1.

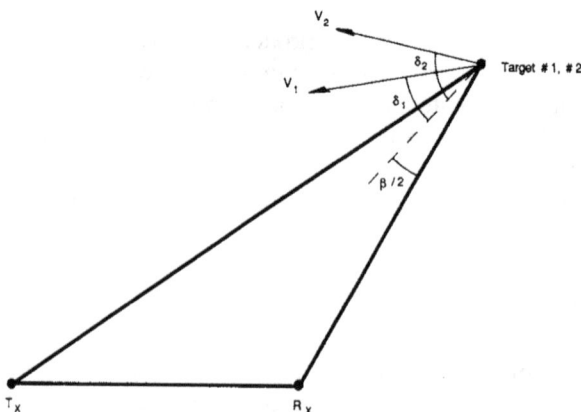

Figure 7.2 Geometry for bistatic doppler resolution.

7.3 ANGLE RESOLUTION

In a monostatic radar, angle resolution is conventionally taken as $\geq \Delta\theta_M$, the 3-dB (one-way) antenna beamwidth [153, 155]. That is, the amplitude returns of two targets can be observed to separate when the targets are separated in angle by at least the 3-dB (one-way) beamwidth. This angular separation is equivalent to a physical target separation of $\geq \Delta\theta_M R_M$, the cross-range dimension of the monostatic beam, where R_M is the monostatic target range.

This monostatic convention invokes two-way antenna directivity, in which both the transmitting and receiving beams are used for resolution. In this case a plot of the target's received signal strength versus angle is narrower and more sharply peaked than that from the receiver's one-way directive pattern. As a result the targets need only be separated by the 3-dB (one-way) beamwidth to achieve resolution, rather than the more intuitive 6-dB, or even null-to-null (one-way) beamwidth ($\approx 2\Delta\theta_M R_M$).

In a bistatic radar, the cross-range dimensions of the transmitting and receiving beams at the target, $\Delta\theta_T R_T$ and $\Delta\theta_R R_R$, respectively, are usually different. This difference is caused by different ranges to the target, i.e., $R_T \neq R_R$, and by the usually different-sized transmitting and receiving apertures, such that $\Delta\theta_T \neq \Delta\theta_R$. Hence, the two-way antenna directivity advantage of a monostatic radar cannot usually be applied to the bistatic case. Furthermore, geometry degrades angle resolution in all but some special cases.

Figure 7.3 shows the geometry for two targets lying on the same isorange contour, and $\Delta\theta_T R_T > 2\Delta\theta_R R_R$; that is, the cross-range dimension of the 3-dB transmitting beam is larger than the receiver's null-to-null beam cross-range dimension. Thus, it is too large to contribute to angle resolution. This geometry typically occurs when the target is near the receiver site.* In this case an adequate (but possibly conservative) degree of angle separation between the two targets is the receiver's null-to-null beamwidth, $2\Delta\theta_R$, and a corresponding, required cross-range dimension of $2\Delta\theta_R R_R$, as shown in Figure 7.3.

To generate a $2\Delta\theta_R R_R$ cross-range separation, the two targets must be physically separated by $(\Delta R_\theta)_u$, where

$$(\Delta R_\theta)_u \approx 2\Delta\theta_R R_R/\sin(90 - \beta/2) \tag{7.6a}$$

$$\approx 2\Delta\theta_R R_R/\cos(\beta/2) \tag{7.6b}$$

and the subscript u denotes unequal cross-range dimensions of the transmitting and receiving beams.

---

*This development also applies to the reciprocal case when $\Delta\theta_R R_R > 2\Delta\theta_T R_T$.

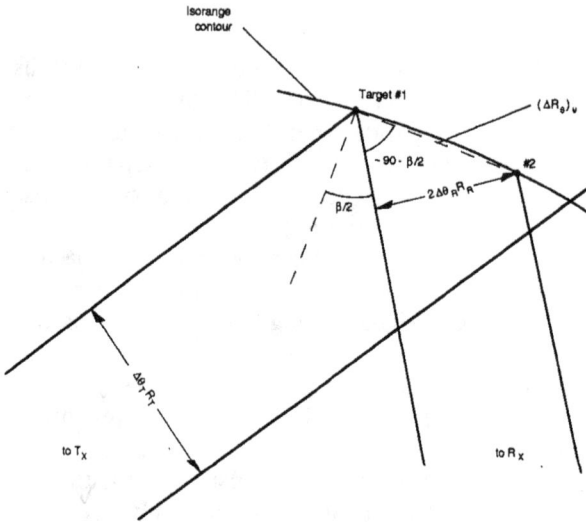

Figure 7.3 Geometry for angle resolution with only one antenna beam contributing to resolution, $\Delta\theta_T R_T > 2\Delta\theta_R R_R$.

Figure 7.4 shows the geometry, again for two targets lying on the same iso-range contour, but now for the special case of $\Delta\theta_T R_T = \Delta\theta_R R_R$; that is, equal cross-range dimensions for the transmitting and receiving beams at the target, such that both beams contribute to angle resolution. In this case, an adequate degree of cross-range separation between the two targets is $\Delta\theta_R R_R$ $(= \Delta\theta_T R_T)$. Note that an angle separation criterion cannot be used because $R_R \neq R_T$ in most geometries. To generate a $\Delta\theta_R R_R$ cross-range separation, the two targets must be physicially separated by $(\Delta R_\theta)_e$, where

$$(\Delta R_\theta)_e \approx \Delta\theta_R R_R/\cos(\beta/2) \tag{7.7a}$$

$$\approx \Delta\theta_T R_T/\cos(\beta/2) \tag{7.7b}$$

and the subscript e denotes equal cross-range dimensions of the transmitting and receiving beams. When $\beta = 0$, $(\Delta R_\theta)_e = \Delta\theta_R R_R = \Delta\theta_T R_T$, the monostatic or pseudomonostatic case. Note, however, that when $\Delta\theta_R R_R = \Delta\theta_T R_T$ at $\beta = 0$, $\Delta\theta_R R_R \neq \Delta\theta_T R_T$ at other β; thus, (7.6a) must be used.

There are regions on the bistatic plane where $\Delta\theta_R R_R > \Delta\theta_T R_T < 2\Delta\theta_R R_R$.† In these transition regions, the transmitting beam contributes an incremental amount

---

†Again, for the reciprocal case, $\Delta\theta_T R_T > \Delta\theta_R R_R < 2\Delta\theta_T R_T$.

137

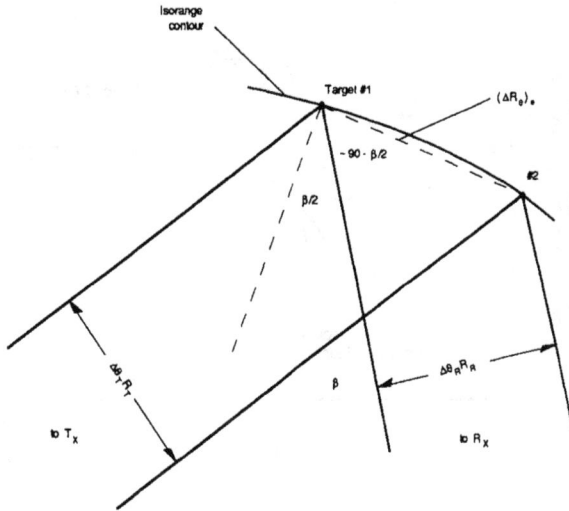

Figure 7.4 Geometry for angle resolution with both antenna beams contributing equally to resolution ($\Delta\theta_T R_T = \Delta\theta_R R_R$).

of angle resolution, depending on its beam shape and the geometry. The general case is shown in Figure 7.5, which is a plot of the required physical separation between two targets to achieve angle resolution, normalized to $\Delta\theta_R R_R/\cos(\beta/2)$, as a function of the ratio of the cross-range dimension of the transmitting and receiving beams. For $\Delta\theta_T R_T/\Delta\theta_R R_R > 2$, (7.6b) applies. For $\Delta\theta_T R_T/\Delta\theta_R R_R = 1$, (7.7b) applies. For $\frac{1}{2} < \Delta\theta_T R_T/\Delta\theta_R R_R < 2$, the transition region is defined, where the curve approximates the transmitting mainbeam Gaussian roll-off. For the reciprocal case, a similar curve is obtained, where the curve approximates the receiving mainbeam roll-off. In general, a conservative estimate for bistatic angle resolution is (7.6b), unless special geometries can be invoked.

A final geometry is shown in Figure 7.6, where two targets lie on different isorange contours, separated by a bistatic range cell width ΔR_B. The physical separation between targets ΔR_ψ required to achieve angle resolution is also the practical limit for target physical separation in the bistatic range resolution case (Section 7.1). It is developed as follows, assuming $\Delta\theta_T R_T > 2\Delta\theta_R R_R$, such that (7.6b) applies. From the law of cosines:

$$\Delta R_\psi^2 \approx \frac{\Delta R_B^2}{\cos^2(\beta/2)} + \frac{(2\Delta\theta_R R_R)^2}{\cos^2(\beta/2)} - \frac{2(\Delta R_B)(2\Delta\theta_R R_R)}{\cos^2(\beta/2)}\cos(90-\beta/2) \quad (7.8a)$$

$$\Delta R_\psi \approx \frac{1}{\cos(\beta/2)}(\Delta R_B^2 + 4\Delta\theta_R^2 R_R^2 - 4\Delta\theta_R R_R\Delta R_B\sin(\beta/2))^{1/2} \quad (7.8b)$$

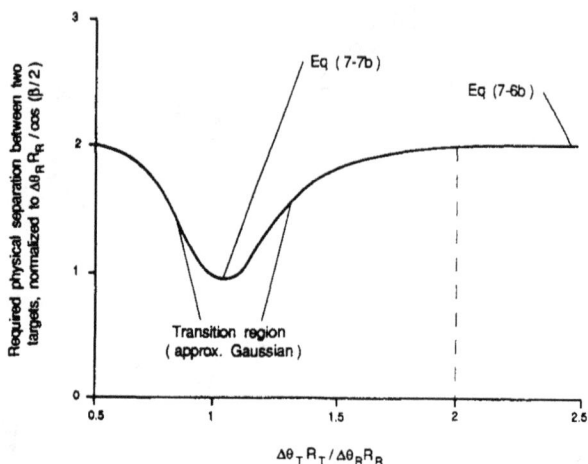

Figure 7.5 General case for angle resolution.

When the cross-range resolution is much larger than the range resolution, i.e., $2\Delta\theta_R R_R \gg \Delta R_B$, the typical case,

$$\Delta R_\psi \approx 2\Delta\theta_R R_R/\cos(\beta/2) \tag{7.9a}$$

$$\approx (\Delta R_\theta)_u \tag{7.9b}$$

In developing expressions for angle resolution, three approximations were made: (1) parallel rays for respective transmitting and receiving beams, (2) straight lines for isorange contours, and (3) a null-to-null beamwidth, $2\Delta\theta_R$ (or $2\Delta\theta_T$), angle separation between two targets required for angle resolution, (7.6b). Limits associated with the first two approximations are discussed in Section 7.1. However, errors associated with these approximations are typically smaller than that associated with the conservative beamwidth assumption and usually can be ignored.

7.4 SYNTHETIC APERTURE RADAR ISORANGE RESOLUTION

In a monostatic radar, the SAR cross-range resolution τ_c in a squint mode is given as [1]:

$$\tau_c = \frac{\lambda R_M}{2V_M T \sin|\delta_M - \theta_M|} \tag{7.10}$$

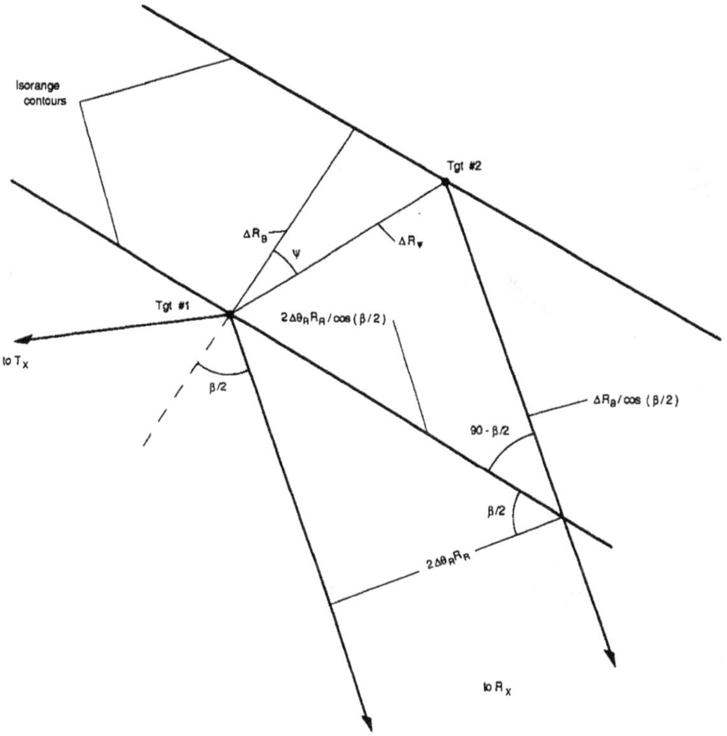

Figure 7.6 Geometry for angle resolution when two targets lie on different isorange contours.

where

$$R_M = \text{radar-to-target range,}$$
$$V_M = \text{velocity of radar,}$$
$$|\delta_M - \theta_M| = \text{angle between the radar-to-target LOS and the radar's velocity}$$
vector (always taken as a positive value in the North-referenced
coordinate system),
$$T = \text{array, or coherent integration, time,}$$
$$\lambda = \text{wavelength.}$$

In practical SAR implementations, T is extended by a factor K_s to account for weighting to reduce azimuth sidelobes. For -30-dB sidelobes, $K_s \approx 1.2$ [133].

Because the angle of rotation, ρ, of the monostatic radar about the target area in time, T, is

---

$$\rho \approx (V_M T/R_M) \sin |\delta_M - \theta_M| \tag{7.11}$$

(7.10) becomes

$$\tau_c \approx \frac{\lambda}{2\rho} \tag{7.12}$$

The factor of 2 accounts for the two-way propagation path, with each path contributing to the cross-range resolution. In bistatic SAR, the transmitting and receiving propagation paths and their respective velocity vectors are usually different. Thus, for the bistatic case, (7.12) becomes

$$\tau_i \approx \frac{\lambda}{\rho_T + \rho_R} \tag{7.13}$$

where

τ_i = bistatic radar isorange resolution,
ρ_T = angle of rotation of the transmitter about the target area (referenced to the bistatic bisector),
ρ_R = angle of rotation of the receiver about the target area (referenced to the bistatic bisector).

Note that τ_i is the resolution along a bistatic isorange contour, an ellipse, and that the rotation angles are referenced to the common bistatic bisector.

The terms ρ_T and ρ_R are now developed. Figure 7.7 shows the geometry for ρ_T. The transmitter travels a distance $V_T T$ in T, starting at an angle $\delta_T - \theta_T$ with respect to the target. This distance is translated to the target position as line $\overline{AB}$. When $\overline{AB}$ is projected onto the cross-range dimension of the transmitter-to-target LOS, the projection has a length $\overline{CD} = V_T T \sin(\delta_T - \theta_T)$.

Although not shown in Figure 7.7, the angle subtending this cross-range dimension is approximately $(V_T T/R_T) \sin(\delta_T - \theta_T)$, which is ρ in (7.11) when the transmitter operates as a monostatic radar. A further projection of the vector onto the isorange contour $\overline{EF}$ yields

$$\overline{EF} \approx V_T T \sin(\delta_T - \theta_T) \cos(\beta/2) \tag{7.14}$$

The angle subtending $\overline{EF}$ on the bistatic bisector, ρ_T, is

$$\rho_T \approx (V_T T/R_T) \sin(\delta_T - \theta_T) \cos(\beta/2) \tag{7.15}$$

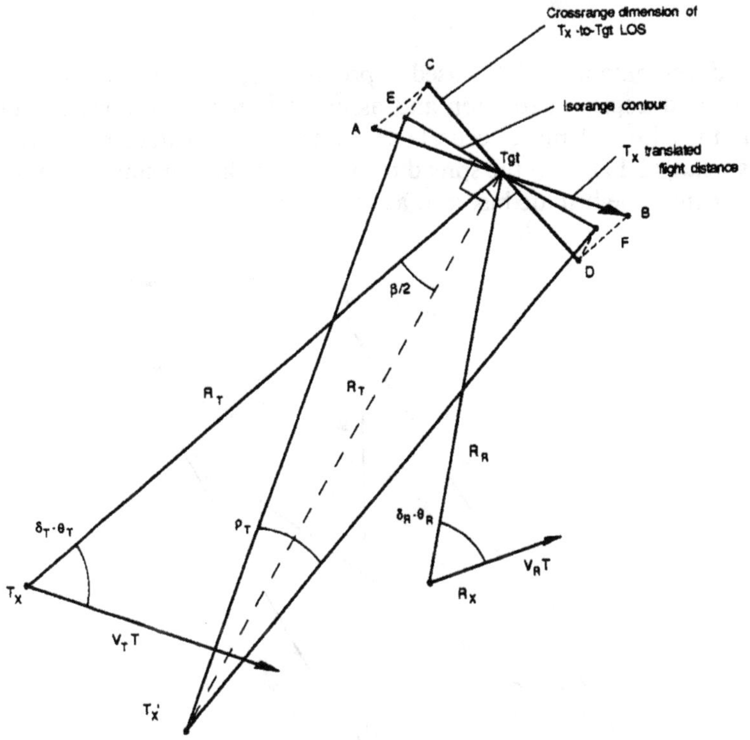

Figure 7.7 Geometry for determining transmitter's angle of rotation, ρ_T, about the target, referenced to the bistatic bisector, for bistatic SAR isorange resolution.

A similar development is made for ρ_R:

$$\rho_R \approx (V_R T/R_R)\sin(\delta_R - \theta_R)\cos(\beta/2) \tag{7.16}$$

Combining (7.13), (7.15), and (7.16) yields

$$\tau_i \approx \frac{\lambda}{T\cos(\beta/2)}\left(\left|\frac{V_T\sin(\delta_T - \theta_T)}{R_T} + \frac{V_R\sin(\delta_R - \theta_R)}{R_R}\right|\right)^{-1} \tag{7.17}$$

The two sine terms in parentheses are the respective transmitter and receiver angular rates about the target [133]. When the transmitter and receiver rotate about the target in the same direction, their angular rates add; when they counter-rotate about the target, their angular rates subtract. Following the monostatic convention, the resultant angular rate is taken positive, which yields a positive value for τ_i. When $V_T = V_R$, $\delta_T = \delta_R$, $\theta_T = \theta_R$, $R_T = R_R$, and $\beta = 0$, (7.17) collapses to (7.10), the sanity check.

The development of (7.17) used approximate expressions for velocity vector projections and angles of rotation and assumed that the isorange contour was a tangent to the ellipse. A more accurate development of τ_i can be made using clutter doppler spread, (6.11), with the spread referenced to the transmitting and receiving sites, rather than the bistatic bisector, as shown in Figure 7.8.

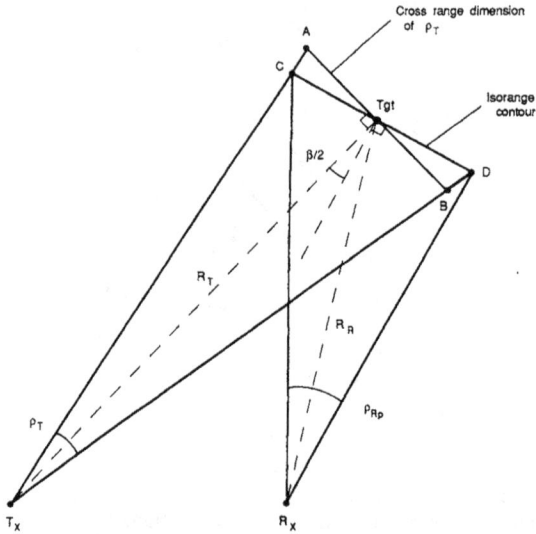

Figure 7.8 Geometry for clutter doppler spread calculations of bistatic SAR isorange resolution.

In the monostatic SAR case, clutter doppler spread, Δf_M, is given as [1]:

$$\Delta f_M = (2T/\lambda R_M)V_M^2 \sin^2(\delta_M - \theta_M) \tag{7.18}$$

Combining (7.10), (7.11), and (7.18) yields

$$\tau_c \approx \frac{\rho R_M}{T\Delta f_M} \tag{7.19}$$

where ρR_M is the cross-range dimension of the "SAR beam" defined by ρ. For example, in Figure 7.8, if the bistatic transmitter operates as a monostatic radar, $\overline{AB} = \rho_T R_T$.

In the bistatic SAR case, resolution across the isorange contour, $\overline{CD}$ in Figure 7.8, is required. Thus, τ_i is defined as

$$\tau_i \approx \frac{\overline{CD}}{T \Delta f_{TR}} \tag{7.20}$$

where Δf_{TR} is the bistatic clutter doppler spread, (6.12). Because $\overline{CD} = \overline{AB}/\cos(\beta/2) = \rho_T R_T/\cos(\beta/2)$, (7.20) becomes

$$\tau_i \approx \frac{\rho_T R_T}{T \Delta f_{TR} \cos(\beta/2)} \tag{7.21}$$

which assumes that $\rho_T < \rho_R$, such that ρ_T establishes the length of the isorange contour over which the bistatic SAR operates. The term ρ_{R_p} in Figure 7.8 is the pseudoreceiving rotation angle, analogous to the pseudoreceiving beamwidth $\Delta\theta_{R_p}$, of Section 6.3. When ρ_T is established by (7.15), it becomes the term $\Delta\theta_T$ in the bistatic clutter doppler spread equation, (6.11), which can now be solved, given the receiver's velocity vector (V_R, δ_R).

By invoking the approximations, $\theta_{R_p} \approx \theta_R$ and $\Delta\theta_{R_p} \approx (R_T/R_R)\Delta\theta_T$ (or the analogous approximation $\rho_{R_p} \approx (R_T/R_R)\rho_T$), (6.17a) and (6.18a), respectively, the spread becomes

$$\Delta f_{TR} \approx (\rho_T/\lambda) |\, V_T \sin(\delta_T - \theta_T) + (V_R R_T/R_R) \sin(\delta_R - \theta_R)| \tag{7.22}$$

where again only a positive value of spread is taken. Combining (7.21) and (7.22) yields (7.17), the expression for bistatic SAR isorange resolution developed from rotation angles about the target. Also note that combining (7.15) and (7.22), and assuming $\beta = 0$ and the other usual equalities for monostatic operation, yields (7.18), the sanity check. Because $\rho_T \ll 1°$ in most cases of interest, the approximations for θ_{R_p} and $\Delta\theta_{R_p}$ are reasonably accurate for ellipse eccentricites, $e < 0.99$, allowing (7.17) to be used for most geometries.

Chapter 8
TARGET CROSS SECTION

The bistatic radar cross section of a target, σ_B, is a measure as is the monostatic RCS, σ_M, of the energy scattered from the target in the direction of the receiver. Bistatic cross sections are more complex than monostatic cross sections because σ_B is a function of aspect angle and bistatic angle.

Three regions of bistatic RCS are of interest: pseudomonostatic, bistatic, and forward-scatter (sometimes called near-forward-scatter). Each region is defined by the bistatic angle. The extent of each region is set primarily by physical characteristics of the target.

8.1 PSEUDOMONOSTATIC RCS REGION

The Crispin and Siegel monostatic-bistatic equivalence theorem applies in the pseudomonostatic region [36]: For vanishingly small wavelengths, the bistatic RCS of a sufficiently smooth, perfectly conducting target is equal to the monostatic RCS measured on the bisector of the bistatic angle. Sufficiently smooth targets typically include spheres, elliptic cylinders, cones, and ogives. Figure 8.1 shows the theoretical bistatic RCS of two perfectly conducting spheres as a function of bistatic angle [1, 92–95]. For the larger sphere (near the optics region), the pseudomonostatic region extends to $\beta \approx 100°$, with an error of 3 dB. Even for the smaller sphere (in the resonance region), the pseudomonostatic region extends to $\beta \approx 40°$.

For targets of more complex structure, the extent of the pseudomonostatic region is considerably reduced, and a variation of the equivalence theorem developed by Kell [41] applies to this case: For small bistatic angles, typically less than 5°, the bistatic RCS of a complex target is equal to the monostatic RCS measured on the bisector of the bistatic angle at a frequency lower by a factor of $\cos(\beta/2)$.

Kell's complex targets are defined as an assembly of discrete scattering centers (simple centers such as flat plates, reflex centers such as corner reflectors, skewed reflex centers such as a dihedral with corner not equal to 90°, and stationary phase regions for creeping waves). When the wavelength is small compared to the target

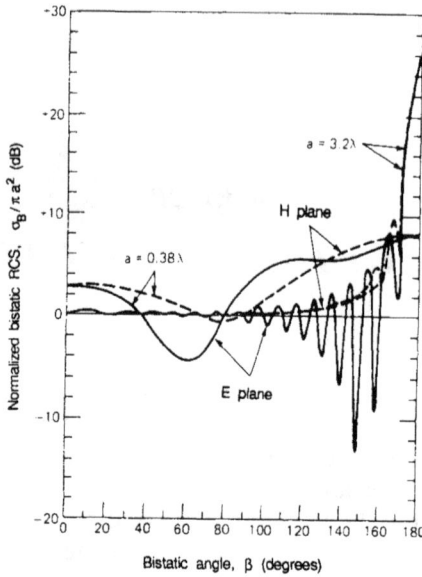

Figure 8.1 Theoretical bistatic RCS for two perfectly conducting spheres [1, 92–95].

dimensions, these complex target models approximate many aircraft, ships, ground vehicles, and some missiles. The targets can be composed of conducting and dielectric materials.

The $\cos(\beta/2)$ frequency reduction term has little effect in Kell's pseudomonostatic region, $0 < \beta \lesssim 5°$ because a 5° bistatic angle corresponds to less than a 0.1% shift in wavelength. At $\beta > 5°$ the change in radiation properties from discrete scattering centers is likely to dominate any $\cos(\beta/2)$ frequency reduction effect [41]. Thus, the $\cos(\beta/2)$ term is often ignored.

Both versions of the equivalence theorem are valid when the positions of the transmitter and receiver are interchanged, given that the target scattering media are reciprocal. Most media are reciprocal. Exceptions are gyrotropic media, such as ferrite materials and the ionosphere [103].

Whenever the equivalence theorem is valid, Kell [41] provides a simple method for deriving bistatic RCS data from monostatic RCS data when plotted as a function of target aspect angle. Bistatic RCS data for the same polarization are obtained by translating along the target aspect angle axis by one-half the desired bistatic angle. If monostatic RCS data are also available as a function of frequency, the monostatic curve for $f\sec(\beta/2)$, where f is the bistatic frequency, is used to estimate the bistatic RCS at f. As outlined earlier, this correction is usually small.

8.2   BISTATIC RCS REGION

The bistatic angle at which the equivalence theorem fails to predict the bistatic RCS identifies the start of the second bistatic region. In this region the bistatic RCS diverges from the monostatic RCS. Kell [41] identified three sources of this divergence for complex targets and for a target aspect angle fixed with respect to the bistatic-bisector. These sources are: (1) changes in relative phase between discrete scattering centers, (2) changes in radiation from discrete scattering centers, and (3) changes in the existence of centers—appearance of new centers or disappearance of those previously present.

The first source is analogous to fluctuations in monostatic RCS as the target aspect angle changes, but now the effect is caused by a change in bistatic angle [104]. For example, when significant coupling occurs between two discrete scattering centers separated by $\Delta R_B \sin\psi$ on an isorange contour as defined in Figure 7.1, σ_B can be reduced in regions near $(\beta\Delta R_B \sin\psi) = (n + 0.5)$, $n = 0,1,2, \ldots$ The reduction can vary from a few decibels to more than 20 dB depending on the magnitude of coupling [104].

The second source occurs when, for example, the discrete scattering center reradiates (i.e., retroreflects) energy toward the transmitter, and the receiver is positioned on the edge of, or outside, the retroflected beamwidth; thus, the received energy is reduced. (The specific geometry that generates this effect is developed in Section 12.3, where the beamwidth of the discrete retroreflector [136] replaces that of the retrodirective system analyzed in Section 12.3.) The third source is typically caused by shadowing, for example, by an aircraft fuselage blocking one of the bistatic paths, the transmitter or receiver LOS to a scattering center.

In general, this divergence results in a bistatic RCS lower than the monostatic RCS for complex targets. Exceptions include (1) some target aspect angles that generate a low monostatic RCS and a high bistatic specular RCS at specific bistatic angles, (2) targets that are designed for low monostatic RCS over a range of aspect angles, and (3) shadowing that sometimes occur in a monostatic geometry and not in a bistatic geometry [92].

Ewell and Zehner [97] measured the monostatic and bistatic RCS of coastal freighters at X-band near grazing incidence for both the transmitter and receiver. Figure 8.2 shows the results for median values of RCS. The ranges of median monostatic RCS measurements are as follows:

Kansas:	23.5–31.4 dB/m²
Hellenic Challenger:	33.0–37.6
Phillipean Rizal:	28.5–35.7
Delaware Sun:	25.3–31.9

Although these ships lacked the extensive superstructure of combat vessels, and thus were less complex targets as defined by Kell, the data match Kell's model: The

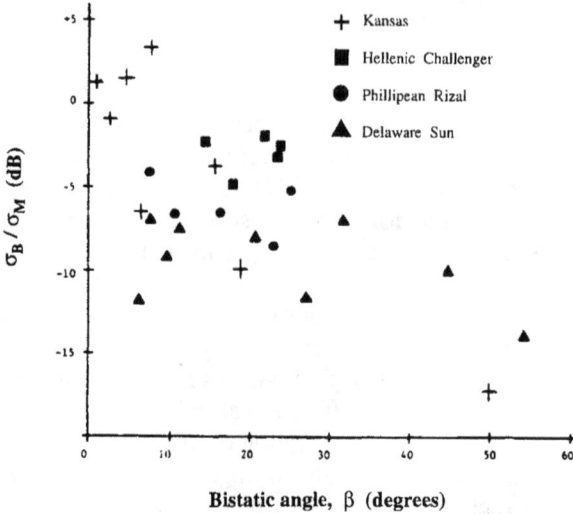

Figure 8.2 Ratio of bistatic to monostatic median RCS for four coastal freighters measured at X-band near grazing incidence. (© 1980 IEEE [97].)

bistatic RCS is reduced starting at $\beta = 5°$ in most cases. The three data points for the Kansas that are larger than the monostatic RCS at $\beta < 10°$ suggest that the bistatic RCS divergence is not always uniformly lower.

This divergence effect also appears, but to a lesser extent, as the wavelength becomes relatively large compared to the target dimension (near the resonance RCS region). Figure 8.3(a) shows the measured maximum monostatic and bistatic RCS of a large jet aircraft at 250 MHz and horizontal polarization [92]. The bistatic RCS is generally lower than the monostatic RCS over most aspect angles, and at some aspect angles 10 dB lower. The measured median RCS values (Figure 8.3b) also exhibited this divergence, except at $\beta = 135°$, which exceeded the monostatic value over about a third of the angles. Thus, again the bistatic divergence is not always uniformly lower. (One is tempted to ascribe this higher bistatic RCS to the onset of forward-scatter enhancement, Section 8.4, even though the effect does not appear at all aspect angles. The data, however, are inconclusive on this assertion.)

8.3 GLINT REDUCTION IN THE BISTATIC RCS REGION

A second effect can occur in the bistatic region. When the bistatic RCS reduction is caused by a loss or attenuation of large, discrete scattering centers (for example, through shadowing), target glint is often reduced. Target glint is the angular displacement in apparent phase center of a target return. It is caused by the phase interference between two or more dominant scatterers within a radar resolution

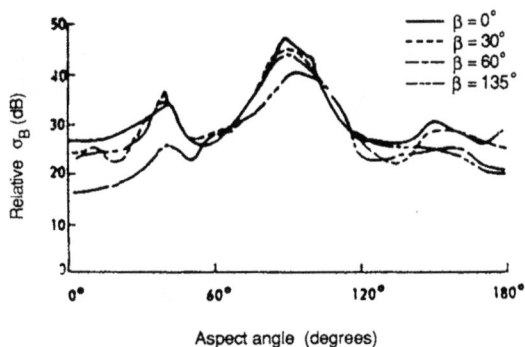

a) 10° Maximum values

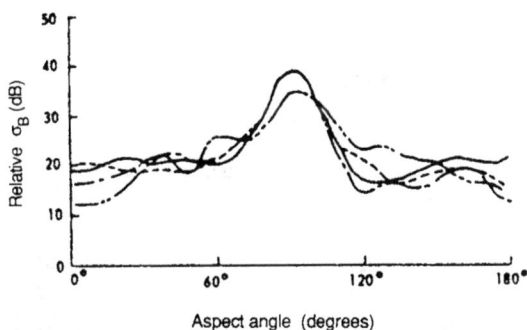

b) 10° Median values

Figure 8.3 Relative monostatic and bistatic RCS for a large jet aircraft at 250 MHz, horizontal polarization, where aspect angle is the angle between the aircraft nose and the bistatic bisector in the horizontal plane. (Courtesy of Academic Press [92].)

cell. As the target aspect angle changes, the apparent phase center shifts, often with excursions beyond the physical extent of the target. These excursions can significantly increase the errors in angle tracking or measurement systems. When the returns from dominant scatterers are reduced in the bistatic region, the source and hence the magnitude of glint excursions are reduced.

Figure 8.4 shows glint measurements in the yaw plane of an aircraft at four bistatic angles: 0° (monostatic), 15°, 30°, and 45° [54]. The radial lines emanating from the approximate geometric center of the aircraft represent the magnitude of

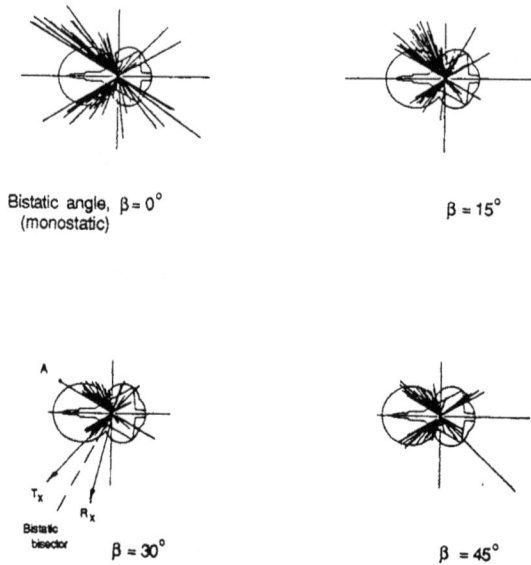

Figure 8.4 Yaw plane, angle glint measurements of a fighter aircraft. (Courtesy of *Microwave Systems News* [54].)

the angular displacement in apparent phase center (i.e., glint excursion) as the bisector of the bistatic angle moves $\pm 60°$ relative to nose aspect. For example, in the plot for $\beta = 30°$, point A represents the glint excursion measured when the bistatic bisector is positioned $60°$ counterclockwise from nose aspect. The aircraft is then rotated $120°$ counterclockwise to give the remaining measurements. Note that the glint excursions can be to the left or right of the bistatic bisector, depending on the phase relationship between dominant scatterers [208]. Figure 8.4 shows that for $\beta \gtrsim 30°$ the magnitude of glint excursions is reduced by roughly a factor of 2 compared to monostatic glint, with some reduction also occurring at $\beta = 15°$.

8.4 FORWARD-SCATTER RCS REGION

The third bistatic RCS region, forward scatter, occurs when the bistatic angle approaches $180°$. When $\beta = 180°$, Siegel *et al.* [33] showed, based on physical optics, that the forward-scatter RCS, σ_F, of a target with shadow (or silhouette) area A is $\sigma_F = 4\pi A^2/\lambda^2$, where λ, the wavelength, is small compared to the target dimensions. The targets can be either smooth or complex structures, and from the application of Babinet's principle, can be totally absorbing [37, 91].

In fact, Babinet's principle provides a simple, intuitive explanation of the forward-scatter effect. Babinet's principle as used in optics states that (1) two diffraction screens are complimentary if the clear regions of the first are opaque (shadow) regions of the second and vice versa and (2) when the two complimentary screens are illuminated by a source, the fields produced on the other side of the screens add to give a field that would be produced with no screens. Specifically, if U_1 is the ratio of the field produced behind the first screen to the field produced with no screen, and U_2 is the same ratio for the complimentary screen, then $U_1 + U_2 = 1$. When applied at radio frequencies, Babinet's principle is modified as follows [207]: The screens must be plane and infinitely thin, and the illumination source for the first screen must be the conjugate source for the complimentary screen, where conjugate is defined as a 90° rotation in polarization of the source. Using an acoustic analogue, Kock [91] demonstrated that the screens could be conducting or absorbing. Kock also elaborated on the infinitely thin condition for the screens:

Babinet's principle applies strictly only for an infinitely thin screen and for an infinitely thin object replacing the aperture. Thick irregularly shaped objects are difficult to treat analytically, and because of this one cannot conclude theoretically that the forward lobe exists in equal magnitude for a thin, flat disk or for a thick, irregularly shaped object having the same shadow area. Straightforward reasoning suggests, however, that the lobes should be similar since, for objects large with respect to the wavelength, the energy striking the front face of the object is not instrumental in creating the forward lobe.

The classic application of Babinet's principle was to show that the radiation patterns of a dipole and a slot are identical, and thus an array of slots could replace an array of more cumbersome dipoles for microwave antennas [207].

Figure 8.5 shows Babinet's principle applied to the forward-scatter case. A "dipole target" and its complimentary "slot target," with opaque and clear areas A, respectively, are illuminated by conjugate sources (transmitters). A receiver is positioned on the other side of the targets at $\beta = 180°$. The dipole target represents the real-world case of a target cutting a hole of area A in the transmitter-to-receiver path, and the slot target represents a Babinet model of the real-world case. From the Babinet principle, the forward-scatter radiation patterns from both targets are identical when illuminated by conjugate sources. Now the transmitter's power "received" by the slot target is proportional to the slot's capture area, A. The power "reradiated" from the slot target is proportional to the slot's antenna gain, $G = 4\pi A/\lambda^2$. Thus, the power at the receiver is proportional to $AG = 4\pi A^2/\lambda^2$, which is the σ_F term in the bistatic radar range equation. As demonstrated by Kock, the hole of area A cut in the transmitter-to-receiver path can be generated by either reflecting or absorbing targets.

152

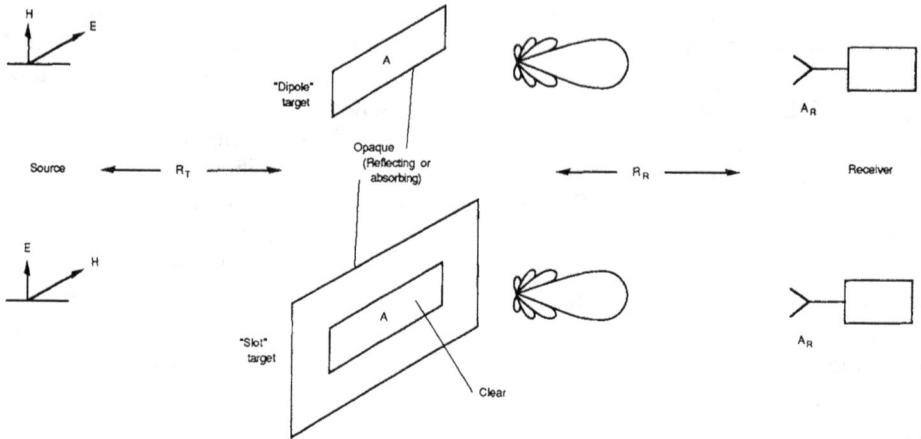

Figure 8.5 Babinet's model for the forward-scatter case with $\beta = 180°$.

For $\beta < 180°$, the forward-scatter RCS rolls off from σ_F. The roll-off is approximated by treating the shadow area A as a uniformly illuminated antenna aperture. The radiation pattern of this "shadow aperture" is equal to the forward-scatter RCS roll-off when $(\pi - \beta)$ is substituted for the angle off the aperture normal. A sphere of radius a will roll off 3 dB at $(\pi - \beta) \approx \lambda/\pi a$, when $a/\lambda \gg 1$ [15]. Although the $a/\lambda \gg 1$ criterion is not satisfied in Figure 8.1, the curve for $a = 3.2\lambda$ still exhibits this phenomenon: 3-dB reduction in σ_F at $\beta \approx 174°$. (The value of σ_F at $\beta = 180°$ also matches $4\pi A^2/\lambda^2$ within 1 dB.) Figure 8.1 shows the roll-off approximating $J_0(x)/x$ down to $\beta \approx 130°$, where J_0 is a Bessel function of zero order. A linear aperture of length D, with aspect angle orthogonal to the transmitter LOS, will roll off 3 dB at $(\pi - \beta) = \lambda/2D$, where $D/\lambda \gg 1$. The forward-scatter RCS roll-off continues, with sidelobes approximating $(\sin x)/x$ over the forward-scatter quadrant $(\beta > 90°)$ [105]. For other aspect angles and targets with complex shadow apertures, calculation of the forward-scatter RCS roll-off usually requires computer simulation.

The forward-scatter RCS of more complex bodies has been simulated and measured; the bodies were both reflecting and absorbing [34, 37, 38, 92, 98, 100–102]. Paddison et al. [100] report both measurements and calculations via computer simulation of forward-scatter RCS for a small (16 × 1.85 cm) right-circular aluminum cylinder at 35 GHz and bistatic angles up to 175.4°. Calculations were made via the method of moments [106], which applies to electrically small targets, i.e., targets with dimensions of the order of several wavelengths. (Measurements were made by Delco [98].) A good match between measurements and calculations was obtained, with typical errors of a few decibels and occasional peak excursions of about 10 dB. A similar match to the Delco measurements was obtained by Cha *et al.* [102], using the physical theory of diffraction methods for targets that are electrically large (i.e., larger than several wavelengths), and the method of moments otherwise.

Figure 8.6 shows calculations of the 16 × 1.85 cm cylinder with 992 facets at 35 GHz, for three fixed transmitter-to-target geometries: (a) near end on, (b) 45° aspect angle, and (c) broadside [100]. The calculations are normalized to λ^2. The broadside geometry shows the classic forward-scattering lobe from a rectangular aperture, with approximate $(\sin x)/x$ sidelobe roll-off out to $\beta \approx 110°$. (Although not shown in the figure, the wavelength-normalized, forward-scatter peak, $\sigma_F/\lambda^2 = 4\pi A^2/\lambda^4$, is 43 dB.) The three bistatic RCS regions are quite distinct: pseudomonostatic at $\beta < 20°$, bistatic at $20° < \beta < 140°$, and forward scatter at $\beta > 140°$. The other two geometries show a similar, but broader, forward-scatter lobe, as expected because the silhouette area, and hence the shadowing aperture, is smaller. The 45° aspect geometry is of interest because the RCS in the bistatic region is larger than the monostatic RCS for most bistatic angles. The large spike at $\beta = 90°$ is the bistatic specular lobe, analogous to the monostatic specular lobe in the broadside geometry. While Figure 8.6 shows the clear dependency of bistatic RCS on both

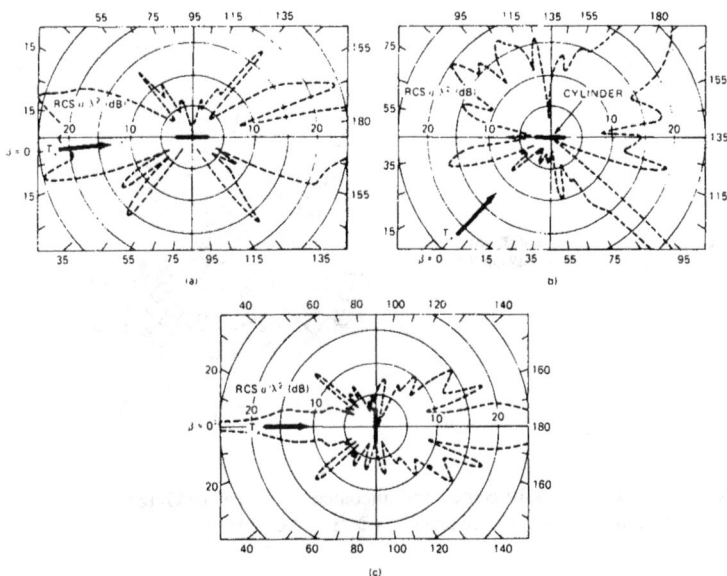

Figure 8.6 Calculated bistatic RCS, replotted as a function of bistatic angle, β, for a conducting cylinder, 16 × 1.85 cm at 35 GHz, HH polarization. (Courtesy of Electromagnetics [100].)

a) Uncoated diameter = 2.54 cm (1 in.)
   Coated diameter = 2.93 cm (1.15 in.)

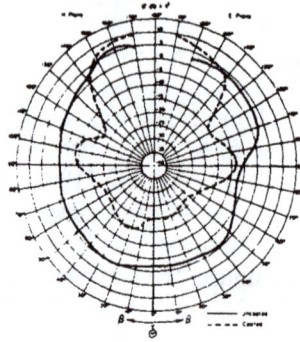

b) Uncoated diameter = 3.33 cm (1.31 in.)
   Coated diameter = 3.72 cm (1.47 in.)

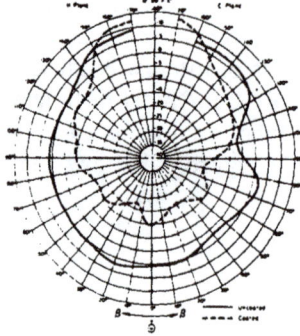

c) Uncoated diameter = 4.03 cm (1.59 in.)
   Coated diameter = 4.42 cm (1.74 in.)

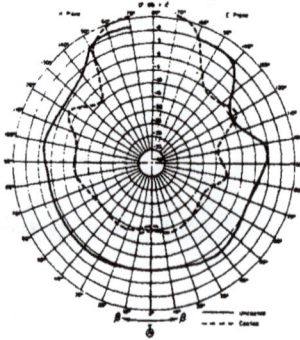

Figure 8.7 Measured bistatic RCS for coated and uncoated spheres at 10 GHz ($\lambda = 3$ cm) in resonance region, as a function of bistatic angle β. (© 1961 IRE [38].)

aspect angle and bistatic angle, it also serves to caution attempts to use oversimplified bistatic RCS models, especially in the bistatic region.

Forward-scatter enhancement diminishes as the wavelength becomes large with respect to body dimensions. For example, a sphere of radius $a = \alpha\lambda$, where α is a constant of proportionality, and an area-normalized bistatic RCS of $\sigma_B/\pi a^2 = \sigma_B/\pi\alpha^2\lambda^2$, will have an area-normalized forward-scatter RCS of $\sigma_F/\pi a^2 = 4\pi^2\alpha^2$. When $\alpha = 3.2$, $\sigma_F/\pi a^2 = 26$ dB; when $\alpha = 0.38$, $\sigma_F/\pi a^2 = 7.6$ dB. Both results match to within 1 dB of the theoretical values shown in Figure 8.1 for $\beta = 180°$.

Measurements of small spheres in the resonance region tend to confirm these predictions. Figure 8.7 shows results of measurements for both coated and uncoated small spheres at 10 GHz [38]. Again the bistatic RCS is normalized to λ^2. The uncoated spheres (solid curves) show only a small forward-scatter enhancement as β becomes large: ≈ 5 dB for $\alpha = 0.42$ [Figure 8.7(a)]; ≈ 8 dB for $\alpha = 0.56$ [Figure 8.7(b)]; and ≈ 11 dB for $\alpha = 0.67$ [Figure 8.7(c)].

The measured values for the solid curve in Figure 8.7(a), with $a = 0.42\lambda$, closely match those of the theoretical curves for $a = 0.38\lambda$ in Figure 8.1. When the normalized terms in both figures are accounted for, σ_M ($\beta = 0°$) matches within 1.5 dB and σ_F ($\beta = 180°$) matches within 2.5 dB. The shapes of the curves are also similar: Both E-plane curves show a dip in σ_B of about 7 dB at β values of 60° to 80°. Both H-plane curves show a smaller dip in σ_B of about 2–3 dB at $\beta = 90°$.

The slightly larger coated spheres (dashed curves) in Figure 8.7 show similar trends for large values of β: σ_B values of from 3–5 dB larger than the uncoated counterparts caused by the larger shadowing area. The coating is effective at $\beta \lesssim 90°$, at which point the forward-scatter enhancement starts to dominate the bistatic RCS, again confirming Babinet's principle.

Chapter 9
CLUTTER

Radar clutter is defined as unwanted echoes, typically from the ground, sea, rain or other precipitation, chaff, birds, insects, and aurora [183]. This chapter covers two bistatic clutter topics: surface clutter, consisting of ground and sea echoes, and chaff.

9.1 SURFACE CLUTTER

The bistatic RCS of surface clutter, σ_c, is a measure, as is the monostatic radar surface clutter cross section, of the energy scattered from a clutter cell area, A_c, in the direction of the receiver. The bistatic RCS is defined as $\sigma_c = \sigma_B^\circ A_c$, where σ_B° is the scattering coefficient, or the clutter cross section per unit area of the illuminated surface. The clutter cell area is given for beam- and range-limited cases in Section 5.6. This section considers measured and estimated values of σ_B°, which vary as a function of the surface composition, frequency, and geometry. Also, in contrast to the monostatic case, little measured data for σ_B° have been reported [42, 43, 107–115].

The available database for ground and sea clutter at microwave frequencies consists of six measurement programs, which are summarized in Table 9.1. The measurement angles shown in Table 9.1 are defined in Figure 9.1, which is a clutter-centered coordinate system. Because ground and sea are reciprocal media, θ_i and θ_s are interchangeable in the subsequent data [103]. The Pidgeon data were analyzed by Domville [109] and Nathanson [116]. Vander Schurr and Tomlinson [117] analyzed the Larson and Cost data.

In addition to this database, bistatic reflectivity measurements have been made at optical [118] and sonic wavelengths [119], and of buildings [120], airport structures [121], and planetary surfaces [66, 122]. In each of these measurements, the reflectivity data are expressed in terms of reflected power, not σ_B° [103].

Table 9.1
Summary of Measurement Program for Bistatic Scattering Coefficient, σ_B°.

Reference (Date)	Organization	Author	Surface Composition	Frequency	Polarization	θ_i	θ_s	ϕ
42 (1965)	Ohio State University (Antenna Laboratory)	Cost, Peakes	Smooth sand, Loam, Soybeans	10 GHz	VV, HH,	5–30, 10–70	5–30, 5–90	0–145, 0, 180
			Rough sand, Loam with stubble, Grass		HV	5–70	5–90	0, 180
43 (1966)	Johns Hopkins University (APL)	Pidgeon	Sea (Sea states 1, 2, 3)	C-band	VV, VH	0.2–3	10–90	180
107 (1967)			Sea (Beaufort, wind 5)	X-band	HH	1–8	12–45	180
108 (1967)			Rural land, Urban land, Sea (20-kt wind)			6–90*	6–180*	180, 165
109 (1968)	G.E.C. (Electronics) Ltd., England	Domville	Sea (20-kt wind)	X-band	VV, HH	≈0–90*	≈0–180*	180, 165
110 (1969)			Semidesert			≈0	?	180, 165

	Organization	Author	Target	Frequency	Polarization			
111 (1977) 112 (1978)	University of Michigan (ERIM)	Larson, Heimiller	Grass with cement Taxiway Weeds and scrub trees	1.3 and 9.4 GHz	HH, HV	10, 40 10, 15, 20	5, 10, 20	0–180 0–105
113 (1982) 114 (1984)	Georgia Institute of Technology (EES)	Ewell, Zehner	Sea (0.9m, 1.2–1.8 m wave heights)	9.38 GHz	VV, HH	≈0	≈0	90–160
115 (1988)	University of Michigan (Dept. of Electrical Engineering & Computer Science)	Ulaby et al.	Visually smooth sand			24	24	0–170
			Visually smooth sand Rough sand Gravel	35 GHz	VV, HH, VH, HV	30 30	30 10–90	0–170 0–90

*Measured and interpolated data ranges.

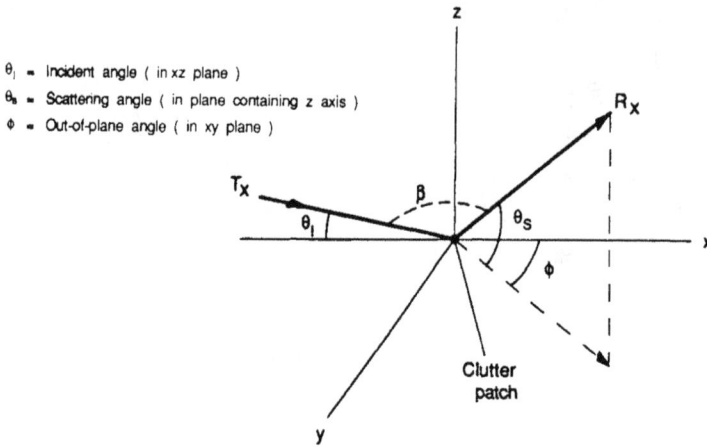

Figure 9.1 Coordinate system for bistatic clutter measurements.

The bistatic angle is calculated from the angles in Figure 9.1 by the use of direction cosines:

$$\beta = \cos^{-1}(\sin\theta_i \sin\theta_s - \cos\theta_i \cos\theta_s \cos\phi) \tag{9.1}$$

When measurements are constrained to the x-z plane, $\phi = 180°$ and $\beta = \theta_s - \theta_i$. This case is called "in-plane." When measurements are taken at low grazing angles, $\theta_i = \theta_s \approx 0°$ and $\beta \approx 180° - \phi$. The monostatic case is given for $\theta_i = \theta_s$ and $\phi = 180°$. Note that $\phi = 180°$ and not $0°$, as might be more intuitively appealing for the monostatic case. This convention was chosen to conform with the coordinate system used in many of the bistatic measurement programs.

Two measurement sets are of interest: in-plane where $\phi = 180°$,* and out-of-plane, where $\phi < 180°$. Most of the data are taken at X-band, with the most substantial in-plane database provided by Domville [108–110]. Because the database is sparse, mean values for σ_B^0 are usually given, with occasional standard deviations and probability distributions calculated.

9.1.1 In-Plane Ground Clutter

Figure 9.2 is a plot of Domville's X-band, vertically polarized data summary for rural land, consisting of open grassland, trees, and buildings [108]. Domville reports that because the data were a composite of different sources and averaged over different terrain conditions, differences of 10 dB in the values sometimes

---

*Again, note that $\phi = 180°$, not $0°$, for in-plane data.

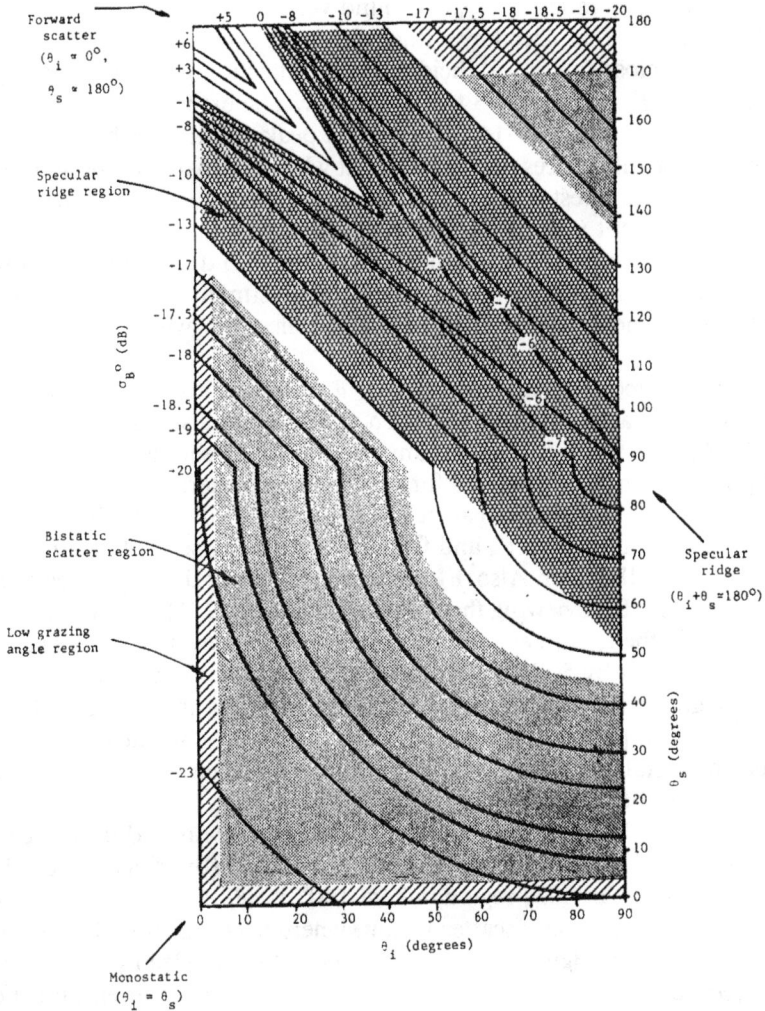

Figure 9.2 *X*-band, vertically polarized, σ_B°, in-plane ($\phi = 180°$) data summary for rural land. (Courtesy of GEC Electronics, Ltd. [108].)

occurred. The spread in raw data within any data set ranged from 1 to 4.5 dB, however. The measured database consists of points near the lines $\theta_i = \theta_s$, $\theta_i = 90°$, and $\theta_s = 90°$, and points along the specular ridge near the forward-scatter region. The remaining data points are interpolations.

Domville also summarized in-plane data for forest and urban areas [108]. The shapes of constant σ_B° contours for all of Domville's terrain types are similar and consist of cones with different slopes around vertical incidence and reflection ($\theta_i = \theta_s = 90°$). The cone is replaced by a ridge about the specular angle ($\theta_i + \theta_s = 180°$). The ridge starts close to vertical incidence and reflection for rural land, Figure 9.2. For urban areas σ_B° is generally 3 to 6 dB higher. The extent of the ridge is smaller, starting at $\theta_i \approx 40°$ and $\theta_s \approx 140°$, but with a peak about 3 dB higher than that of rural land. These differences are insignificant when compared to the spread in measured data. Because forest terrain is a more uniform scatterer, the cones extend into the forward quadrant ($\theta_s > 90°$). The ridge is very small, ranging from $0 < \theta_i < 26°$ and $154° < \theta_s < 180°$, with a peak of about -10 dB, or 16 dB below that of rural land. Other values of σ_B° for forest terrain are similar to those of rural land for $\theta_s < 90°$, except near-vertical incidence and reflection, where σ_B° is about 10 dB lower.

Domville reports [109] that at low θ_i no significant variation in σ_B° was observed for rural and forest terrain when measured at a small out-of-plane angle, $\phi = 165°$. Also at low θ_i, no significant variation between horizontal, vertical, or crossed polarizations was observed for rural and forest terrain.

For semidesert, σ_B° was measured [110] at -40 dB for both horizontal and vertical polarization at $\theta_i \lesssim 1°$, and for all $\theta_s \gtrsim 1°$. Crossed-polarization measurements were 5–10 dB lower. Also, σ_B° is reduced by about 0.3 dB/degree as ϕ moves from 180° to 165°. In reviewing these data, Nathanson [231] observed that "a Land Rover exceeded the terrain echo by almost 10 dB. However, a herd of camels only exceeded the clutter by 5 dB. The number of humps was not recorded."

Although terrain conditions are different, the Cost in-plane data [42] match the Domville data [108] within about 10 dB. The Cost data curves do not always approach the bistatic specular ridge monotonically, even though the terrain conditions appear to be more uniform.

The in-plane Domville ground clutter data can be divided into three regions: a low grazing angle region, where $\theta_i \lesssim 3°$ or $\theta_s \lesssim 3°$ (the single-hatched area in Figure 9.2); a specular ridge region, where $140° \leq (\theta_i + \theta_s) \leq 220°$ (the double-hatched area); and a bistatic scatter region, where (θ_i, θ_s) assume the values shown as the shaded areas in Figure 9.2. Each region can be modeled by a "semiempirical process" (containing arbitrary constants that are adjusted to fit empirical data) as follows.

The low grazing angle and bistatic scatter regions are based on the constant γ monostatic clutter model:

$$\sigma_M^\circ = \gamma \sin\theta_i \tag{9.2}$$

where σ_M° is the monostatic scattering coefficient, θ_i is the monostatic, or incident, angle in Figure 9.1, and γ is a normalized reflectivity parameter. For farm land, $\gamma \approx -15$ dB; for wooded hills, $\gamma \approx -10$ dB [123].

The constant γ, bistatic scatter region model is developed using a variation of the monostatic-bistatic equivalence theorem (Chapter 8), in which $\sin\theta_i$ is replaced by the geometric mean of the sines of the incident and scattering angles, $(\sin\theta_i \sin\theta_s)^{1/2}$, in (9.2) [123]. Hence,

$$(\sigma_B^\circ)_b = \gamma(\sin\theta_i \sin\theta_s)^{1/2} \tag{9.3}$$

where $(\sigma_B^\circ)_b$ is the scattering coefficient in the bistatic scatter region. Now γ can be estimated from Figure 9.2 using monostatic data, which is plotted along the line $\theta_i = \theta_s$. A value of $\gamma = -16$ dB in (9.3) fits the monostatic data within about 2 dB, and yields a match within 3 dB to the bistatic data, including the small triangle in the forward quadrant.

The low grazing angle region is modeled by the sine of the arithmetic mean of the incident and scattering angles, $\sin[(\theta_i + \theta_s)/2]$. Hence,

$$(\sigma_B^\circ)_l = \gamma \sin[(\theta_i + \theta_s)/2] \tag{9.4}$$

where $(\sigma_B^\circ)_l$ is the scattering coefficient in the low grazing angle region. The data match is again ≈ 3 dB for $\gamma = -16$ dB, including the small quadrilateral in the upper right corner of Figure 9.2. Because $(\theta_i + \theta_s)/2 = \theta_i + \beta/2$, (9.4) is an exact application of the monostatic-bistatic equivalence theorem. When $(\sigma_B^\circ)_b$ is calculated by (9.4), the results are equal to or higher than those of (9.3). For regions excluding the specular ridge, (9.4) was found to set an upper bound to measured bistatic scattering coefficients [117].

For very low grazing angles (θ_i or $\theta_s \lesssim 1°$), but excluding the specular ridge region, the calculations for $(\sigma_B^\circ)_l$ in (9.4) must be multiplied by the pattern propagation factors F_T and F_R, and loss terms L_T and L_R. Assuming only $\theta_i \lesssim 1°$, then for a flat terrain with rms surface roughness, $\sigma_h \ll h_T$, where h_T is the transmitting antenna height above the average local plane, $F_T \approx R_1/R_T$ in the reflection-interference region [123]. This region is defined as $R_1 < R_T < R_2$ [123], where R_1 and R_2 are given as

$$R_1 = 4\pi h_T \sigma_h / \lambda, \tag{9.5}$$

$$R_2 \approx 130\sqrt{h_T} \tag{9.6}$$

where h_T and R_2 are in kilometers and R_T is the range from transmitter to clutter patch. When $R_T < R_1$, $F_T \approx 1$; when $R_T > R_2$, propagation is by diffraction, unless mountains or steep hills are present [123]. When $\theta_s \lesssim 1°$, F_R is calculated in the same manner.

The specular ridge region is modeled for values of $(\sigma_B^o)_s \leq 1$ by a variation of the Beckmann and Spizzichino theory of forward scattering from rough surfaces [124, 125]:

$$(\sigma_B^o)_s = \exp[-(\beta_c/\sigma_s)^2] \qquad (9.7)$$

where

$(\sigma_B^o)_s$ = scattering coefficient in the specular ridge region,
σ_s = rms surface slope,
β_c = angle between vertical and bistatic bisector of θ_i and $\theta_s = |90° - (\theta_i + \theta_s)/2|$

For flat terrain $\sigma_s \approx 0.1$ rad. With a value of $\sigma_s = 0.17$ rad, (9.7) matches the specular ridge in Figure 9.2 within 5 dB, for $(\sigma_B^o)_s \leq 1$.

For the special case of very low grazing angles within the specular ridge region, (9.7) must include terms that account for reflection from tilted facet surface elements within the "glistening region" [123, 124]. When $\theta_i \ll 1°$, Barton shows that the glistening region extends over $\approx(180° - 4\sigma_s) \leq \theta_s \leq 180°$ [123]. When $\sigma_s = 0.17$ rad, the region extends from about 140° to 180°. This region matches the specular ridge region of Figure 9.2, and can be viewed as a slice through the ridge at $\theta_i \approx 0°$. A similar slice occurs at $\theta_s \approx 180°$.

For Domville's urban land data, a 3-dB match to $(\sigma_B^o)_b$ and $(\sigma_B^o)_l$ is obtained with $\gamma = -12$ dB. For Domville's forest data, the match is obtained over all cone regions, including those extending into the forward quadrant, for $\gamma = -16$ dB. Note, however, that the monostatic model suggests $\gamma = -10$ dB for wooded hills, a 6-dB difference, but still within Domville's 10-dB data averaging range. The match to $(\sigma_B^o)_s$ for Domville's urban land data is accurate to within 3 dB when $\sigma_s = 0.22$ rad. The $(\sigma_B^o)_s$ model breaks down for the small specular ridge present in forest terrain. Table 9.2 summarizes parameters for the three bistatic in-plane clutter regions.

9.1.2 In-Plane Sea Clutter

Figure 9.3 is a plot of Domville's X-band, vertically polarized, in-plane data summary for the sea [109]. The measured database consists of points near the lines $\theta_i = 0°$ and $\theta_s = 0°$, and points along the specular ridge near the forward-scatter region. The remaining data points are interpolations. A 20-kt wind speed was measured during the tests, but no measure of sea state conditions was made. The Domville data bracket, within ± 5 dB, that of Pidgeon's data [43], taken at C-band for $0.2° \leq \theta_i \leq 3.0°$ and $10° \leq \theta_s \leq 90°$.

Table 9.2
σ_B^o Model of Domville's Bistatic, In-Plane, Ground Clutter Data [108] Using Normalized Reflectivity, γ, and rms Surface Slope, σ_s, Parameters

| | Bistatic Regions* | | |
	Low Grazing Angle	Bistatic Scatter	Specular Ridge
Equation Number	9.4	9.3	9.7
Rural Land	$\gamma = -16$ dB	$\gamma = -16$ dB	$\sigma_s = 0.17$ rad†
Urban Land	$\gamma = -12$ dB	$\gamma = -12$ dB	$\sigma_s = 0.22$ rad
Forest	$\gamma = -16$ dB‡	$\gamma = -16$ dB‡	**

*Except where noted, data match is ≈ 3 dB.
† ≈ 5 dB data match.
**No data match.
‡Monostatic model suggests $\gamma \approx -10$ dB.

The absence of measured sea state is a significant omission in the Domville data. For example, Pidgeon [43] reports that values of σ_B^o at sea state 3 are approximately 10 dB larger than the corresponding values for sea state 1. However, sea state 3 and a 10-kt wind gave 3- to 5-dB larger values than those of sea state 2 and 20- to 30-kt winds. Pidgeon ascribes this apparent anomaly to the fact that a fully developed sea requires a wind blowing with a given velocity over a given distance for a given time. Clearly, the latter situation did not meet these requirements. Thus, the Domville data uncertainty should be considered as ~ 10 dB.

The measured values for σ_B^o in the regions $\theta_i \lesssim 2°$ and $\theta_s \lesssim 2°$ in Figure 9.3 are shown as a constant -50 dB. This result is an average of data, including Pidgeon's C-band data [43], and can vary by approximately ± 10 dB depending on the specific value of θ_i or θ_s, sea state, wind conditions, and ducting. The low values of σ_B^o in these regions are caused by the propagation factor, F_T or $F_R \ll 1$, implicitly included in the measurements. This type of data is sometimes called "effective σ_B^o" [114].

Regions of the in-plane Domville sea clutter data can be modeled using a direct application of the constant γ monostatic clutter model, (9.2), when θ_s is held constant [109]. For $\gamma = -20$ dB, (9.2) matches Domville's data within 5 dB for the shaded areas in Figure 9.3. Similar results are obtained when θ_s replaces θ_i in (9.2) and θ_i is held constant. Because the sea clutter ridge values for σ_B^o are >0 dB in most ridge areas, they cannot be modeled by (9.7).

Data for horizontal polarization, reported by Domville [109] under the same condition as for vertically polarized data, show similar trends for constant σ_B^o contours. The values of σ_B^o are typically 1–5 dB lower than those for vertical polariza-

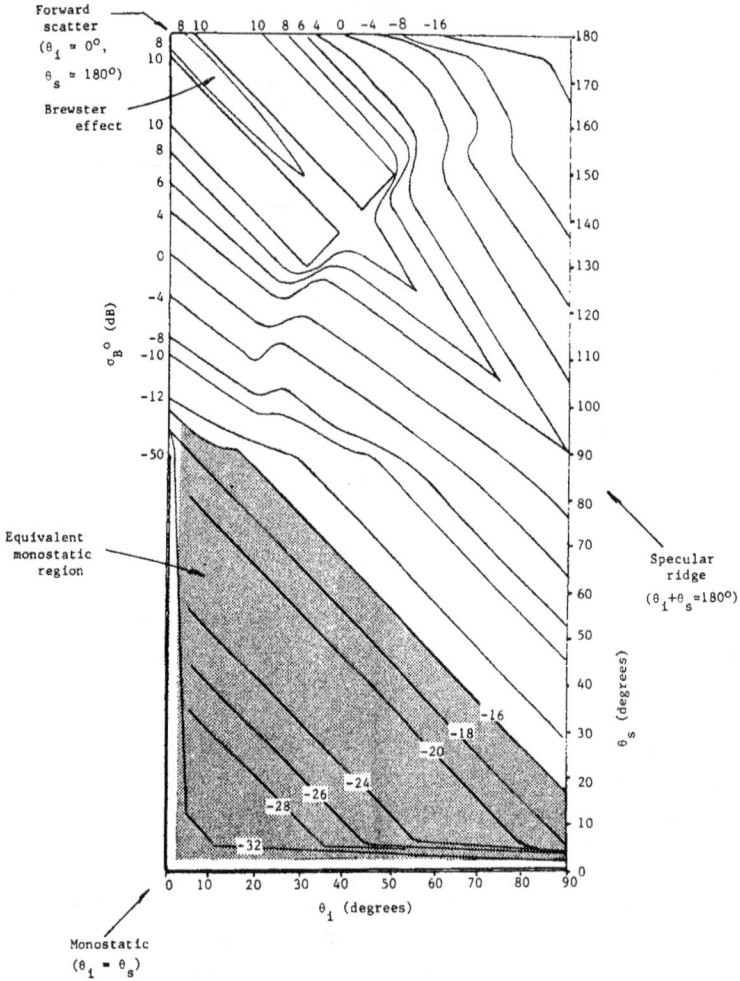

Figure 9.3 *X*-band, vertically polarized, σ_B°, in-plane ($\phi = 180°$) data summary for a 20-kt sea. (Courtesy of GEC Electronics Ltd [109].)

tion, but these differences are insignificant when compared to the spread in measured data. The Brewster effect near forward scatter is absent.

Pidgeon [107] reports *X*-band horizontal polarization measurements for $1° < \theta_i < 9°$ and $12° \leq \theta_s \leq 55°$, as shown in Figure 9.4. A Beaufort 5 sea (19- to 24-kt wind velocity) was measured, which is equivalent to sea state 3 for a fully developed

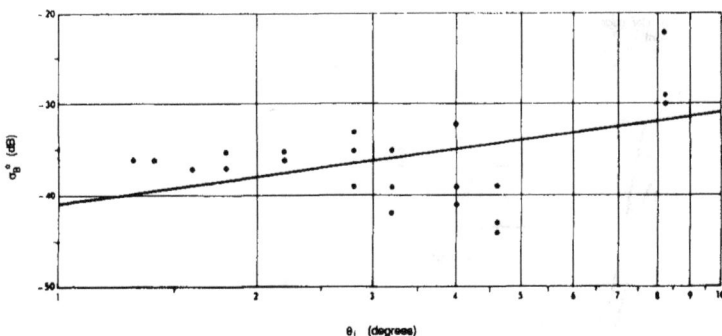

Figure 9.4 *X*-band, horizontally polarized, σ_B°, in-plane (ϕ = 180°) data for a 19- to 24-kt sea, $12° \leq \theta_s$ $\leq 55°$ [107], solid line is (9.2) for $\gamma = -23$ dB [116].

sea. The data spread is again about 10 dB. However, the σ_B° trend as θ_i increases is not as pronounced. Specifically, σ_B° is shown at about -35 dB for $1.5° < \theta_i < 3°$, which is about 10 dB higher than that for vertical polarization. It then decreases to about -40 dB for $3° < \theta_i < 4.6°$. Differences in sea states and wind condition could have caused these variations. In any case the data spread appears to be 10 to 15 dB.

The horizontally polarized data from Domville [109] and Pidgeon [107] can also be modeled using (9.2), but with $\gamma \approx -23$ dB [107, 116]. Domville reports [109] that the use of (9.2) with $\gamma = -35$ dB models wind speeds below about 12 kt, but again with the sea state caveat previously cited by Pidgeon. Measured cross-polarized (VH) values for σ_B° are 10–15 dB lower than those for copolarized (VV) values at $\theta_i < 1°$, but only 5–8 dB less at $\theta_i \approx 3°$ [43].

9.1.3 Out-of-Plane Ground Clutter

Cost [42] and Larson [111, 112] measured out-of-plane σ_B° for ground clutter under similar—and in some cases identical—geometries. Figure 9.5 shows Cost data [42] and Figure 9.6 shows Larson data [112], both for $\theta_i = \theta_s = 10°$. Even when differences in the surface composition are accounted for [sand, loam, and 3-in (8-cm) soybean foliage by Cost and 4-ft (120-cm) weeds and scrub trees by Larson], the data show only limited correlation under identical geometries and polarization, and frequencies scaled to the surface composition (10 GHz for Cost and 1.3 GHz for Larson).

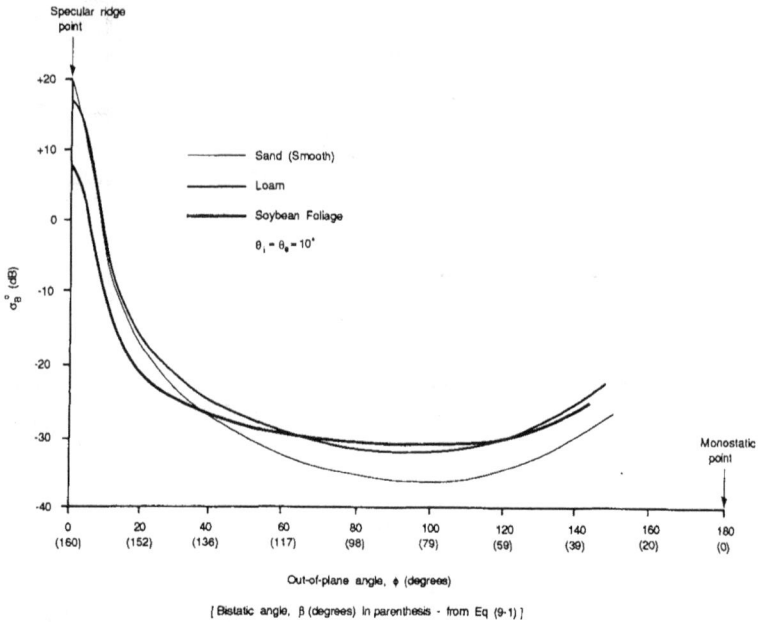

Figure 9.5 Out-of-plane, horizontally polarized, σ_B° data at 10 GHz. (Courtesy of Ohio State University [42].)

Three significant geometries for comparison are: $\phi = 180°$, the monostatic point; $\phi = 0°$, the in-plane specular ridge point; and regions around $\phi = 90°$, where σ_B° often exhibits a minimum. For the monostatic point, the Larson 9.4-GHz data show $\sigma_B^\circ = \sigma_M^\circ = -13$ dB. From (9.2), γ is calculated as -5 dB, roughly 10 dB higher than the Barton monostatic data suggest [123]. While the Cost data only show measurements to $\phi \approx 150°$, an extrapolation to $\phi \approx 180°$ gives $\sigma_B^\circ = \sigma_M^\circ \approx -20$ dB and $\gamma \approx -12$ dB, which is still high, but only by about 3 dB.

As reported by Vander Schurr and Tomlinson [117], a better match to γ for the Cost and Larson data can be made using the Larson data at 1.3 GHz ($\lambda = 23$ cm). Now the ratios of surface composition height-to-wavelength are roughly comparable: 120 cm/23 cm $= 5.2$ for Larson data, and 8 cm/3 cm $= 2.7$ for the Cost data. For the monostatic point, the Larson 1.3-GHz data show $\sigma_B^\circ = \sigma_M^\circ = -17$ dB, and $\gamma \approx -9$ dB, which is again high but only by about 6 dB.

For the specular ridge point ($\theta_i = \theta_s = 10°$, $\phi = 0°$), the Larson X- and L-band data show $\sigma_B^\circ \approx -2$ dB, while the Cost data show σ_B° of $+8$ dB. The Domville data, Figure 9.2, for the same conditions ($\theta_i = 10°$, $\theta_s = 170°$, $\phi = 180°$) show $\sigma_B^\circ = +6$ dB, a reasonable match to the Cost data.

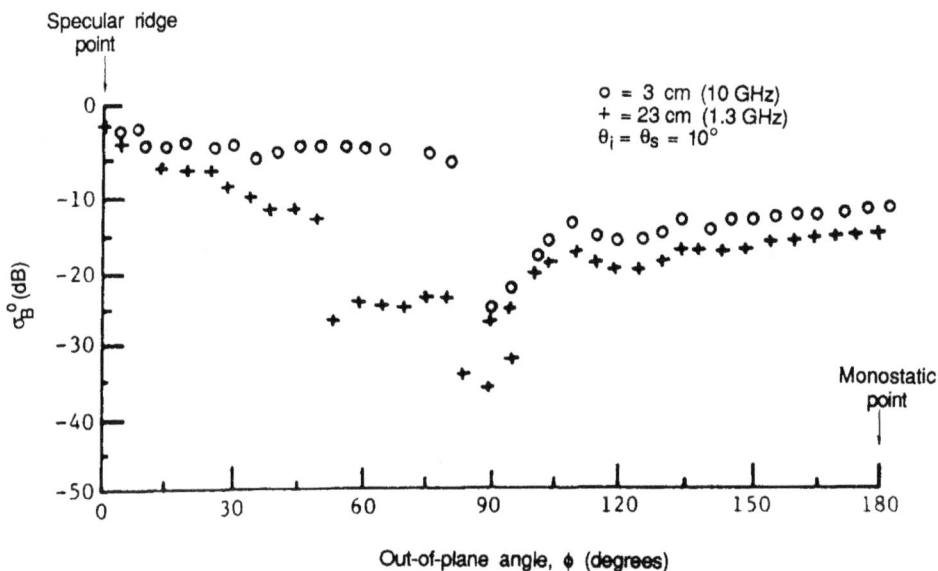

Figure 9.6 Out-of-plane, horizontally polarized, σ_B^o data for tall weeds and scrub trees. (© 1978 IEEE [112].)

In both the Cost and Larson X-band data for θ_i and $\theta_s = 10°$, σ_B^o typically shows a minimum value of $\phi \approx 90°$. However, the Cost data show a steep drop in σ_B^o at $\phi \gtrsim 10°$, and then a broad valley of $\sigma_B^o < -30$ dB for $\approx 60° < \phi < \approx 130°$. In contrast, the Larson data show a dip in σ_B^o only for $80° < \phi < 100°$, with a sharp minimum of -25 dB at $\phi \approx 90°$. The broad reduction in σ_B^o near $\phi = 90°$ appears in most of the Cost data for HH, VV, and VH polarization. It does not appear at all in some of the Larson HH data. When it does appear, it is more pronounced at 1.3 GHz than at 9.4 GHz, as Figure 9.6 shows.

Ulaby [115] measured horizontal, vertical, and crossed-polarized, out-of-plane σ_B^o for sand and gravel at 35 GHz at moderately high incident angles ($\theta_i = 24°$ and $30°$). A typical result for HH polarization is shown in Figure 9.7. The shape of all of Ulaby's HH polarized curves are similar to those of Cost, Figure 9.5, with peaks of $+12$ to $+24$ dB at the specular ridge point and broad minimums of -15 to -24 dB near $\phi = 90°$.

Ulaby found that visually smooth sand (rms surface roughness < 0.1 cm) is electromagnetically smooth in the specular direction ($\theta_i + \theta_s = 180°$, $\phi = 0°$) at 35 GHz. Under these conditions the reflected signal is totally coherent. As the surface departs from visually smooth, the coherent scattering component decreases and the diffuse scattering component increases. For sand with rms surface rough-

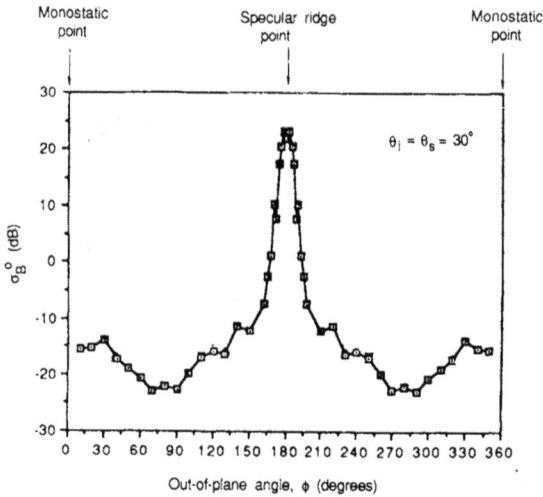

Figure 9.7 Out-of-plane, horizontally polarized, σ_B° data for smooth sand at 35 GHz. (© 1988 IEEE [115].)

ness of 1.67 cm, about twice the wavelength, the peak of the specular ridge was ~ 10 dB less (+13 dB versus +24 dB) than that for visually smooth sand, and the width of the ridge had broadened from $\phi \approx 15°$ to $\phi \approx 30°$ at 20 dB below the peak. Ulaby also observed that VV values for σ_B° in the specular ridge region were about 12 dB smaller than those for HH values, while outside the ridge, VV values tended to be slightly larger. The lower VV values in the specular ridge region were not analyzed by Ulaby, but are at the geometry where the Brewster effect occurs. Hence, lower values would be expected.

Although some inconsistencies between out-of-plane σ_B° data for ground clutter exist and cannot be resolved, general trends are apparent. First σ_B° usually approaches a minimum as ϕ approaches 90°, with values 10 to 20 dB below the monostatic value ($\theta_i = \theta_s$, $\theta = 180°$). Second, out-of-plane σ_B° values are not significantly different (within $\approx$5 dB) from in-plane σ_B° values for $\phi \lesssim 10°$ and $\phi \gtrsim 140°$, i.e., angles close to in-plane conditions. The $\phi \lesssim 10°$ limit is based on Cost, Ulaby, and Domville data; the $\phi \gtrsim 140°$ limit, on Ulaby and Larson data.

9.1.4 Out-of-Plane Sea Clutter

Ewell [113, 114] measured horizontally and vertically polarized, out-of-plane σ_B° for sea clutter at θ_i and θ_s near grazing incidence (θ_i, $\theta_s \ll 1°$). Visual estimates of sea conditions ranged from 0.9- to 1.8-m wave height. Ratios of bistatic-to-monostatic scattering coefficients (median values) were calculated, with bistatic angles, $\beta \approx$

$180° - \phi$, ranging from 23° to 85°. A typical result is shown in Figure 9.8. The data implicitly included pattern propagation factors, F_T and F_R. Because antenna heights were different, F_T and F_R are expected to be different, but were not measured. In all cases the measured bistatic-to-monostatic ratios were less than unity. In two cases they ranged from -2 to -12 dB, and in the third case, shown in Figure 9.8, they dropped from about -5 dB at $\beta = 23°$ to -20 to -25 dB at $\beta = 60°$. The trend was generally downward as β increased. Values for horizontal and vertical polarization showed no significant differences. For the most part, both monostatic and bistatic data exhibited nearly log-normal amplitude distributions. Ewell showed that selected data sets could be modeled by assuming locally correlated scattering areas, similar to the facets of some sea clutter theories [148]. However, general application of this model requires detailed measurements of sea conditions (wave angular spectra and slope, scatterer size distributions) that are not usually available. Thus, no general model is available.

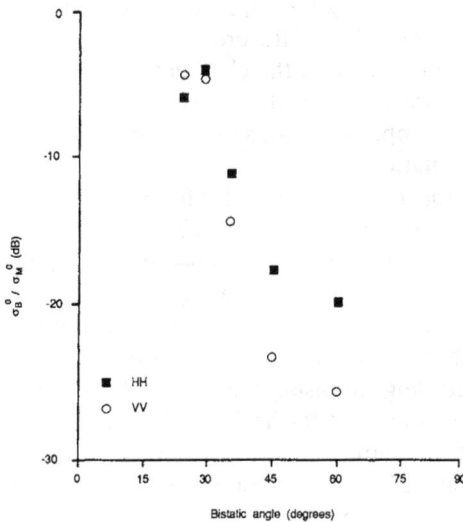

Figure 9.8 Ratio of bistatic-to-monostatic radar scattering coefficients, σ_B^0/σ_M^0, for horizontal (HH) and vertical (VV) polarizations, with visual estimates of significant wave heights of 1.2–1.8 m. (Courtesy of IEE [113], and Artech House [114].)

9.2 CHAFF

Chaff is defined as an airborne cloud of lightweight reflecting objects typically consisting of strips of aluminum foil or metal-coated fibers, which produce clutter echoes in a region of space [183]. Peebles [221] has reviewed the open literature on the bistatic RCS of chaff and developed analytical expressions for the bistatic RCS of a cloud of randomly positioned and randomly oriented resonant dipoles for all bistatic geometries and transmitting and receiving antenna polarizations. The expressions apply for dipole lengths $L = \lambda/2$, λ, and $3\lambda/2$, where λ is the wavelength. In simplest terms,

$$\sigma_{ch} = N \bar{\sigma}_1 \tag{9.8}$$

where σ_{ch} is the bistatic RCS of the chaff cloud, $\bar{\sigma}_1$ is the spherically averaged bistatic RCS of a single dipole, which is modified for a specific transmitting and receiving polarization and a specific geometry, and N is the number of chaff dipoles that are both illuminated by the transmitter and observed by the receiver.

Typically, a chaff cloud is described in terms of dipole density, or number of dipoles per unit volume. In this case the chaff cell volume must be calculated for a specific bistatic radar configuration and geometry. A first-order estimate of this volume can be made by multiplying the expressions for clutter cell area [Section 5.6, and in particular (5.31) or (5.32)] by the cross-range dimensions of either the transmitter or receiver elevation beam at the chaff cloud range, whichever is smaller. In this case $\Delta\theta_T$ and $\Delta\theta_R$ are taken as the transmitter and receiver azimuth beamwidths, respectively. The approximations used in developing (5.31) and (5.32) obviously apply to this calculation.

Expressions developed by Peebles [221] for modifying $\bar{\sigma}_1$ to account for polarization and geometry, while analytically tractable, are particularly complex. Rather than replicating them here, only special—and elegantly simple—cases are extracted. They are elegantly simple because $\bar{\sigma}_1$ can be characterized by dipole length L, polarization referenced to the bistatic plane and the bistatic angle β. Figure 9.9 extracts Peebles's results [221] for a transmitter and receiver with vertical polarization in the scattering (or bistatic) plane, and for $\bar{\sigma}_1$ normalized with respect to λ^2. Figure 9.10 is a similar plot for horizontal polarization. The scattering plane is identical to the bistatic plane defined in Chapter 3. The monostatic values for $\bar{\sigma}_1$ at $\beta = 0°$ are calculated by an expression developed by Chu in an unpublished work cited by Peebles:

$$(\bar{\sigma}_1)_{\beta=0°} = \lambda^2 \frac{1.178(L/\lambda) - 0.131 + 0.179 \ln(22.368L/\lambda)}{[\ln(22.368L/\lambda)]^2} \tag{9.9}$$

This expression applies only to resonant dipoles, i.e., $L = M\lambda/2$ where $M = 1,2, \ldots$. Thus, $(\bar{\sigma}_1)_{\beta=0°} = 0.153\lambda^2$, $0.166\lambda^2$, and $0.184\lambda^2$ for $L = \lambda/2$, λ, and $3\lambda/2$, respectively, which are the $\beta = 0°$ points on Figures 9.9 and 9.10. Note that the average bistatic RCS of a single dipole is less than, and in one case equal to, that for the monostatic case.

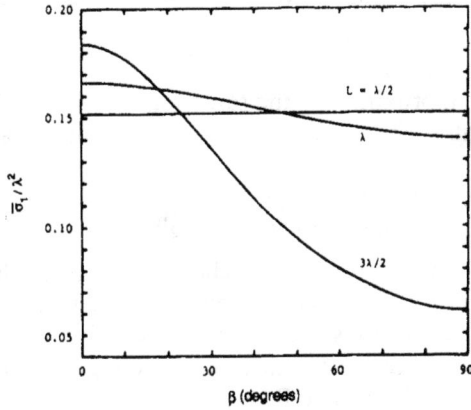

Figure 9.9 Spherically averaged bistatic cross sections for linear transmitting and receiving polarizations perpendicular to scattering plane. Cross sections have even symmetry about $\beta = \pi/2$ (90°) and $\beta = \pi$ (180°). (© 1984 IEEE [221].)

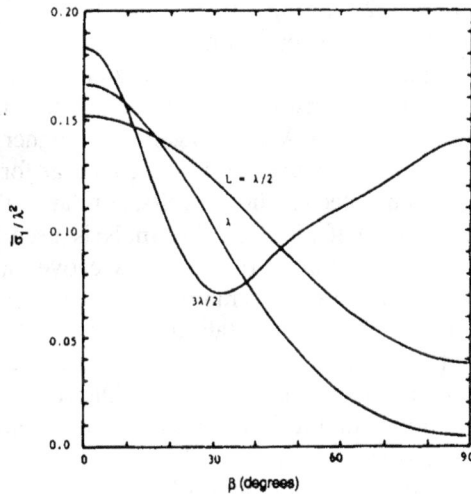

Figure 9.10 Spherically averaged bistatic cross sections for linear transmitting and receiving polarizations parallel to scattering plane. Cross sections have even symmetry about $\beta = \pi/2$ (90°) and $\beta = \pi$ (180°). (© 1984 IEEE [221].)

The following assumptions were made by Peebles for the results shown in Figures 9.9 and 9.10:

- Incident wave is planar.
- Incident field is approximately the same for all dipoles in the cloud.
- Dipoles are infinitely conducting wires, resonant at the transmitting frequency.
- Scattering between dipoles is negligible.
- Mutual coupling between dipoles is negligible.

The first assumption is satisfied when the chaff lies in the far field of the transmitting antenna beam. The second assumption is satisfied when the cloud extent is small with respect to R_T. The last assumption is satisfied when dipoles are spaced at least two wavelengths apart in any direction. Average spacing down to 0.4λ can produce up to a 3-dB loss in bistatic RCS [223].

Peebles also calculated $\overline{\sigma}_1$ for a cloud of horizontally oriented, resonant chaff dipoles [222]. Specifically, the dipole axes were assumed to lie in a horizontal plane with respect to the earth's surface, with random and uniform distribution of directions within the plane. This case is of interest for some chaff deployments, for example, in the atmosphere. Equation (9.8) again applies to this case. Unfortunately, $\overline{\sigma}_1$ was found to be a function of specific geometries, and thus could not be expressed analytically or graphically. Tabular results are given in [222], which shows values bracketing those shown in Figures 9.9 and 9.10. Specifically, for vertical polarization and $L = \lambda/2$, $\overline{\sigma}_1$ ranges from a factor of 2 higher in some geometries to vanishingly small values whenever either the transmitter or the receiver is located in the horizontal plane containing the chaff dipoles. (In this case the dipole plane is orthogonal to either the transmitting or receiving polarization.) In contrast, for horizontal polarization and $L = \lambda/2$, $\overline{\sigma}_1$ is generally higher, by a factor of 2 or less, and does not show very low values. Results are similar for $L = \lambda$ and $3\lambda/2$.

Dedrick *et al.* [224] generated (earlier) results, similar to those of Peebles, but for a cloud of nonresonant chaff dipoles. The analysis used Stokes's parameter description of the general radiation field. An average over near-resonant ($L = 0.5694\lambda$ and 1.213λ) wire orientation was made by a Monte Carlo method, and an average over long ($L = 6.06\lambda$) wire orientation was made analytically. Dedrick's and Peebles's [221] data plotting methods are identical and show similar data trends for the resonant and near-resonant case, with Dedrick showing lower values because the dipoles were nonresonant. Dedrick's long wire results show large values in the forward-scatter direction ($\beta \rightarrow 180°$), with $\overline{\sigma}_1 = 3.0$ at $\beta = 180°$ for all polarizations, which is one to two orders of magnitude higher than in the backscatter ($\beta = 0°$) direction. These results are not unexpected because forward-scatter RCS values for targets large with respect to wavelength exhibit this phenomenon, as discussed in Section 8.4.

Chapter 10
ELECTRONIC COUNTERMEASURES AND COUNTER-COUNTERMEASURES

All radars must be able to operate in the presence of naturally occurring interference resulting from transmissions from other sources. Military radars must also operate in hostile environments, which consist of deliberate interference designed to degrade their performance [1]. This deliberate interference is called electronic countermeasures (ECM), and is traditionally divided into two categories: active ECM, consisting of noise and deception waveforms, and passive ECM, typically consisting of chaff and decoys.

Methods for mitigating the effects of naturally occurring interference are the subject of electromagnetic compatibility, which is treated elsewhere [196, 198]. Methods for countering ECM are called electronic counter-countermeasures (ECCM). The goal of ECCM is to raise the cost of ECM to the point where it is prohibitive [1] or, failing that goal, to provide an equal cost penalty for all types of ECM, so that the radar is not vulnerable to a specific type of ECM.

The topics of ECM and ECCM for monostatic radars are now treated extensively in the open literature [1, 105, 136, 192–197]. Specifically, ECM and ECCM techniques are listed and described by Schleher [193], Van Brunt [192], and Johnston [195]. More importantly, their operating principles and effects are analyzed by Skolnik [1], Blake [136], Barton [105], Maksimov *et al.* [196], Leonov and Fomichev [197], with the first, classic analysis given in Boyd *et al.* [195]. The purpose of this chapter is to extend the monostatic ECM and ECCM analyses to the bistatic radar case, with emphasis on the differences between monostatic and bistatic ECM and ECCM. The chapter concentrates on active ECM and its counters because the effects of passive ECM, such as chaff (Section 9.4), are very similar for monostatic and bistatic radars and can be countered by doppler processing. One exception occurs when a decoy uses a passive retroreflector to enhance its monostatic RCS; this case is treated in Section 12.3.

10.1 NOISE JAMMING

A noise jammer typically radiates continuous, random noise with bandwidth B_J that is wider than the radar receiver's noise bandwidth B_n. The effect is to raise the total power spectral density of the background noise in the receiver from N_0 to $N_0 + J_0$, where J_0 is the jamming PSD at the radar receiver [105]. Following the method of Blake [136] and Barton [105], J_0 is defined as

$$J_0 = \frac{P_J G_J G_R F_J^2 \lambda^2}{(4\pi)^2 B_J R_J^2 L_J} \tag{10.1}$$

where

$\quad P_J G_J$ = jammer transmitted ERP,
$\qquad F_J$ = jammer pattern propagation factor = $F_J' f_J f_{RJ}$,
$\qquad F_J'$ = propagation factor to account for multipath, diffraction, and refraction,
$\qquad f_J$ = jammer antenna pattern factor, referenced to the receiver site,
$\qquad f_{RJ}$ = radar receiving antenna pattern factor, referenced to the jammer site,
$\qquad G_R$ = radar receiving antenna power gain,
$\qquad B_J$ = jammer bandwidth,
$\qquad R_J$ = jammer-to-radar receiver range,
$\qquad L_J$ = jammer system losses, including atmospheric attenuation; bandwidth mismatch, $B_J < B_n$, where B_n is the noise bandwidth of the radar receiver's predetection filter; noise quality factor (non-Gaussian noise); and polarization mismatch.

Note that under typical noise-jamming conditions $B_J > B_n$, in which case no bandwidth mismatch loss results. When $B_J \approx B_n$, the jammer is called a spot noise jammer; when $B_J \gg B_n$, the jammer is called a barrage noise jammer. Also, when G_J is small, the jammer is called a broadcast jammer.

The various antenna pattern factors require clarification. Two are defined with (10.1): f_J, the jammer antenna pattern factor referenced to the receiver site, and f_{RJ}, the radar receiving antenna pattern factor referenced to the jammer site. A third is defined with the basic, thermal noise-limited range equation, (4.1a): f_R, the radar receiving antenna pattern factor referenced to the target site. Their values change as the geometry changes.

Four geometries are usually of interest and are shown in Figure 10.1. In the simplest geometry, Figure 10.1(a), the jammer and target are collocated, with the receiving beam pointed directly at the jammer and target, and with the jammer beam pointed directly at the receiver. In this case $f_R = f_{RJ} = f_J = 1$. It is usually

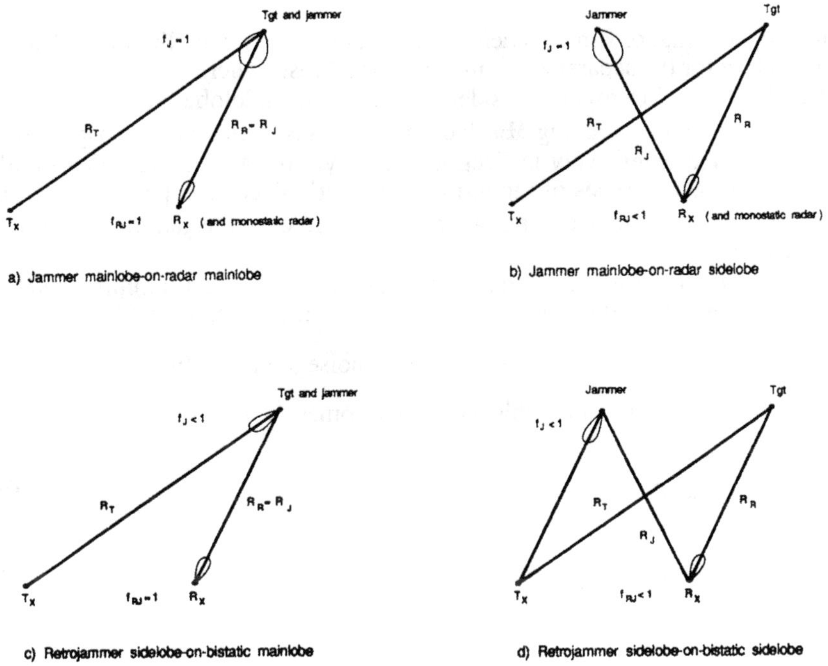

Figure 10.1 Typical geometries defining jammer and radar antenna pattern factors; in all cases $f_R = 1$ (receiving antenna pointed at target).

called mainlobe jamming, and is analogous to the monostatic case in which the monostatic radar is located at the receiver site. Figure 10.1(b) shows the geometry when the jammer and target are separated, with the receiving beam pointing directly at the target and with the jammer beam pointed directly at the receiver. In this case, $f_R = f_J = 1$, but $f_{RJ} < 1$. It is usually called sidelobe jamming and again is analogous to the monostatic case in which the monostatic radar is located at the receiver site. In both cases, the jammer antenna gain, G_J, can assume all possible values.

The last two geometries are bistatic variants of the first two, where a high-gain jammer beam is now pointed directly at the transmitting site, rather than at the receiving site. As outlined later in this chapter, and also in Section 12.3, a jammer operating with a high-gain antenna typically points its antenna using AOA measurements of the radar's transmitting signal, which of course comes from the transmitting site. Figure 10.1(c) shows the geometry for the collocated jammer and target case, where $f_R = f_{RJ} = 1$ but $f_J < 1$. It is called retrojammer sidelobe-on-bistatic mainlobe jamming, or simply sidelobe-on-mainlobe jamming. Figure 10.1(d) shows the geometry for the separated jammer and target case, where $f_R = 1$, but $f_{RJ} < 1$ and $f_J < 1$. It is called retrojammer sidelobe-on-bistatic sidelobe jamming, or simply sidelobe-on-sidelobe jamming. Each of these cases is treated in this chapter, with the last case treated extensively in Section 12.3, Hybrid Radars. Note that in all four cases $f_R = 1$. Some types of jamming, particularly deception jamming, attempt to generate receiving antenna pointing errors. The effect of this jamming can often be characterized by $f_R < 1$.

When multiple noise jammers are present, each with a jamming PSD at the radar receiver $(J_0)_i$, the background noise in the receiver is increased from N_0 to $N_0 + \sum_{i=1}^{n} (J_0)_i$ where n is the number of noise jammers. In terms of the receiver's system noise temperature, this increase becomes

$$T_s' = T_s + \sum_{i=1}^{n} (J_0)_i / k \qquad (10.2a)$$

$$= T_s + \sum J_0 / k \qquad (10.2b)$$

where

T_s' = effective system noise temperature (under jamming conditions),
T_s = system noise temperature (under thermal noise conditions),
k = Boltzmann's constant,

and for convenience,

$$\sum J_0 = \sum_{i=1}^{n} (J_0)_i$$

Substituting T_s' for T_s in (4.1a) yields

$$(R_T R_R)_J = \left[\frac{P_T G_T G_R \lambda^2 \sigma_B F_T^2 F_R^2}{(4\pi)^3 (k T_s + \Sigma J_0) B_n (S/N)_{min} L_T L_R} \right]^{1/2} \qquad (10.3a)$$

$$= \kappa_J \qquad (10.3b)$$

where $(R_T R_R)_J = \kappa_J$ is the bistatic maximum range product under noise-jamming conditions. Substituting (10.1) into (10.3a) and assuming a single jammer with J_0

$\gg kT_s$, a typical case of interest, yields:

$$(R_T R_R)_{J1} = \left[\frac{P_T G_T}{P_J G_J} \cdot \frac{F_T^2 F_R^2}{F_J^2} \cdot \frac{L_J}{L_T L_R} \cdot \frac{B_J}{B_n} \cdot \frac{\sigma_B R_J^2}{4\pi (S/N)_{min}} \right]^{1/2} \qquad (10.4a)$$

$$= \kappa_{J1} \qquad (10.4b)$$

Equations (10.3) and (10.4) are the general bistatic burn-through equations for multiple and single noise jammers, respectively, and are equivalent to the monostatic burn-through equations when $R_T = R_R$ and $\sigma_B = \sigma_M$. Similar to the monostatic case, burn-through or target detection occurs when $R_T R_R \leq (R_T R_R)_J$; when $R_T R_R > (R_T R_R)_J$ the target is said to be screened by jamming.

A more compact form of (10.3) can be written in terms of the thermal noise-limited bistatic maximum range product κ, (4.1b):

$$(R_T R_R)_J = \kappa_J = \kappa \left(\frac{kT_s}{kT_s + \Sigma J_0} \right)^{1/2} \qquad (10.5)$$

where $[kT_s/(kT_s + \Sigma J_0)]^{1/2}$ represents the reduction in κ under noise-jamming conditions. Equation (10.5) is correct for all values of ΣJ_0.

For convenience the terms comprising J_0 can be separated into two groups: geometry-dependent, J_D, and geometry-independent, J_I. Thus,

$$J_0 = J_D J_I \qquad (10.6)$$

where

$$J_D = F_J^2 / R_J^2 L_J \qquad (10.7)$$

$$J_I = \frac{P_J G_J G_R \lambda^2}{(4\pi)^2 B_J} \qquad (10.8a)$$

$$= \frac{P_J G_J A_R}{4\pi B_J} \qquad (10.8b)$$

and A_R is the effective aperture of the bistatic receiving antenna.

Consider an example using the geometry of Figure 10.2 and the parameters given in Table 10.1.

The maximum bistatic range product κ is assumed constant over all geometry, which is not usually the case because σ_B, F_T, and F_R, vary with geometry. The

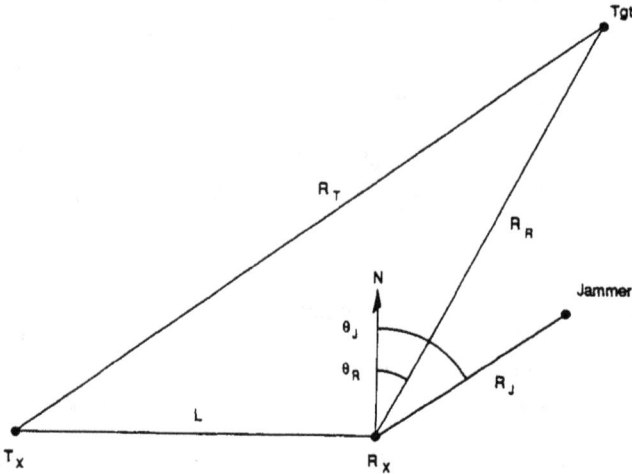

Figure 10.2 Geometry for standoff jamming of a bistatic radar.

constant κ assumption is used to illustrate basic jammer-radar interactions. Inserting parameters from Table 10.1 into (10.8a) yields the geometry-independent term, $J_1 = 5.6 \times 10^{-8}$ W-m²/Hz.

The remaining parameter to be specified is F_J, the jammer's geometry-dependent propagation factor, defined in (10.1), and consisting of three parts, F'_J, f_J, and f_{RJ}. Assume no multipath, diffraction, and refraction, so $F'_J = 1$. The jammer is assumed to broadcast jamming uniformly over its forward hemisphere, which includes the bistatic receiving (and transmitting) site. Thus, the jammer's antenna pattern factor, $f_J = 1$, to generate the geometry of Figure 10.1(a) and (b). The bistatic receiver's antenna pattern factor, f_{RJ}, depends on the jammer's angular posi-

Table 10.1
Parameters for Bistatic Radar Burn-Through against a Single, Standoff, Broadcast, Barrage Noise Jammer

$P_J = 500$ W
$G_J = 3$ dBi
$B_J = 1$ GHz
$L_J = 3$ dB
$R_J = 100$ km
$\lambda = 0.1$ m
$G_R = 29$ dBi ($A_R \approx 0.7$ m²)
$kT_s = 6 \times 10^{-21}$ W/Hz (≈ 2 dB noise figure)
$\kappa = 10{,}000$ km² (constant)

tion with respect to the target, $\theta_J - \theta_R$ in Figure 10.2, and the bistatic receiving antenna beamwidth, $\Delta\theta_R$. When $|\theta_J - \theta_R| < \Delta\theta_R/2$ the jammer's transmission will be received through the mainlobe of the bistatic receiving antenna, Figure 10.1(a). When $|\theta_J - \theta_R| \gtrsim \Delta\theta_R$ reception will be through the sidelobes, Figure 10.1(b). This geometry is analogous to the monostatic situation, where the former case is called standoff mainlobe jamming and the latter case is called standoff sidelobe jamming. It is identical when the monostatic radar is located at the bistatic receiving site. Because G_R is assumed to be 29 dBi, $\Delta\theta_R \approx 6°$. Thus, mainlobe reception occurs when the jammer is positioned with $\pm 3°$ of the target. Sidelobe reception occurs at angles beyond $\pm 6°$. Assume two cases: mainlobe reception where $f_{RJ}^2 = 1$ and sidelobe reception where $f_{RJ}^2 = -35$ dB (–6 dBi average power sidelobes).

Inserting these and the Table 10.1 parameters into (10.7) yields the geometry-dependent term, $J_D = 5 \times 10^{-11}$ m^{-2} and 1.6×10^{-14} m^{-2} for mainlobe and sidelobe reception, respectively. Multiplying by J_I yields the jammer's PSD at the bistatic receiver, $J_0 = 2.8 \times 10^{-18}$ W/Hz and 8.9×10^{-22} W/Hz, respectively. Inserting these and the Table 10.1 parameters into (10.5) yields $(R_T R_R)_J = \kappa_J = 0.046$ $\kappa = 460$ km^2 and 0.93 $\kappa = 9300$ km^2, respectively. Thus, for standoff mainlobe jamming, the maximum range product under jamming conditions is reduced to about 5% of the maximum range product under thermal noise conditions, whereas it is approximately 93% for standoff sidelobe jamming. Note that for the sidelobe case, $J_0 < kT_s$, and thus the exact expression, (10.5), must be used. [If the approximate expressions, (10.4) or $(kT_s/J_0)^{1/2}$, were used, $\kappa_J = 2.6\kappa$, which shows that the jamming enhances target detection range — clearly impossible.] Both the approximate and exact expressions give the same results for the mainlobe case because $J_0 \gg kT_s$.

Given a transmitter-receiver separation, or baseline L, the two values for κ_J can be plotted as "angle-gated" ovals of Cassini. Figure 10.3 shows the example for $L = 100$ km, using the methodology detailed in Section 4.3. The large oval is a plot of $\kappa_J = 9300$ km^2, the sidelobe jamming case. The small dotted oval is a plot of $\kappa_J = 460$ km^2, the mainlobe jamming case. Using (12.3), this small oval can be approximated by a circle centered on the receiving (and transmitting) site with radius $R_0 \approx \kappa_J / L = 4.6$ km. The jammer is at 100 km and at an arbitrary angle with respect to the receiving site. The jammer cuts out a mainlobe wedge of width $\Delta\theta_R$ (and near-sidelobe wedge of width $2\Delta\theta_R$) in the large oval, representing the mainlobe jamming sector.

In the monostatic jamming case, the ovals reduce to circles. When the monostatic radar with equivalent thermal noise-limited monostatic range $\sqrt{\kappa}$ is located at the receiving site, the circles have a radius $\sqrt{\kappa_J}$, or 96 km and 21 km for sidelobe and mainlobe jamming, respectively. These circles are also shown in Figure 10.3. The mainlobe jamming sector is identical to that of the bistatic case, but is not shown. Ignoring the mainlobe jamming wedge, the monostatic radar has a coverage area against sidelobe jamming of $A_M = \pi\kappa_J = 2.92 \times 10^4$ km^2, and the bistatic

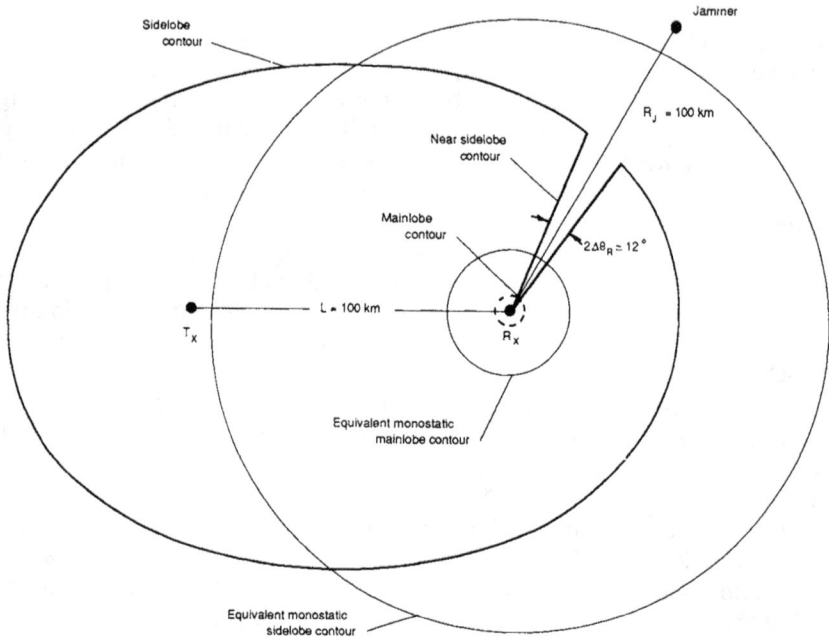

Figure 10.3 Target detection contours for the jamming conditions of Table 10.1 (and subsequent text), approximately to scale.

radar of $A_{B1} = 2.87 \times 10^4$ km^2, from (5.20), just slightly less than the monostatic coverage. Note that the shapes of the coverage areas are different, which can work to the bistatic radar's advantage or disadvantage, depending on the direction of target attack.

Against mainlobe jamming, while both monostatic and bistatic burn-through ranges and coverage areas are significantly reduced, bistatic performance is much lower than monostatic performance. This reduction occurs when $L > 2\sqrt{\kappa_J}$ and the oval of Cassini breaks into two small ovals. In this case, $L = 4.8\sqrt{\kappa_J}$ and, from (5.21), $(A_{B2})/2 = 68.4$ km^2, with average radius $R_0 = 4.6$ km. For the monostatic case, the radius is 21 km and the coverage area is 1385 km^2. Thus, the monostatic radar has an advantage over mainlobe jamming when the bistatic baseline is large. Usually, this advantage is not operationally significant because the burn-through ranges are small in either case. We again emphasize that the ovals and circles are artificial because κ was assumed invariant with geometry and target RCS, which is not usually the case.

For the example of Table 10.1, mainlobe jamming severely constrains radar performance, whereas sidelobe jamming has a minimal effect on radar performance. A standoff jammer operating through the receiver's sidelobes has three primary options to reduce the bistatic (and monostatic) coverage regions: closer jammer-to-receiver range, R_J; reduced jamming bandwidth, B_J; and increased effective radiated power, $P_J G_J$. Each option increases J_0 at the bistatic receiver and thus reduces the bistatic maximum range product, κ. Moving the jammer closer to the receiver increases the risk of jammer engagement, for example, by location via triangulation and subsequent attack. As a consequence, standoff jammers are usually located outside weapon system engagement ranges, which limits the reduction in R_J. Reducing the jammer's bandwidth is usually implemented with an intercept receiver designed to measure the radar's transmitting frequency. The jammer's bandwidth is then tailored to this frequency. This operation is called *spot noise jamming.* Obvious counters to spot noise jamming are radar transmitter frequency diversity or frequency agility, which might force the jammer back to barrage jamming to cover all diverse or agile frequencies [105].

Increasing the effective radiated power can be implemented by increasing P_J or G_J. In the example, $P_J = 500$ W and $G_J = 3$ dBi. A jammer designer would consider both, and might conclude that a 3- to 6-dB increase in P_J and a 10- to 30-dB increase in G_J were possible, given adequate prime power and platform real estate. The assumption of a 22-dB increase in G_J yields a 25-dBi jammer antenna gain with an 11° beamwidth, which in turn requires *pointing the beam at the receiving site.* When operating against a monostatic radar, this pointing is typically done by measuring the transmitting signal's AOA, and using this measurement to point the jammer's beam. The process is called *retrodirective operation,* or smart jamming [105]. Against a bistatic radar, retrodirective operation is useful only when the jammer's main beam covers both sites. Otherwive, J_0 is reduced and κ_J is increased. Two generic examples are shown in Figures 10.1(c) and (d).

Consider an example again using the geometry of Figure 10.2 in which a monostatic radar with equivalent monostatic range $\sqrt{\kappa}$ is now located at the transmitting site. The parameters of Table 10.1 are again used except that $G_J = 25$ dBi rather than 3 dBi. When this retrojammer operates through the monostatic radar's sidelobes at $R_J = 100$ km, the monostatic radar's detection range is reduced from $\sqrt{\kappa_J} = 96$ to 7.6 km. When the bistatic receiver is located outside the retrojammer's main beam, Figure 10.1(d) applies. For example, when the bistatic receiver is in a −3-dBi average sidelobe of the retrojammer, $f_J = -28$ dB. Thus, the bistatic radar's performance is increased from $\kappa_J = 9300$ to 9820 km², which is 98% of κ, the maximum range product under thermal noise conditions.

The usual monostatic radar response to this level of retrodirective sidelobe jamming is sidelobe cancellation, with a canceller loop for each jammer [105]. When 22-dB cancellation is achieved, monostatic performance is restored to 96 km; with 28-dB cancellation, the equivalent monostatic range-squared matches that of the bistatic maximum range product, κ_J. Thus, a design and cost trade-off can be made between a (multiple) monostatic sidelobe cancellation system and a properly sited bistatic system—without sidelobe cancellers—when operating against standoff sidelobe retrodirective jamming. In most cases, the trade-off favors the monostatic configuration.

However, when many retrodirective jammers are present and ΣJ_0 is very high, the performance of a single monostatic radar, in particular a search radar, can be unavoidably degraded. In such cases Barton [105] asserts that "only the use of netted radars, having considerable overlap in the benign environment, can preserve search coverage in the presence of a concerted ECM attack." When the concerted ECM attack includes retrodirective jammers that are capable of multiple-beam operation, such as is possible with a Rotman lens system [192], each monostatic radar in the net can be jammed simultaneously by each retrojammer. In this particularly ugly ECM attack scenario, properly designed bistatic receivers with noise sidelobe cancellers have the potential of enhancing net performance through (quiet) bistatic spatial diversity. This configuration is called a *hybrid radar system* and is analyzed in Section 12.3.

When the target and (single) jammer are collocated $R_J = R_R$, which is the self-screening case, Figures 10.1(a) and (c). The bistatic self-screening, or crossover, range is established by substituting (10.6) into (10.5) with $R_T = R_R$, and again assuming $J_0 \gg kT_s$:

$$(R_T)_{J1} = \kappa \sqrt{kT_sL_J/F_J^2J_1} \tag{10.9}$$

Equation (10.9) is analogous to the monostatic self-screening range when the monostatic radar is located at the bistatic transmitting site. Assuming that $F_J = 1$ and κ and L_J are invariant with geometry, $(R_T)_{J1}$ is a constant. That is, the spatial region over which a target with a self-screening jammer can be detected is a circle of radius $\kappa \sqrt{kT_sL_J/J_1}$ centered on the transmitting site. The bistatic receiver can be located anywhere as long as $J_0 \gg kT_s$, such that the detection range is jammer noise-limited, rather than thermal noise-limited.

Under these assumptions, and again using the parameters of Table 10.1, $(R_T)_{J1} = 4.6$ km. That is, the target with self-screening jammer must be within 4.6 km of the transmitting site to be detected by the receiver. In this case the comparable detection region around the receiving site does not exist because the transmitter-to-self-screening target range sets the burn-through or crossover range. Note that $(R_T)_{J1} = 4.6$ km is also the burn-through range for a monostatic radar with equivalent monostatic range $\sqrt{\kappa}$ located at the transmitting site. Coincidentally, the value is the same as $R_0 = 4.6$ km, the radius of the small oval of Cassini generated by a standoff mainlobe jammer shown in Figure 10.3. By equating the expressions for R_0 and $(R_T)_{J1}$ [which again are $R_0 \approx \kappa_J/L$ and $(R_T)_{J1} = \kappa \sqrt{kT_sL_J/J_1}$, where κ_J is defined by (10.5) with $J_0 \gg kT_s$], the condition for equality becomes $J_0L^2L_J = 1$. Because $J_0 = 5 \times 10^{-11}$ m^{-2}, $L = 100$ km, and $L_J = 3$ dB in the example, the equality is satisfied, quite by accident.

In the self-screening scenario, retrodirective jamming with a high-gain jamming antenna ($G_J = 25$ dBi in the previous example) is neither needed nor desired because (1) $(R_T)_{J1}$ is already very small, (2) it would again be ineffective against a properly sited bistatic or hybrid monostatic-bistatic system, Figure 10.1(c), and (3) jammer system complexity and cost would be increased unnecessarily.

In general, bistatic and monostatic radar burn-through performance against self-screening jammers is not adequate to support system operation, such as target engagement. This assertion usually applies to escort and penetrating jammers as well. Barton [154] outlines the alternative air defense system response under these conditions:

> In the case of penetrating jammers, the radar system response is not to await burnthrough, but to identify the angles of the jammers and to attempt triangulation with data from adjacent radar sites. The jammers are then tracked passively in angle, and missile guidance is accomplished by using track-on-jam or a home-on-jam seeker, when the target is judged to be within missile range.

10.2 DECEPTION JAMMING

Deception jamming, or deception ECM (DECM), uses specialized waveforms to degrade radar performance. These waveforms comprise a set of DECM techniques, such as range gate pull-off (RGPO), crosseye, and false targets, which are described elsewhere [1, 105, 136, 192–197]. Radar performance degradation can take many forms, including target acquisition delays, increased tracking errors, loss of track, and overload or saturation of the radar's processor. In many cases, DECM acts to exacerbate a problem naturally present in a radar, such as target fluctuation, depolarization, glint, and multipath [105].

Deception ECM is designed to be more efficient than noise jamming in that DECM bandwidths and waveforms closely match radar bandwidths and waveforms. This matching requires that the DECM system intercept the radar waveform, and if radar waveform is varied frequently, the intercepts must be nearly continuous. To meet these requirements, DECM is implemented with transponders or repeaters. A transponder is defined as a system that receives and detects the radar signal, and then generates, upconverts, amplifies, and transmits a deception signal, with the detection and generation stages usually done at video frequencies. A repeater is defined as a system that receives, modulates, amplifies, and retransmits the radar signal, with all operations usually done at the radio frequency (rf) of the radar signal. It is also called a straight-through repeater [215].

The analysis of DECM requires two separate and distinct steps: (1) determination of the jamming-to-signal power ratio *required* at the radar receiver, $(JSR)_{req}$, to cause performance degradation, and (2) calculation of the jamming-to-signal power ratio, JSR, *available* at the radar receiver as a function of the engagement geometry. When JSR $\geq$ (JSR)$_{req}$, a necessary condition for DECM effectiveness is established. [Sufficiency is established only when other conditions, such as a minimum time duration of JSR $\geq$ (JSR)$_{req}$, are satisfied.] The first step can sometimes be accomplished analytically, but to generate high confidence estimates of (JSR)$_{req}$, simulation and laboratory or field measurements of a specific DECM technique and a specific radar are usually required. Depending on the monostatic radar and DECM technique, values of (JSR)$_{req}$ can vary from less than 0 dB for false targets to greater than 10 dB for angle deception [215]. For cross-polarized jamming, (JSR)$_{req}$ typically ranges from 22 to 30 dB [105]. To a first order, (JSR)$_{req}$ for monostatic and properly designed bistatic radar receivers should be similar, although (JSR)$_{req}$ for bistatic radar receivers has not been documented, if indeed it has ever been established.* When the performance of a radar against DECM is analyzed, the signal-to-jamming ratio required at the radar receiver, (SJR)$_{req}$, to achieve acceptable radar performance is used, where (SJR)$_{req}$ $\geq$ $(JSR)_{req}^{-1}$. The second DECM analysis step is accomplished by using link equations, which of course are different for monostatic and bistatic radars.

Two differences between noise jamming and DECM link equations are the signal reference and the signal bandwidth. For noise jamming, the target signal, S, is referenced to background noise, which has been increased by the jamming PSD, J_0, (10.1). For DECM, the target signal is referenced to the DECM signal power, J, at the radar receiver. Secondly, DECM typically radiates a signal with bandwidth B_J that is matched to the radar receiver's noise bandwidth B_n. When there is a bandwidth mismatch, such that $B_J > B_n$, the mismatch is treated as a loss, and is included in the jammer system loss term, L_J. In some cases, DECM can be implemented such that $B_J < B_n$, which does not generate a loss. This case is different from that of a noise jammer, where $B_J < B_n$ usually causes a loss, as defined for the L_J term in (10.1). In all cases DECM link equations are written in terms of DECM signal power at the radar receiver, rather than spectral density as is done for the noise-jamming case.

A significant difference between DECM transponders and repeaters is that a transponder generates the DECM signal internally, and thus can transmit power independent of the signal power level it receives (detects) from the radar. However, a repeater generates the DECM signal by operating on (time, phase, or amplitude modulating) the received radar signal. Thus, DECM-transmitted power is proportional to the radar's signal power level received by the repeater, which usually allows the repeater to operate in its linear amplification region. When the received

---

* As outlined in Chapters 1 and 2, few bistatic radars have been deployed or reached the field test stage since World War II; those that have are often used for test range instrumentation in which ECM testing against the instrumentation system is not a consideration. Related work [105, 189, 200, 201] on semiactive homing missiles does not discuss (JSR)$_{req}$ for DECM.

power level is strong enough, the transmitted power will reach its maximum output level, so that the repeater operates in the saturated region. To distinguish these two cases, $(P_J)_{max}$ represents the maximum effective transmitted power of a saturated repeater (or transponder) and $(P_J)_{lin}$ represents the effective transmitted power of a linear repeater, or a transponder that purposely reduces its power to implement a specific DECM technique. This latter case is sometimes called ECM or DECM power management.

The following link equations are used for the general DECM case. From (4.1) the power, S, of the target signal that the bistatic radar receives at ranges R_T and R_R is

$$S = \frac{P_T G_T G_R \lambda^2 \sigma_B F_T^2 F_R^2}{(4\pi)^3 R_T^2 R_R^2 L_T L_R} \tag{10.10}$$

where the terms are defined with (4.1) and the max and min subscripts have been omitted.

From (10.1), the DECM signal power, J, which the bistatic radar receives from the DECM at range R_J, is

$$J = \frac{P_J G_J G_R F_J^2 \lambda^2}{(4\pi)^2 R_J^2 L_J} \tag{10.11}$$

where the terms are defined with (10.1), and any DECM bandwidth mismatch $B_J > B_n$ is included in L_J. Thus,

$$\text{SJR} = S/J = \frac{P_T G_T}{P_J G_J} \cdot \frac{F_T^2 F_R^2}{F_J^2} \cdot \frac{L_J}{L_T L_R} \cdot \frac{\sigma_B}{4\pi} \cdot \frac{R_J^2}{R_T^2 R_R^2} \tag{10.12}$$

When (10.12) is solved for $R_T R_R$, it has the same form as the bistatic burn-through range for a single noise jammer, (10.4a), and becomes identical to (10.4a) when $B_J = B_n$ and $S/J = (S/N)_{min}$. In this sense, the terms "burn-through" and "self-screening" can also be used for DECM analysis when $\text{SJR} = (\text{JSR})_{req}^{-1}$.

Equation (10.12) applies to a DECM transponder that transmits either maximum or variable power, and also to a DECM repeater that operates in its saturated power region. When a DECM repeater operates in its linear power region, $(P_J)_{lin} \propto P_T G_T G_j G_e / R_T^2$, where G_e is the internal gain of the repeater and G_j is the gain of the repeater's receiving antenna. When $(P_J)_{lin}$ is substituted for P_T in (10.12) with $G_J = G_j$, the usual case, $(\text{SJR})_{lin} \propto R_J^2 \sigma_B / R_R^2 G_j^2 G_e$. In the monostatic case $(\text{SJR})_{lin} \propto R_J^4 \sigma_M / R_M^4 G_j^2 G_e$. When DECM is carried by the target for self-defense, $R_J = R_R$ for the bistatic case and $R_J = R_M$ for the monostatic case. Thus, $(\text{SJR})_{lin} \propto \sigma_i / (G_j^2 G_e)$ for both bistatic ($i = B$) and monostatic ($i = M$) radars, with appropriate attention paid to possibly different pattern propagation factors, as discussed next.

In the general case when DECM is carried by the target for self-defense, $R_J = R_R$ and $F'_R = F'_J$. That is, the bistatic (and monostatic) propagation factors for the target and jammer are equal, assuming reciprocity. Thus, (10.12) becomes

$$(SJR)_c = (S/J)_c = \frac{P_T G_T}{P_J G_J} \cdot \frac{F_{TJ}^2 f_R^2}{f_{IJ}^2 f_{RJ}^2} \cdot \frac{L_J}{L_T L_R} \cdot \frac{\sigma_B}{4\pi R_T^2} \tag{10.13}$$

where the subscript c denotes a collocated target and DECM.

When the values of S and J are taken at the peak of the receiving beam, $f_R = f_{RJ} = 1$. However, when the DECM system is designed to generate J in the skirts or sidelobes of the receiving beam, $f_{RJ} \ll 1$. Equation (10.13) also represents the monostatic case when $\sigma_B = \sigma_M$, and $R_T = R_M$; that is, when the monostatic radar is located at the bistatic transmitting site. (When the monostatic radar is located at the bistatic receiving site, $R_T \neq R_M$ and the pattern propagation factor, F_T, can be different.)

Consider the following example in which (10.13) is solved for the DECM self-screening range, R_T, and plotted as a function of the radar-to-DECM ERP ratio, $P_T G_T / P_J G_J$, where the parameters are listed in Table 10.2.

Table 10.2
Parameters for Bistatic (and Monostatic) Radar Range Performance Against Deception ECM (DECM)

$$(JSR)_{req} = (SJR)_c^{-1} = 10 \text{ dB}$$
$$f_J^2 = 1$$
$$F_T^2 = 1$$
$$f_R = f_{RJ}$$
$$L_J = 6 \text{ dB}$$
$$L_T L_R = 10 \text{ dB}$$

The L_J term might consist of 3-dB transmission line losses and a 3-dB bandwidth mismatch. Thus, (10.13) becomes

$$R_T = \left(\frac{\sigma_B}{\pi} \frac{P_T G_T}{P_J G_J} \right)^{1/2} \tag{10.14}$$

Figure 10.4 plots (10.14), and shows that a modest $P_J G_J$ is required to achieve $(JSR)_{req} = 10$ dB at very small values of R_T. For example, if DECM operation is required at a minimum range from the transmitter, $(R_T)_{min} = 5$ km, for $\sigma_B (= \sigma_M)$

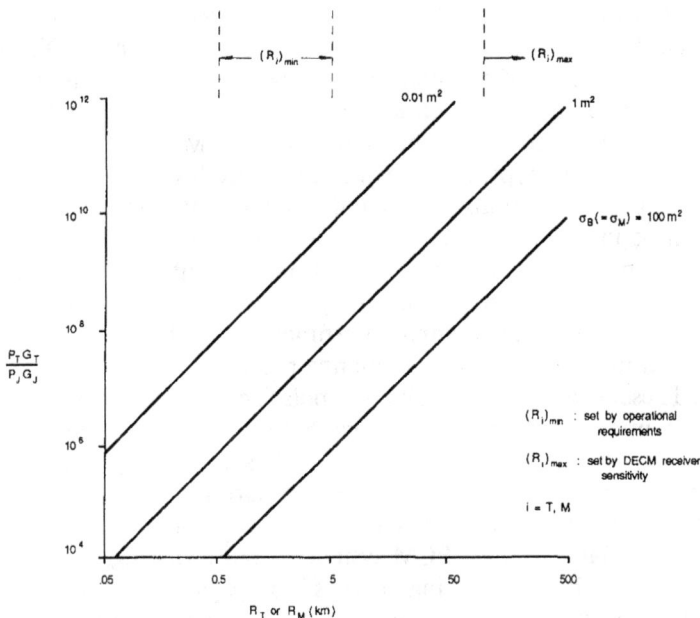

Figure 10.4 Self-defense DECM relative effective radiated power requirements, $P_T G_T / P_J G_J$, when operating at self-screening ranges from monostatic radar (R_M) and bistatic transmitter (R_T), with parameters given in Table 10.2.

$= 100$ m^2, then $P_T G_T / P_J G_J = 8 \times 10^5$. For an airborne fire control radar with $P_T = 10$ kW and $G_T = 35$ dBi, then $P_J G_J = 40$ W. Note that $P_J G_J$ is a maximum at $(R_T)_{min}$. Conversely, a minimum $P_T G_T$ occurs at $(R_T)_{max}$, which is often set by the sensitivity of the DECM receiver. This situation is analogous to the self-screening noise jammer case, (10.9), in which a self-screening noise jammer also requires modest $P_J G_J$. The form of (10.9) and (10.14) is similar, and in both cases performance is independent of the bistatic receiver-to-jammer or target range. Specifically, for a semiactive homing missile, the SJR for both DECM and noise jamming varies only as the illuminator (transmitter)-to-target range, R_T, varies during the engagement.

In contrast, the SJR for an active homing missile—with a monostatic radar seeker—rapidly increases (R_M^{-2}) as the missile closes the DECM or jamming target. The value of $P_T G_T$, however, for the missile will nearly always be smaller than that of the fire control radar illuminator on a launch platform, thus offsetting the SJR increase. For example, if DECM must be effective at $(R_M)_{min} = 0.5$ km against an active homing missile with $P_T = 1$ kW and $G_T = 25$ dBi, $P_J G_J = 40$ W, as before. (The author admits to "backing into" this calculation.)

In view of these modest DECM effective radiated power requirements, with the power battle nearly always favoring on-board self-screening DECM (or jamming), the monostatic (and bistatic) radar designer is driven to alternative system responses. For jammers, track-on-jam and home-on-jam are typical responses [154]. For DECM, receiver modifications, or "ECCM fixes" to counter specific DECM techniques are typical responses. For retrodirective DECM techniques, such as crosseye, separate transmitting and receiving antennas, usually with a small baseline, can be invoked. As discussed in Section 12.3, a bistatic radar configuration is a sufficient, but not necessary, condition to counter crosseye.

Because all bistatic systems, including semiactive homing missiles, inherently use only the silent receiving antenna for estimating target angular position, DECM techniques that measure and exploit antenna polarization and scan parameters are countered. These techniques include cross-polarized jamming, some types of scan-rate modulation and sidelobe false targets that are designed to emulate a credible target trajectory. In the latter case, the DECM false target generator must either measure or predict the receiving beam position in order to position the false target in angle. Unless the bistatic receiver uses pulse chasing, Section 13.2, this information is not available to the DECM system. In pulse chasing, the receiving beam is slaved in angle to the transmitting pulse as it propagates in the transmitting main beam. Thus, by estimating the geometry and the transmitting mainbeam position and measuring the transmitting pulse arrival time, the DECM system can calculate the receiving beam position using the same pulse-chasing algorithm as used by the bistatic receiver. However, if the bistatic transmitter uses a phased array with a random scan pattern (and covertly sends the scan pattern to the bistatic receiver), the DECM system can no longer estimate the transmitting mainbeam position, and thus cannot calculate the receiving beam position.

Because modern monostatic radars are usually designed for monopulse or silent lobing operation, they, too, will counter scan-rate modulation techniques that require measurement of the antenna scan pattern. Further, when the monostatic radar uses a phased-array antenna with random scan pattern, the DECM system cannot estimate the transmitting mainbeam position. Thus, sidelobe false targets that are designed to emulate a credible trajectory are countered, as in the bistatic case. Finally, Barton [105] reports that monostatic radar ECCM to cross-polarization jamming includes use of an arbitrary polarization for reception, unknown to the jammer, which is easily implemented in most types of antennas. The effect of this ECCM is identical to that inherent in silent bistatic antenna operation. Consequently, while bistatic radar operation may complicate DECM operation, it provides no obvious DECM counter-countermeasure advantage over a properly designed monostatic radar.

Chapter 11
MULTISTATIC RADARS

As defined in Chapter 1, a multistatic radar uses two or more receiving sites with overlapping coverage and combines target data coherently or noncoherently at a central location. Examples of noncoherent data combining are SPASUR [1, 7, 184, 185] and the MMS [13, 14] although for both systems, each transmitting-receiving pair operates coherently; that is, each receiver establishes phase coherence with the transmitter and coherently processes the bistatic echo. Examples of coherent data combining are thinned, random, and distorted arrays [3–6], the Radio Camera [11, 12], and interferometers, such as SPASUR when it operates in a calibration mode by cross-correlating radio star signals. Multiple transmitters can also be used, but they are not essential to the definition. When multiple transmitters are used and the transmissions from each site are coherently phased so that the multiple transmitters operate as a single transmitting array, the configuration is called a distributed array radar (DAR) [3, 5].

When the multistatic receiving sites are widely separated, the bistatic echo from complex targets at each receiving site can become spatially decorrelated; that is, a different phase and amplitude of the bistatic echo will be received by each site. When the multistatic radar combines data noncoherently, the system's S/N is improved by noncoherent integration of the decorrelated signals [225–228]. This process is analogous to signal decorrelation by frequency agility in a monostatic radar. For example, a target exhibiting Swerling 1 or 3 fluctuation statistics at one frequency (or site) might exhibit Swerling 2 or 4 statistics at multiple frequencies (or sites), which in turn requires a smaller per-pulse (or per-site) S/N for high probabilities of detection. In contrast, this spatial decorrelation is a disadvantage for multistatic radars that combine data coherently. Consequently, these systems are constrained to limited site separations, which to a first order are established by target complexity, frequency of operation, and the geometry, or bistatic angle. In all implementations, LOS requirements must be satisfied, as outlined in Section 5.5.2.

The noncoherent combining of data from multiple receiving sites is identical to the straightforward, but complex, radar netting process, and is treated

---

extensively by Farina and Studer [209, 210]. This chapter concentrates on the performance issues of coherent data combining by a multistatic radar.

11.1 RECEIVING APERTURE CHARACTERISTICS

A feature common to most multistatic radars that combine target data coherently is the large receiving aperture, and hence narrow receiving beamwidth, generated by the multiple receiving sites. For a linear array of periodic, aperiodic, or random elements distributed over an effective aperture of length L_A, the one-way, half-power beamwidth $\Delta\theta_R$ is approximately 0.886 λ/L_A rad [145]. For n_r identical elements, each with gain G_e, the linear array gain, G_R, is $\approx n_r G_e$. When the elements are isotropic, $G_e \approx \pi$ and $G_R \approx \pi n_r$. If the average spacing between elements is s, $L_A = n_r s$ and $G_R \approx G_e L_A/s \approx (0.886 G_e/\Delta\theta_R)(\lambda/s)$. The effective aperture extends beyond the center of the end elements by $s/2$. As the element gain increases, the element beamwidth decreases, which in turn decreases the angle over which the array can be scanned. For example, $G_e = 20$ dB permits angle scan over approximately $\pm 10°$ from array normal.

When the element spacing s is periodic and $s > \lambda/2$, grating lobes are generated, each with amplitude and width equal to the main lobe. They are spaced $\sin\theta_{R1} - \sin\theta_{R2} = \lambda/s = \lambda n_r/L_A$ apart, where θ_{Ri} is the pointing angle of the ith grating lobe referenced to array normal [145]. When the linear array is colinear with the baseline, θ_{Ri} is referenced to the North coordinate system, Figure 3.1. When the element spacing s is periodic and $s = \lambda/2$, the array is fully filled and no grating lobes are present. In this case $G_R \approx 2G_e L_A/\lambda \approx 1.732 G_e/\Delta\theta_R$, the usual form for one-way linear array gain. When the element spacing is random and $s > \lambda/2$, grating lobes are suppressed, but the average sidelobe level rises to $1/n_r$. When the element spacing is aperiodic, the average far sidelobe level also approaches $1/n_r$. The close-in sidelobes approximately follow the radiation pattern for the underlying distribution [4]. Table 11.1 summarizes these relationships for element spacing greater than $\lambda/2$.

For periodic element spacing, the array pattern can be steered to an angle θ_{Rj} by applying linearly progressive phase increments from element to element, such that the phase between adjacent elements differs by $2\pi(s/\lambda)\sin\theta_{Rj}$ [145]. Beamforming and array weighting are done in the usual manner. For aperiodic or random element spacing, methods for both adaptive beamforming and array pattern steering have been developed [11, 146]. Similar developments are available for planar arrays (i.e., two-dimensional linear arrays) [145].

Some inherent advantages of narrow receiving beam operation include accurate cross-range location of targets at long ranges, enhanced ECCM capability to detect targets between jamming strobes, and resolution of closely spaced targets in

Table 11.1
Summary of Characteristics for Linear Arrays of Identical Elements with $s > \lambda/2$ Spacing

Element Spacing	$\Delta\theta_R$	G_R	Grating Lobes	Average Sidelobe Levels Near	Far
Periodic	$0.886\lambda/L_A$	$n_r G_e$	Equal to main lobe in amplitude and width	Same as grating lobes	
	$= 0.886s/n_r$	$= G_e L_A/s$ $= (0.866 G_e/\Delta\theta_R)(\lambda/s)$			
Aperiodic	Same as above	Same as above	Suppressed	Follows radiation pattern	$1/n_r$
Random	Same as above	Same as above	Suppressed	$1/n_r$	$1/n_r$

where

$\Delta\theta_R$ = one way, half-power beamwidth,
G_R = linear array gain,
λ = wavelength,
L_A = length of effective aperture = $n_r s$,
s = average spacing between elements,
n_r = number of elements,
G_e = gain of one element.

angle [5]. Limitations include ambiguities in target location estimates when grating lobes are present and increased clutter levels through grating lobes or higher sidelobe levels [3].

11.2 SYSTEM S/N

For a multistatic radar operating coherently with n_r receivers and n_t transmitters, S/N for the multistatic radar system is given as a variation of (4.2) and (4.3) [3].

$$S/N = \frac{\sigma_B \Sigma_i^2 \Sigma_t^2}{(4\pi)^2 k} \tag{11.1}$$

where

$$\Sigma_t = \sum_{i=1}^{n_t} \left(\frac{P_{Ti} G_{Ti} F_{Ti}^2}{R_{Ti}^2 L_{Ti}} \right)^{1/2} \tag{11.2}$$

and

$$\Sigma_r = \frac{\sum_{j=1}^{n_r} \left(\frac{A_{Rj}F_{Rj}^2}{R_{Rj}^2 L_{Rj}}\right)^{1/2}}{\left(\sum_{j=1}^{n_r} T_{sj}B_{nj}\right)^{1/2}} \qquad (11.3)$$

where the subscript i denotes the value for the ith transmitting element and the subscript j denotes the value for the jth receiving element. The receiving aperture A_{Rj} is related to the receiving antenna gain G_{Rj} in the usual manner:

$$A_{Rj} = G_{Rj}\lambda^2/4\pi \qquad (11.4)$$

When the n_t transmitter elements are identical, with each having parameters equal to the $i = 1$ element, and each element is at the same range from the target,

$$\Sigma_t = \left(\frac{n_t^2 P_{T1}G_{T1}F_T^2}{R_T^2 L_{T1}}\right)^{1/2} \qquad (11.5)$$

When the n_r receiver elements are identical and at the same range to the target,

$$\Sigma_r = \left(\frac{n_r A_{R1}F_R^2}{R_R^2 T_{s1}B_{n1}L_{R1}}\right)^{1/2} \qquad (11.6)$$

Thus, in this special case,

$$S/N = \frac{n_t^2 n_r P_{T1}G_{T1}A_{R1}\sigma_B F_T^2 F_R^2}{(4\pi)^2 R_T^2 R_R^2 kT_{s1}B_{n1}L_{T1}L_{R1}} \qquad (11.7)$$

Equation (11.7) is identical to that of an active or passive array antenna with n_t identical transmitting elements and n_r identical receiving elements. (The n_r term is not squared because the signal voltages add coherently and the noise voltages add noncoherently in the combining or phasing network [136]. Both (11.1) and (11.7) assume that σ_B does not decorrelate over all pairs of transmitting and receiving elements. This assumption typically is valid for $\beta \lesssim 5°$ for complex targets at microwave frequencies, and for $\beta \lesssim 100°$ for smooth targets, as defined in Section 8.1.

Equation (11.7) shows that the multistatic S/N is equal to bistatic S/N when $(n_t^2 P_{T1}G_{T1})(n_r A_{R1}) = P_T G_T A_R$, assuming identical receiver characteristics and losses. Also, assuming $n_r A_{R1} = A_R$, the effective radiated power of each multistatic transmitting element is reduced by n_t^2. When n_t is large, say, greater than 10, the detectability of each multistatic transmitting element is reduced by >20 dB compared to the bistatic transmitter. This argument assumes an adequate LOS from each transmitting element to the target.

11.3 IMPLEMENTATION REQUIREMENTS

Both signal synchronization and spatial coherence among transmitting and receiving elements must be established. Signal synchronization is treated in Sections 13.4 and 13.5. Establishing amplitude and phase coherence between each transmitting element and the coherent combining of signals from all receiving elements are problems unique to coherent multistatic radar operation. Self-cohering or adaptive beam-forming techniques are potential solutions. For example, the thinned, adaptive synthetic aperture radio camera (TASARC) program at the University of Pennsylvania [5, 11] used adaptive beamforming to correct phase distortions between randomly located transmitting and receiving elements on the ground. A corner reflector located between the antenna array and target area generates a space point phase reference and focuses the beam at the corner reflector. Then a second set of phase corrections is added to refocus the beam at a specific target. This technique is similar to that used for phase correction in airborne SAR systems. The corner reflector can be replaced by naturally occurring clutter discretes, when available.

A self-cohering process for transmitting and receiving elements randomly located on an aircraft is described by Steinberg [4], and a search algorithm for obtaining the reference signal is also described by Steinberg [6, 12]. Other multistatic radar implementation issues include the self-cohering design (methods of phase conjugation and reference phase distribution for the adaptive transmitting-receiving array) and the establishment of suitable data links between the elements and the central signal processor. The effects of multipath and array scattering, along with the aforementioned spatial decorrelation of the target echo as the bistatic angle increases, must also be considered.

In general, the implementation of a multistatic radar is more easily accomplished when the positions of all transmitting and receiving elements are fixed with respect to each other, for example, when located on the ground [147]. Because of the LOS restrictions, they are often designed for operation with high-altitude targets, such as satellites and ballistic missiles.

Chapter 12
SPECIAL CONCEPTS AND APPLICATIONS

Throughout this book, bistatic radar concepts such as forward-scatter fences, clutter tuning, hitchhiking, passive situation awareness, over-the-shoulder operation, range extension, and counter-retrodirective jammer operation have been discussed briefly, but not analyzed in any detail. Many of these concepts are summarized in Table 4.1, which defines their operating regions (ovals of Cassini) and transmitter type (dedicated, cooperative, and noncooperative). The purpose of this chapter is to assemble available information on these concepts and to analyze their utility using range, doppler, and area relationships, as well as target radar cross section and clutter models developed in previous chapters.*

The element common to all of these bistatic radar concepts is, again, geometry: In each concept a particular transmitter-receiver geometry is selected to

---

*Author's Note: Some of the bistatic radar concepts analyzed in this chapter are new; that is, they have not been published in professional journals or trade magazines. Thus, while they may appear here for the first time, I do not take credit for their invention. I know of some that were developed earlier, but either were not documented or if documented, the reports were not published. Often a "proprietary" stamp is used to bury the report in a company's files forever. Possibly others fall into this category. In any case, the concepts analyzed here are developed from first (bistatic) principles and a motivation to find special niches for bistatic radars. Clearly, these bistatic radar concepts do not comprise a complete set. Additional concepts will undoubtedly be identified, developed, and possibly tested, and the current set will certainly be revised and extended over time.

Finally, in this chapter when I assume that the performance of a bistatic receiver matches that of a host monostatic receiver, I do so without proof. The performance match assumes that the bistatic receiver generates matched filtering, integration gain and losses identical to that of the monostatic receiver; however, receiving antenna gains and radar cross sections are excluded from the assumption. The assumption is fairly robust when the monostatic radar is either cooperative or uses predictable waveforms, and time and phase synchronization can be implemented with sufficient fidelity. The assumption becomes more difficult to justify when the monostatic radar is either noncooperative or uses unpredictable waveforms. The difficulty is compounded when noncooperative radar characteristics are not completely known, and synchronization must be implemented through the monostatic radar's sidelobes. A "not always the case" flag is included wherever the assumption is made. In any event the reader is cautioned to assess the performance match assumption on a case-by-case basis.

develop some unique performance capability. In the case of clutter tuning, both geometry and the separate velocity vectors of the transmitter and receiver are exploited. Some concepts require a dedicated transmitter; others hitchhike off a monostatic radar. Some operate autonomously from the monostatic radar; others must be netted with the monostatic radar.

A convenient way to group these bistatic radar concepts is into four categories: hitchhiking, forward-scatter fences, hybrid radars, and clutter tuning. Concepts in the hitchhiking category exploit transmissions from other emitters, usually cooperative or noncooperative monostatic radars (called host radars), operate autonomously from the host radars, and detect targets in either the transmitter- or the receiver-centered ovals of Cassini. They often exploit the covert nature of a bistatic receiver for survivability enhancement in forward-based regions near the battle area. They include range extension and its variant over-the-shoulder operation, passive situation awareness, monitoring, and launch alert. One variant of the launch alert concept operates in the forward-scatter RCS enhancement region, but otherwise has requirements and characteristics of a hitchhiker. Thus, this variant is included here.

Concepts in the forward-scatter fence category exploit RCS enhancement of both normal and stealth targets in regions near $\beta = 180°$. These fences are an extension of the 1930s bistatic radar fences described in Chapter 2 and use fixed or moving sites. The transmitter is dedicated and the receiver detects targets in the cosite operating region, which is typically constrained by contours of constant bistatic angle, β. The concepts include a single fence and multiple, netted fences in a grid configuration.

Concepts in the hybrid radar category use a cooperative monostatic radar as the bistatic transmitter. In this sense, they are hitchhikers, but, in contrast to hitchhikers, they operate exclusively to improve performance of the monostatic radar by passing spatially diverse and covert bistatic data on common targets to the monostatic radar. The bistatic receiver is netted directly with the monostatic radar and can be a monostatic radar that operates in the receive-only mode. Because common targets must be viewed simultaneously by the monostatic radar and the bistatic receiver, operation is again in the cosite region. The latter two conditions define these concepts as multistatic radars, and indeed these hybrid monostatic-bistatic radars are a special case of multistatic radars. Their operating regions are also constrained by contours of constant β. These concepts are designed to counter retrodirective jammers, augmented decoys, and active cancellers. The MMS, described in Section 2.3.5, is another example of a hybrid radar, with the exception that a covert receiver is not required for its range instrumentation operation.

Concepts in the clutter-tuning category manipulate both the geometry and the separate velocity vectors of the airborne- or space-based transmitter and receiver to control doppler shift and doppler spread of ground clutter, Chapter 6. The transmitter is, in concept, a dedicated bistatic transmitter although, in practice, it is usually a monostatic radar dedicated to the operation. The concepts include a forward-looking synthetic aperture radar operating in the receive-centered oval of Cassini region, and a broadside moving target indication radar, operating in the cosite region.

12.1 HITCHHIKING

As outlined in Chapter 1, a bistatic hitchhiking receiver uses a cooperative or non-cooperative monostatic radar as the bistatic transmitter and operates at the pleasure of the host radar. Specifically, the radar must be on, illuminating regions of interest to the bistatic receiver and transmitting a useful waveform at adequate power levels. Furthermore, the bistatic receiver must locate the host radar to solve the bistatic triangle, operate with adequate target LOS and establish time synchronization for range measurements and phase synchronization for coherent operation and clutter suppression. Both time and phase synchronization are often established by receiving host radar sidelobe transmissions over the direct path. While these requirements are more difficult to satisfy and cannot be satisfied all the time, when compared to operation with a dedicated transmitter (a transmitter that is designed for use with, and under the operational control of, the bistatic receiver), potentially useful applications for this concept exist. In fact, the concept has been deployed at least once as the German Klein Heidelberg, which is detailed in Chapters 1 and 2.

A utility assessment of the hitchhiking concept typically starts with the calculation of its coverage area, the region of thermal noise-limited target detections. This area is calculated via the usual range equations and ovals of Cassini, Chapter 4, for selected host radar-to-bistatic receiver baseline ranges, L. The coverage area is then truncated by (1) the LOS-constrained coverage area, Section 5.5.2; (2) spatial nulling, time gating, or doppler filtering of the direct path signal to prevent receiver saturation, Sections 1.2 and 12.3; and (3) any scanning limits of the host radar's transmitting and the bistatic receiving antennas. Solutions to the transmitting-receiving beam scan-on-scan problem, Section 13.1, may also truncate or otherwise limit the coverage area.

A convenient way of referencing bistatic parameters is to the host radar's maximum detection range, $(R_M)_{max}$, or simply R_M. This range and associated parameters in the monostatic range equation are assumed to be known or can be estimated. Thus, the baseline separation, L, between the hitchhiking receiver and host radar is specified in units of R_M, and the bistatic maximum range product, κ, is specified in units of R_M^2. For example, if all bistatic and host radar parameters are identical, $\kappa = R_M^2$. If, however, the bistatic receiving antenna gain is 20 dB lower than the host receiving antenna gain, κ is reduced by 10 dB because, from (4.1), $\kappa \propto \sqrt{G_R}$. Thus, $\kappa = 0.1\, R_M^2$.

Consider the simplest example where $\kappa = R_M^2$ and the bistatic coverage area is not truncated by any of the restrictions listed previously. Figure 12.1 plots bistatic coverage areas at $L = 3 R_M$ and a coverage area for the host monostatic radar, which has radius R_M and area $A_M = \pi R_M^2$, assuming a 360° azimuth scan. For these parameters, two identical ovals of Cassini are generated, one around the bistatic receiver site and one around the host radar site. These ovals represent the hitchhiking receiver's coverage area.

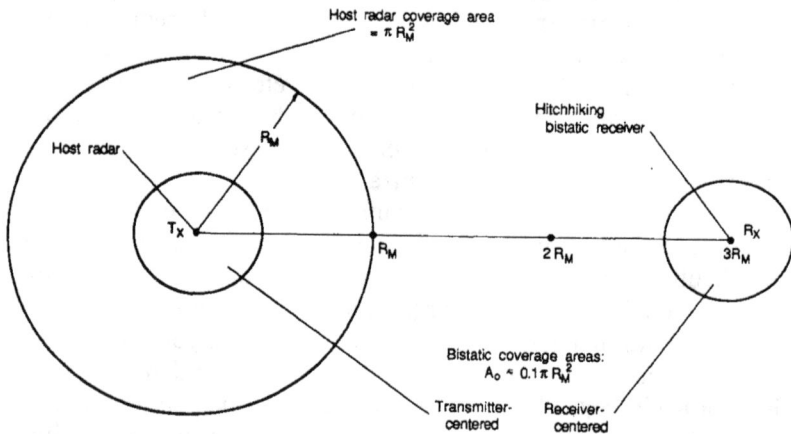

Figure 12.1 Bistatic hitchhiking coverage areas, assuming: adequate lines-of-sight; smooth 4/3 earth's radius; host monostatic radar maximum range, R_M; bistatic maximum range constant $\kappa = R_M^2$; baseline range $L = 3R_M = 3 \sqrt{\kappa}$.

In the general case, two site-centered ovals will be generated when, from (4.8),

$$L > 2\sqrt{\kappa} \tag{12.1}$$

In the special case when all bistatic and host radars are identical, $\kappa = R_M^2$ and $L > 2R_M$.

Whenever (12.1) is satisfied, the oval of Cassini breaks into two parts, one small oval around the host radar and an identical one around the bistatic receiver. These types of ovals are usually generated in bistatic hitchhiking operations. The coverage area, A_o, under each oval is $(A_{B2})/2$, where A_{B2} is given by (5.21). Thus,

$$A_o = \frac{\pi \kappa^2}{L^2}\left(1 + \frac{2\kappa^2}{L^4} + \frac{12\kappa^4}{L^8} + \cdots\right) \tag{12.2}$$

These ovals are roughly in the shape of a circle. The receiver-centered oval is slightly compressed on the side away from the host radar, and slightly extended on the side toward the host radar, as shown in Figure 12.2. The transmitter-centered oval has an identical, mirror-image shape. Definition of an average radius, R_o, of these ovals is useful, where R_o is the radius of a circle with area A_o. Using the first term of (12.2) yields

$$R_o = \kappa/L \tag{12.3}$$

For the receiver-centered oval, $R_o \approx R_R$, the bistatic receiver-to-target detection range. The maximum deviation of R_o from R_R occurs on the baseline and extended baseline; specifically, R_R is a minimum, $(R_R)_{min}$, on the extended baseline away from the host radar, and a maximum, $(R_R)_{max}$, on the baseline toward the host radar, as shown in Figure 12.2. From Table 4.2,

$$(R_R)_{min} = (L^2/4 + \kappa)^{1/2} - L/2 \tag{12.4}$$

Also, $(R_R)_{max} = L/2 - r$, where r is obtained by solving (4.5) for r with $\theta = 0°$, $R_T R_R = \kappa$, and $L > 2\sqrt{\kappa}$. Thus,

$$(R_R)_{max} = L/2 - (L^2/4 - \kappa)^{1/2} \tag{12.5}$$

The value of R_o falls between $(R_R)_{max}$ and $(R_R)_{min}$. For the example of Figure 12.1, $L = 3R_M$ and $\kappa = R_M^2$, so that $L = 3\sqrt{\kappa}$. Thus, $(R_R)_{max} = 0.127L$, $R_o = 0.111L$,

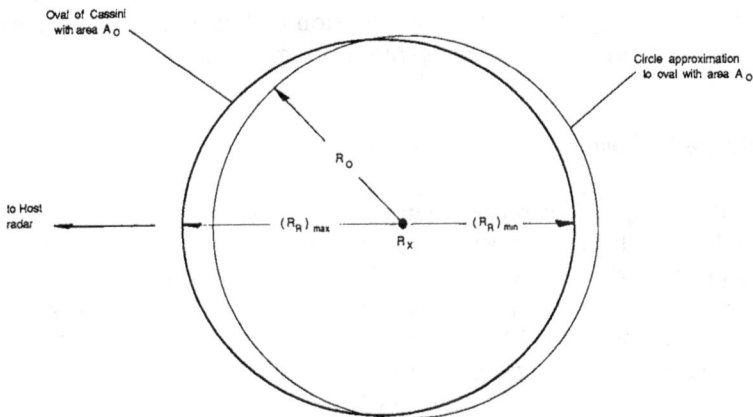

Figure 12.2 Details of receiver-centered oval of Cassini for $\kappa = R_M^2$ and $L = 3R_M$ (and $L = 3\sqrt{\kappa}$). An identical, mirror-image oval exists around the host radar (T_x) site.

and $(R_R)_{min} = 0.101L$. For $L = 5\sqrt{\kappa}$, the values are $0.042L$, $0.040L$, and $0.039L$, respectively. The average radius, R_o, is used for subsequent analysis with the observation that it is slightly optimistic directly away from the host radar, slightly pessimistic directly toward the host radar, and fairly accurate elsewhere.

The distance, $(R_T)_{max}$, from the host radar to the outer edge of the receiver-centered oval is, from Table 4.2,

$$(R_T)_{max} = (L^2/4 + \kappa)^{1/2} + L/2 \tag{12.6}$$

Equation (12.6) is exact and applies to all ovals of Cassini.

The angle, ϕ_o, subtended by the receiver-centered oval at the host radar site, and by the transmitter-centered oval at the receiver site, is approximately

$$\phi_o = 2 \tan^{-1}(R_o/L) \tag{12.7}$$

When operation in the receiver-centered oval is of interest, and the host radar's transmitter beam scans across the receiver-centered oval, the host radar operates as a time-gated transmitter. When the transmitter beamwidth, $\Delta\theta_T$, is on the order of ϕ_o, the host radar operates as a time-gated floodlight transmitter. In either case, when the receiver beamwidth, $\Delta\theta_R$, is narrow, beam scan-on-scan problems arise (Section 13.1). Conversely, when operation in the transmitter-centered oval is of interest and the receiver beamwidth $\Delta\theta_R \approx \phi_o$ is continuously pointed at the transmitter-centered oval, no beam scan-on-scan problems arise. The former case typically occurs with range extension operation; the latter with monitoring and launch alert.

This development can be used in a first-order utility assessment of the following bistatic hitchhiking concepts: range extension and its variant over-the-shoulder operation, passive situation awareness, monitoring, and launch alert.

12.1.1 Range Extension

The hitchhiking concept of range extension typically operates with a cooperative but nondedicated, host, monostatic radar. As the name suggests, the bistatic receiver is positioned at a distance L such that the host radar's coverage area is increased by the area, A_o, of the receiver-centered oval. The concept is also called forward coverage extension [149], which is usually considered for air defense operations, particularly when operating against low RCS, "stealth-type" aircraft, where the host radar's coverage area is significantly reduced. In this case the hitchhiking concept can recover some of the lost coverage. While the bistatic receiver operates autonomously from the host radar, its target reports are often netted with the host radar and other sites in an air defense system.

Figure 12.1 shows a typical example for the range extension hitchhiking concept. In this concept only the oval of Cassini around the receiving site is needed because the transmitter-centered oval provides redundant data in the monostatic coverage area. In Figure 12.1 the bistatic and host monostatic radar parameters are assumed to be identical (not always the case); thus, $\kappa = R_M^2$. Because $L = 3R_M = 3\sqrt{\kappa}$, then $R_o = 0.33R_M$ and $(R_T)_{max} = 3.3R_M$, from (12.3) and (12.6), respectively. That is, the distance from the host radar to the outer edge of the extended range coverage area is 3.3 times that of the host radar's maximum detection range. The coverage area of the receiver-centered oval, A_o, is $0.114\pi R_M^2$, roughly one-tenth of the host radar's coverage area. These calculations apply to both conventional and stealth targets, given that monostatic RCS, σ_M, equals the bistatic RCS, σ_B. When $\sigma_M \neq \sigma_B$, R_M and κ must be modified, which yields distorted bistatic coverage areas. These calculations are usually done via computer modeling.

In this range extension concept, both the host radar and the bistatic receiver are assumed to use high-gain antennas, which raises the beam scan-on-scan problem, Section 13.1. Because the transmitting and receiving beams are not colinear, a single receiving beam must scan many beam positions during the time, T_H, in which a host radar's beam dwells on the oval. This operation is analogous to time-gated illumination or transmission. When a single receiving beam scans n beam positions during T_H, its dwell-per-beam position, T_R, must be reduced such that $T_R = T_H/n$. Consequently, the number of pulses integrated per receiving beam dwell is reduced by n compared to the host radar, and assuming perfect integration, κ is reduced by $\sqrt{n}$. Thus, the oval radius, R_o, shrinks by approximately $\sqrt{n}$ in each receiving beam position. Because n can be large, approaching $180°/\Delta\theta_R$ for transmitting beam positions near the baseline, and $360°/\Delta\theta_R$ for positions straddling the baseline, R_o can be reduced by an order of magnitude or more. As a result, the oval coverage area, A_o, is reduced by two orders of magnitude or more.

Measures that can be taken to increase the oval coverage area include (1) increasing T_H, which is analogous to the step scan remedy (Section 13.1), and incidentally increases the host radar's maximum detection range, R_M; (2) reducing the receiver's scan sector; (3) using multiple, simultaneous, receiving beams (and receivers); and (4) implementing pulse chasing (Section 13.2). Note that increasing the receiving beamwidth, $\Delta\theta_R$, is not a viable measure because the increase in beam dwell, T_R, is offset by the reduction in beam gain, G_R.

Any increase in T_H is subject to target data rate limitations and will probably be modest because T_H is maximized for the host radar operation in the first place. Some reduction in receiver scan sector might be possible, for example, by constraining the sector to the region near the extended baseline. This configuration also simplifies target location calculations, as outlined in Section 5.1, and reduces direct path interference from the host radar because the receiving beam is always pointed away from the transmitting site. This configuration is called over-the-shoulder operation. However, these two measures used together probably cannot recover the lost area, so the more complex and costly measures of multiple, simultaneous beams or pulse chasing must be invoked. A cluster of n receiving beams covering the receiver's scan sector, with each beam having the gain of the single beam configuration, will recover the lost area. However, because a single transmitted pulse will traverse multiple receiving beams in the scan sector, a beam-pulse scheduling algorithm is required. This algorithm is similar to the pulse-chasing algorithm, but requires no beam scanning. The bistatic alerting and cueing hitchhiker, Section 2.3.3, apparently operated this way.

Pulse chasing with a single beam, or in the leapfrog configuration, can also be considered when (1) the receiving antenna is a phased array electronically scanning in at least the azimuth dimension, and (2) the host radar's maximum PRF allows one pulse to traverse the coverage area at one time. As detailed in Section 5.7, this latter requirement is satisfied when operation of the host radar is unambiguous in range and $\sqrt{\kappa} < 2R_M$. In the example $\sqrt{\kappa} = R_M$, so the requirement is satisfied. Either solution is possible, although pulse chasing has not been fully developed.

LOS constraints on the range extension hitchhiking concept are set by $(R_T)_{max}$, the distance from the host radar to the outer edge of the receiver-centered oval, (12.6). First assume that the host radar has been designed to detect conventional targets at $R_M = \sqrt{\kappa} = 100$ km. In this normal monostatic case and for a smooth 4/3 earth's radius, the host radar's antenna would need to be elevated to $h_T = 4$ m to detect a target at altitude $h_t = 500$ m, from (5.24). For range extension, with $L = 3R_M = 300$ km and $(R_T)_{max} = 3.3R_M = 330$ km, the host radar's antenna would need to be elevated to $h_T = 3.36$ km to illuminate the same target, from (5.30). Alternatively, if h_T remained at 4 m, only targets at altitude $h_t = 6.1$ km could be illuminated. Thus, unless R_M is small, L is not too large, or only high-altitude targets are of interest, range extension for conventional targets requires an airborne host radar.

LOS constraints are significantly reduced when operating against stealth targets because detection ranges are reduced. Assume that both σ_M and σ_B are reduced by 24 dB [211] and ignore bistatic RCS enhancements caused by specular reflections and forward scatter. With R_M reduced from 100 to 25 km and, again, $L = 3R_M$, then $(R_T)_{max} = 3.3R_M = 82.5$ km. That is, the bistatic receiver can detect the stealth target at 82.5% of the host radar's maximum detection range for a conventional target *along the baseline*. Because $(R_T)_{max}$ for stealth targets is less than R_M for conventional targets, an adequate host radar LOS for conventional targets will also satisfy range extension LOS requirements for stealth targets flying at the same minimum altitude.

Multiple bistatic receiving sites can be used to broaden the coverage area. In the previous example, adjacent sites would be located on a circle of radius $3R_M$ and separated from each other by slightly less than $2R_o$, or $0.66R_M$, as shown in Figure 12.3. The single host radar can service each site by scanning in azimuth across all

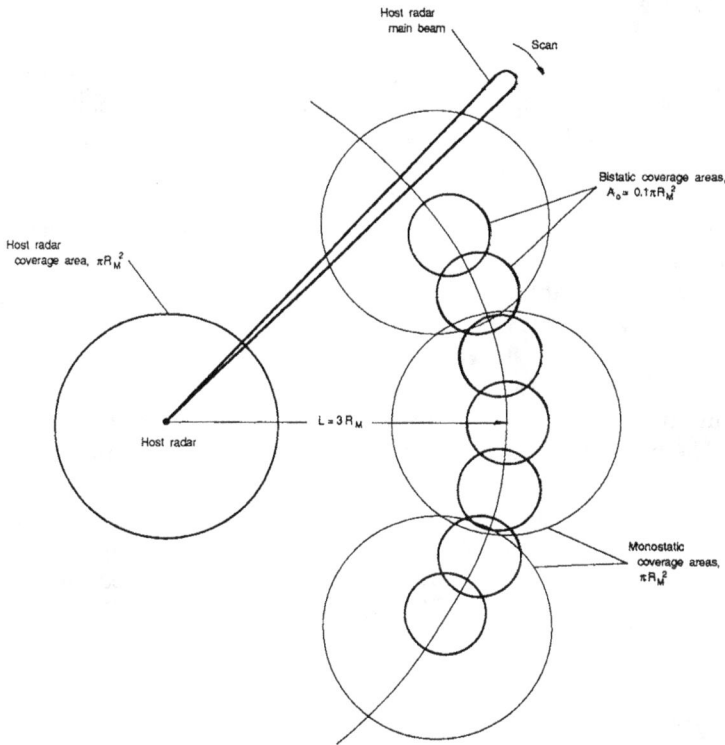

Figure 12.3 Multiple bistatic hitchhiking receiving sites configured for a range extension fence around the host radar, with a monostatic fence configuration shown for comparison, approximately to scale.

the coverage areas, which are configured as a fence annulus surrounding the host radar.

If monostatic radars with performance equal to that of the host radar were located on the extended circle, their spacing would be much greater: $\approx 2R_M$ versus $\approx 0.66R_M$ for the $L = 3R_M$ case, as shown in Figure 12.3. The LOS requirements would also be much easier to satisfy, and no beam scan-on-scan problems would exist. Consequently, bistatic range extension is constrained to scenarios in which some combination of size, weight, cost, and vulnerability favor a forward-based, passive receiving site. For example, when the forward-based monostatic radar must be smaller or must be designed for low probability of intercept operation, such that its detection range is significantly reduced, the bistatic alternative can be attractive. A cost-performance trade-off is usually required and should include vulnerability factors in counter-stealth and counter-antiradiation missile scenarios.

12.1.2 Passive Situation Awareness

The PSA hitchhiking concept typically operates with either a cooperative or a non-cooperative host monostatic radar. The bistatic receiver is usually located on an aircraft to warn a pilot of an intruding threat aircraft or missile in sectors not covered by other sensors, or when the pilot's primary sensor, the monostatic radar, is off. PSA is also called situation awareness, and operates autonomously in the receiver-centered oval of Cassini. Because the PSA hitchhiker is usually airborne, LOS constraints are not of concern, unless the airborne platform is at a very low altitude. Because available real estate on an aircraft is limited and because near-spherical PSA coverage may be required, many small receiving antennas with wide beamwidths are usually installed around the aircraft. This configuration is similar to radar warning receiver (RWR) antenna configurations. Consequently, the receiving antenna power gain, G_R, and the bistatic maximum range product, κ, will be reduced. The basic PSA requirement is to calculate the maximum aircraft-to-host radar range, L, that provides the pilot with a warning at a given range, R_o, from the aircraft, as shown in Figure 12.4. Note that R_o is the average radius of the receiver-centered oval of Cassini. The baseline L can be calculated using (12.3).

Consider the following base case example, in which the required R_o = 20 km and the host radar operates at 3 GHz with G_T = 38 dBi and R_M = 200 km. Also

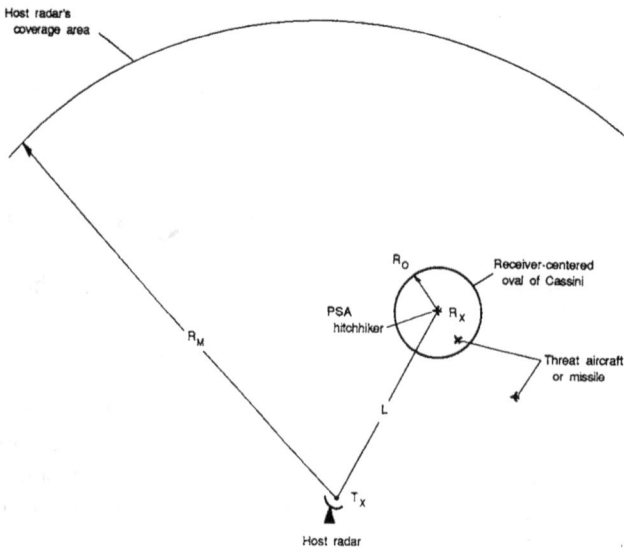

Figure 12.4 Geometry for PSA hitchhiking concept.

assume that each PSA antenna has a diameter of 0.15 m with $G_R = 11$ dBi, and that all other bistatic receiver and host radar parameters are identical (again not always the case). When $G_R = 11$ dBi, the receiving beamwidth, $\Delta\theta_R = 45°$. This wide beam usually mitigates beam scan-on-scan problems that were of concern in the range extension hitchhiking concept. In this example the bistatic maximum range product κ is reduced by a factor of $(G_R/G_T)^{1/2}$, assuming the host radar has equal transmitting and receiving antenna gains. Thus,

$$\kappa = R_M^2(G_R/G_T)^{1/2} \tag{12.8}$$

and for $G_R = 11$ dBi and $G_T = 38$ dBi, $\kappa = 1787$ km^2. From (12.3), $L = 90$ km for $R_o = 20$ km. In other words, the aircraft must be within 90 km of the host radar to give the pilot a 20-km warning range of a threat that *has the same RCS that the host radar is designed to detect.* Figure 12.4 shows this example, approximately to scale.

If the host radar were noncooperative, for example, a threat radar, the PSA aircraft usually requires warning whenever it enters the host radar's coverage area, which in the Figure 12.4 example is a circle with radius $R_M = 200$ km. In this way the PSA aircraft can be warned of threats that are vectored by the noncooperative host radar. Therefore, $L = R_M = 200$ km and, from (12.3), κ must be increased to ≈ 4000 km^2, which in turn requires G_R to be increased. Substituting (12.3) into (12.8) and solving for G_R yields

$$G_R = R_o^2 L^2 G_T/R_M^4 \tag{12.9}$$

Thus, G_R must be increased from 11 to 18 dBi. Now $\Delta\theta_R$ is reduced from 45° to 20°, which raises the beam scan-on-scan problem. Potential solutions are outlined in Section 13.1. The most straightforward solution is to add more PSA antennas to the aircraft, which is analogous to the single-aperture, multiple, simultaneous beam solution. Complexity and cost are obviously increased.

The use of stealth platforms, i.e., platform with low monostatic RCS, σ_M, significantly changes PSA performance. Three cases are of interest as shown in Table 12.1.

Table 12.1
Stealth Platform Combinations for PSA Scenarios

Case	PSA Aircraft RCS	Threat RCS
1	Low	Conventional
2	Conventional	Low
3	Low	Low

In all cases, the host monostatic radar parameters are held constant, except for σ_M, and the required warning range, R_o, is held constant at 20 km. When operating with a noncooperative host radar, the PSA aircraft must be warned whenever $L \leq R_M$. The constant R_o assumption, in turn, assumes that the performance of weapons carried by the threat is invariant with threat RCS. Finally, σ_M is assumed to be reduced by 24 dB at all platform aspect angles; i.e., conventional RCS = low RCS + 24 dB [211].

Case 1: Stealth versus Conventional

The host radar's maximum detection range, R_M, assumes two values, one for detection of the stealth PSA aircraft, $(R_M)_{PSA}$, and one for detection of the conventional threat, $(R_M)_T$, which is also the target to be detected by the PSA aircraft. Using parameters of the previous example, $(R_M)_{PSA}$ is reduced from 200 to 50 km, while $(R_M)_T$ remains at 200 km. The corresponding PSA maximum range product, κ, remains at 4000 km^2. For a noncooperative host radar, L can now be reduced to 50 km. Thus, either R_o can be increased or its receiving antenna gain, G_R, can be reduced. For the former case, $2\sqrt{\kappa} = 126$ km, which is greater than L. Thus, from (12.1), a single, cosite oval is generated around the host radar and PSA aircraft. It can be evaluated by (5.20) if required. More important, however, is the latter case, in which G_R can be reduced from 18 dBi when operating at 200 km to 6 dBi at 50 km, just the R^2 path change. The corresponding receiving beamwidth is 80°, which considerably simplifies the PSA aircraft configuration because fewer antennas are now required for spherical coverage. Hence, case 1 becomes a viable PSA configuration, given that (1) the PSA receiver can emulate performance of the noncooperative host receiver, particularly in terms of matched filtering, processing gain, and losses, and (2) RCS variations with platform aspect angle can be predicted and tolerated.

Case 2: Conventional versus Stealth

Performance is now reversed: $(R_M)_{PSA} = 200$ km, $(R_M)_T = 50$ km, $\kappa = 250$ km^2, assuming $G_R = 18$ dBi, and $L = 200$ km for a noncooperative host radar. Thus, $R_o = 1.3$ km. The value of G_R would have to be increased from 18 to 42 dBi to restore R_o to 20 km, clearly an impossible task for a tactical aircraft aperture. Therefore, case 2 is not viable, which of course is the conventional-versus-stealth penalty.

Case 3: Stealth versus Stealth

Now $(R)_{PSA} = (R_M)_T = 50$ km, $\kappa = 250$ km^2, assuming $G_R = 18$ dBi, and $L = 50$ km for the noncooperative host. Thus, $R_o = 5$ km. The value of G_R would have to be increased from 18 to 30 dBi to restore R_o to 20 km, a very difficult task for tactical aircraft. The basic problem with this scenario is that all ranges except for the required PSA detection range, R_o, are reduced. The PSA detection range must remain at R_o = 20 km, which requires a 12-dB increase in G_R. Note that when the PSA aircraft is allowed to overfly the host radar so that $L \approx 0$, $R_o \approx \sqrt{\kappa}$, the equivalent monostatic range, which yields $R_o \approx$ 16 km, the maximum possible R_o with κ = 250 km². Thus, unless a significant reduction in R_o can be tolerated, or a significant increase in G_R can be achieved, case 3, while possible, is not particularly viable.

When the PSA aircraft operates with a cooperative (friendly) host radar, the requirement for $L = (R_M)_{PSA}$ can be eliminated. For the base case parameters, R_o = 20 km can be achieved with G_R = 11 dBi at L = 93 km, and with G_R = 18 dBi at L = 200 km, which yields useful capability at some increase in cost and complexity. However, for case 3 the maximum R_o remains at 16 km. In all cases when both the PSA aircraft and the threat are within the cooperative host radar's detection range, the host radar itself can pass warning information to the PSA aircraft. If so, the PSA configuration is not required. Table 12.2 summarizes results from these PSA examples.

12.1.3 Monitoring

The threat monitoring hitchhiking concept operates autonomously with a noncooperative host monostatic radar in the transmitter-centered oval of Cassini, Figure 12.1. The bistatic receiver is usually located on a surveillance platform at a considerable distance from the host radar, possibly on a satellite. Its purpose is to detect noncooperative targets operating with the noncooperative host radar, for example, during test range operations or air defense exercises. Because the baseline is large, at least one of the three sites (transmitter, receiver, or target) will probably need to be elevated to satisfy LOS requirements (Section 5.5.2).

The basic monitoring needs require the calculation of the receiving antenna power gain needed to detect targets in the transmitter- (host radar-)centered oval with specified average radius R_o and a host radar-receiver separation, or baseline L. The geometry is shown in Figure 12.5 for two bistatic receiver locations, airborne and space-based, with parameters shown for the airborne case. Assuming all other bistatic receiver and host radar parameters are identical (not always the case), (12.9) is again used to calculate a first-order approximation to G_R:

$$G_R = R_o^2 L^2 G_T / R_M^4 \tag{12.9}$$

For ideal monitoring operations, the bistatic receiver should be able to detect the same "traffic" as the host radar; thus, $R_o \approx R_M$. Because the bistatic receiver is usually at a considerable distance from the host radar, $L \gg R_M$. For example, if R_o

Table 12.2
Results for PSA Hitchhiking Examples

Case	PSA Aircraft RCS	Threat RCS	Host Radar		κ (km²)	L (km)	R_o (km)	G_R (dBi)	Comments
			$(R_M)_{PSA}$ (km)	$(R_M)_T$ (km)					
Base	Conventional	Conventional	200	200	1787	[90]	20	11	Aircraft risk for noncooperative host radar
1	Stealth	Conventional	50	200	4000	200	20	[18]	PSA cost and complexity
					4000	50	>20 (cosite)	[18]	PSA cost and complexity
					1787	50	>20 (cosite)	11	Viable
					1000	50	20	6	Viable
2	Conventional	Stealth	200	50	250	200	[1.3]	[18]	Not viable
3	Stealth	Stealth	50	50	4000	200	20	42	Not viable
					250	50	[5]	[18]	Aircraft risk, cost and complexity
					1000	50	20	[30]	Probably not viable

Notes:

(1) All cases with cooperative or noncooperative host radar, except where noted.

(2) Host radar G_T = 38 dBi at S-band.

(3) $(R_M)_{PSA}$ = detection range on PSA aircraft; $(R_M)_T$ = detection range on PSA's threat; and κ = PSA bistatic maximum range product.

(4) Conventional RCS = stealth RCS + 24 dB, all aspect angles.

(5) PSA receiver matched to host radar receiver, except for G_R.

(6) Critical parameters are boxed.

(7) $L \approx \kappa/R_o$.

(8) $G_R = R_o^2 L^2 G_T/(R_M)_T^4$.

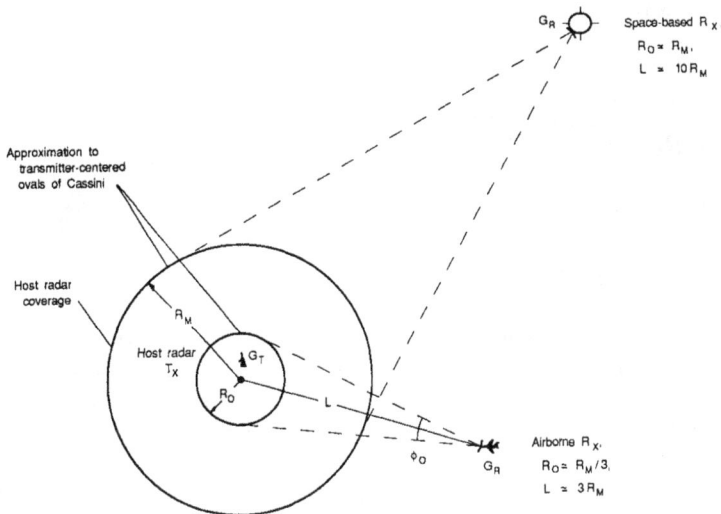

Figure 12.5 Two bistatic hitchhiking concepts for monitoring activity near a noncooperative host radar site, T_x, not to scale.

$= R_M$ and $L = 10 R_M$, $G_R = 100 G_T$. If $G_T = 38$ dBi at S-band, $G_R = 58$ dBi, which requires a circular receiving aperture diameter of about 35 m. Alternatively, if $R_o = R_M/3$ and $L = 3R_M$ are acceptable, $G_R = G_T = 38$ dBi, which reduces the diameter to 3.5 m. The former case might be useful for a space-based receiver when $R_M = 200$ km and thus $L = 2000$ km; the latter, for an airborne receiver operating beyond the host radar's detection range. Both of these cases are shown in Figure 12.5. Note that multiple aircraft, each with a bistatic receiver hitchhiking off the same host radar, cannot increase the coverage area because all ovals are centered on the host radar. In any case $G_R \geq G_T$ is typical for monitoring operations.

Beam scan-on-scan problems can be avoided when the receiving antenna beamwidth $\Delta\theta_R$ covers at least half the oval and the receiver beam dwells on the oval during the time the host radar scans that part of the oval. The angle subtended by the transmitter-centered oval at the receiving site, ϕ_o, is given by (12.7). When $\Delta\theta_R = \phi_o/2$, the receiving beam can dwell first on one side of the baseline and then on the other as the host radar's beam scans each half of the oval. Note that if the host radar's scan pattern is not regular, as might be the case for a phased-array radar, the receiving beam must cover the entire oval; thus, $\Delta\theta_R = \phi_o$.

For a uniformly scanning transmitting antenna, the average oval radius, R_o, that satisfies both (12.9) and $\Delta\theta_R = \phi_o/2$ can be developed as follows. Because antenna beamwidth is inversely proportional to the square root of antenna gain, (12.9) can be rewritten as $\Delta\theta_R = \Delta\theta_T R_M^2/R_o L$, assuming the host radar and bistatic receiver use similar types of antennas. Thus, $\Delta\theta_T R_M^2/R_oL = \phi_o/2$. Substituting (12.7) into this expression yields

$$\Delta\theta_T = \frac{R_o}{R_M} \cdot \frac{L}{R_M} \cdot \tan^{-1}(R_o/L) \tag{12.10}$$

For small angles $\tan^{-1}(R_o/L) \approx R_o/L$, so that $\Delta\theta_T \approx R_o^2/R_M^2$. Thus,

$$R_o \approx R_M\sqrt{\Delta\theta_T} \tag{12.11}$$

Equation (12.11) is independent of the baseline L. That is, for any host radar-bistatic receiver separation L such that $L > 2\sqrt{\kappa}$ is satisfied, the bistatic receiver can monitor an oval with average radius $R_o = R_M\sqrt{\Delta\theta_T}$ and avoid beam scan-on-scan problems. In the previous example, $G_T = 38$ dBi, which corresponds to $\Delta\theta_T = 2°$ for a circular aperture. Thus, $R_o = 0.19R_M$. If this coverage is not adequate, the monitoring receiver must implement beam scan-on-scan remedies. The use of pulse chasing with a noncooperative host radar is only possible when the radar's scan rate and PRF are predictable—an unlikely prospect if the host radar suspects that a monitoring hitchhiker may be present. Hence, multiple, simultaneous receiving beams (and receivers) become the remaining scan-on-scan remedy, but with cost and complexity penalties.

12.1.4 Launch Alert

The launch alert hitchhiking concept operates autonomously with a noncooperative host monostatic radar in the transmitter-centered oval of Cassini, Figure 12.1. Two launch alert configurations are possible. In one launch alert configuration, the bistatic receiver is located on a surveillance aircraft that is in the vicinity of air combat operations, Figure 12.6. Its purpose is to detect missiles or other weapons launched from the host radar platform and possibly nearby platforms that are illuminated by the host radar, and then warn friendly aircraft of the launch. In a second launch alert configuration, the bistatic receiver is carried by the combat aircraft itself, to provide self-warning of a missile launch, Figure 12.7. Because the missile is positioned near the baseline, the geometry for this second configuration is a forward-scatter fence, with the target flying down rather than across the fence. In either launch alert configuration, LOS requirements are usually satisfied with existing geometry.

A launch alert assessment must first consider the type of threat system that is launched at the target aircraft. If the threat is a semiactive homing missile, continuous or near-continuous illumination of the target is required. While this operation can be exploited for a bistatic alert, a simpler and probably more reliable

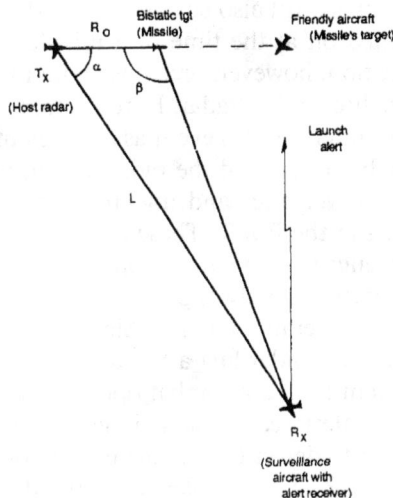

Figure 12.6 Geometry for missile launch alert from a standoff surveillance aircraft.

exploitation is to identify the illumination signal via an on-board radar warning receiver (RWR) and use that signal as the alert. If the threat is an autonomous missile using an active radar seeker, the seeker signal can be used in a similar way. However, if the threat is an autonomous missile using a passive (for example, infrared) seeker, then an RF type of RWR cannot provide alert. In this case, if the launch platform fire control radar remains on directly after launch and if the missile flies in or near the host radar's beam, a bistatic alert solution is possible. Thus, both the threat configuration and its operation are critical to a utility assessment of the bistatic hitchhiking alert concept. Usually an engagement simulation of a specific threat system, its operation, and both missile and target kinematics is required to determine a useful bistatic alert geometry.

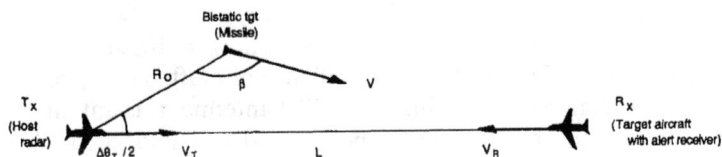

Figure 12.7 Geometry for missile launch alert when the alert receiver is carried by the target aircraft, not to scale.

Threat countermeasures must also be considered. If the launch platform's fire control radar can be turned off at the time of a missile launch, the bistatic alert solution is lost. This operation, however, precludes both midcourse missile updates and kill assessment by the fire control radar. Furthermore, turn-off can be detected by the RWR, which might interpret this event as a sufficient alert. The launch platform might then counter by turning off the radar at other times, if that operation does not compromise its surveillance and tracking performance. The effect is to generate false launch alerts at the RWR. These countermeasure options must also be assessed, usually in an engagement simulation.

So much for the operations analysis part of the alert assessment. If a bistatic alert solution is found to be operationally feasible, the next step is to determine if the performance is feasible. Consider first a bistatic alert receiver carried by a surveillance aircraft in the vicinity of air combat operations. The geometry is shown in Figure 12.6. The basic alert requirement is again to calculate the receiving antenna power gain needed to detect the missile in the host radar-centered oval of Cassini, with average radius R_o. Launch alert is required early in the engagement so that the friendly target aircraft has time to react. Because the target will be within R_M, the host radar's maximum detection range, R_o will be a small fraction of R_M, for example, $R_o = 0.1 R_M$. The surveillance aircraft should remain outside the host radar's coverage, so that $L > R_M$. Finally, the bistatic target, the missile, is physically smaller than an aircraft; therefore, its monostatic and bistatic RCS will usually be smaller, for similar geometries.

The bistatic maximum range product, κ, is now modified by changes in both the antenna gain and the bistatic RCS:

$$\kappa = R_M^2 \left(\frac{\sigma_B}{\sigma_M} \frac{G_R}{G_T} \right)^{1/2} \tag{12.12}$$

where σ_B is the bistatic RCS of the missile and σ_M is the monostatic RCS of the missile's target (i.e., the friendly aircraft). Substituting (12.3) into (12.12) and solving for G_R yields

$$G_R = (R_o^2 L^2 / R_M^4)(\sigma_M / \sigma_B) G_T \tag{12.13}$$

For example, assume $G_T = 32$ dBi, $R_o = 0.1\ R_M$, $\sigma_M/\sigma_B = 10$, and $L = 2R_M$. With those parameters, $G_R = 28$ dBi. If the surveillance aircraft is equipped for ELINT operations using a relatively high-gain (≈ 28-dBi) antenna, this antenna could possibly be shared with the bistatic alert function and use associated ELINT information on host radar type and location, or at least angular direction from the surveillance aircraft, for the alert broadcast. The viability of this concept is a result of the early launch requirement, which in turn requires a small value of R_o. Beam scan-on-scan is not a problem in this scenario because the receiving antenna can be continuously pointed at the forward half of the small transmitter-centered oval, i.e., to the right of the baseline in Figure 12.6.

Note that if the surveillance aircraft is positioned near the host radar-to-friendly aircraft LOS, the geometry becomes forward scatter, with $\beta \Rightarrow 180°$. Special detection measures are usually required to operate in this region and are detailed in subsequent paragraphs. The onset of forward-scatter geometry can be defined by a cone with its apex at the host radar, axis on the host radar-to-friendly aircraft LOS, and a cone half-angle of $\alpha = 90° - \theta_T$ (Figure 12.6), defined as

$$\alpha = \sin^{-1}(R_o \sin\beta/L) - \beta \tag{12.14}$$

with the bistatic angle, β, establishing the forward-scatter region. For example, if $\beta > 160°$ for the forward-scatter region, $R_o = 0.1R_M$, and $L = 2R_M$, as in the previous example, $\alpha = (180° - 1°) - 160° = 19°$. In tactical air operations, orchestration of this geometry may not be possible. Without special detection measures, alert operations are precluded when the surveillance platform is within this cone.

Next consider a bistatic alert receiver carried by the target aircraft, the aircraft against which the missile is launched. The geometry now becomes forward scatter, as shown in Figure 12.7. The missile's bistatic RCS will be enhanced, with a forward-scatter RCS, σ_F, directly on the baseline. Because the airborne host fire control radar's operating frequency will be high, probably near 10 GHz, σ_F will be high at $\beta = 180°$ and, depending on missile aspect angle, will roll off rapidly at $\beta < 180°$, as outlined in Section 8.4. Thus, the RCS enhancement diminishes as the missile is displaced from the baseline. A second constraining aspect of this geometry is the host radar's main beam, which is assumed to be pointed at the target aircraft, and which must also illuminate the missile for a bistatic alert solution. Figure 12.7 shows the geometry for a missile located at the edge of the host radar's beam, with beamwidth $\Delta\theta_T$. From the law of tangents,

$$\beta = \tan^{-1}\left[\frac{R_o + L}{R_o - L} \cdot \tan\frac{\Delta\theta_T}{4}\right] - \frac{\Delta\theta_T}{4} \tag{12.15}$$

Assuming $\Delta\theta_T = 4°$, which corresponds to $G_T = 32$ dBi, and $R_o = 0.01L$, which corresponds to the missile location directly after launch, $\beta = 178°$. When $R_o = 0.5L$, roughly at the midpoint of the missile's trajectory, $\beta = 176°$. Thus, missile alert operations are constrained to $\beta > 176°$ for host radar mainbeam illumination of the missile.

Two design aspects should be considered for this alert configuration: required receiving antenna power gain and the detection mechanism at very large β. Equation (12.13) is again used for receiving antenna power gain, G_R. In the example for

(12.13), G_T = 32 dBi and R_o = $0.1R_M$, which also applies to this example. Now assume that missile launch occurs when the target with the bistatic receiver is at half the host radar's maximum range, or L = $R_M/2$. Consequently, R_o = $0.2L$, and from (12.15) β = 177.5° for this alert geometry.

Next the missile's bistatic RCS, σ_B, in the forward-scatter region must be esti-mated for β = 177.5°. When the missile is on the baseline, β = 180° and σ_B = σ_F = $4\pi A^2/\lambda^2$, where A is the missile's shadow aperture and λ is the wavelength. Assume that the missile has a tail-on, circular shadow aperture A = 1 m² and λ = 0.03 m; thus, σ_B = σ_F = 1.4 × 10⁴ m². As outlined in Section 8.4, a validated computer model [100, 102] is required to estimate σ_B for β < 180° and specific vehicle configurations and aspect angles. Absent this data, σ_B can be approximated for large β ($\beta \gtrsim$ 176°) as follows. As shown in Section 8.4, the forward-scattered pattern of the sphere with radius a will roll off 3 dB at $(\pi - \beta)$ = $\lambda/\pi a$. Assuming the missile's circular shadow aperture approximates a sphere with a = 0.56 m, and λ = 0.03 m, then $(\pi - \beta)$ = 1°. (Note that this value is one-half the 3-dB beam-width of a circular aperture with A = 1 m² at λ = 0.03 m.) Because β = 177.5° in this alert geometry, $\pi - \beta$ = 2.5°. Thus, the alert receiver will be somewhere near the first sidelobe of the forward-scattered lobe. Assuming σ_B = σ_F − 20 dB = 140 m² at β = 177.5° and σ_M for the missile target (friendly aircraft) = 3 m² yields σ_M/σ_B = −17 dB. With R_o = $0.1R_M$, L = $R_M/2$, and G_T = 32 dBi, G_R = −11 dBi from (12.13). Thus, with a modest RWR-type antenna of G_R = 3 dBi, the alert receiver has a 14-dB excess SNR or a 14-dB link margin. This margin allows the geometry of Figure 12.7 to open somewhat. For example, the missile might now be farther from the baseline, with illumination of the missile through the first sidelobe of the host antenna and reception of the missile echo through the second sidelobe of its forward-scatter lobe. In any case, receiving antenna gain is not an issue in this forward-scatter geometry, which is not unexpected because σ_B is large.

A detection mechanism for use in this high-β geometry has been developed and successfully tested at short range with a cooperative transmitter using a pulsed waveform. The test bed, called the Aircraft Security Radar, was designed to detect moving intruders near parked aircraft, and is described in Section 2.3.4. It used a variation of the beat frequency detection method, where the time-overlapping direct path transmitter pulse and the doppler-shifted target echo are mixed in the receiver by a square law detector. The resultant beat frequency, or beat note, f_N, appears as AM of the direct path transmitter pulse. Thus, the detection problem becomes one of calculating the AM frequency and amplitude referenced to the host radar's carrier frequency and amplitude. The frequency is the beat note, which appears as a sideband of the detected carrier at baseband, and is displaced from the carrier by f_N. Amplitude is expressed as the degree of modulation M = $(S_B/S_{DP})^{1/2}$, where S_B is the power received over the bistatic path, S_{DP} is the power received over the direct path, and M is the amplitude of the AM sideband referenced to the car-rier amplitude.

In this high-β geometry, the missile doppler will be low. Furthermore, the host radar and bistatic warning receiver are moving, which causes a doppler shift over both the direct and the missile path. The expression for f_N is given by (6.25), with parameters defined in Figure 6.1. In this case V is the missile velocity, V_T is the host radar velocity, and V_R is the bistatic receiver velocity.

Again using the previous example, and assuming that both the host radar and the missile are flying toward their target (bistatic receiver), and that the bistatic receiver is flying toward the host radar, as shown in Figure 12.7, $\delta_T = 90°$, $\delta_R = -90°$, and $\delta = \beta/2$. Thus, (6.25) becomes

$$ f_N = \frac{1}{\lambda}[2V\cos^2(\beta/2) + V_T(\sin\theta_T - 1) - V_R(\sin\theta_R + 1)] \tag{12.16} $$

With $\lambda = 0.03$ m and $\beta = 177.5°$, and assuming that $V = 670$ m/s, $V_T = V_R = 335$ m/s, $\theta_T = 90 - \Delta\theta_T/2 = 88°$, then $\theta_R = -89.5°$, and the beat frequency $f_N = 14$ Hz. From (6.5) the mainbeam clutter doppler at the missile location is 22.3 kHz, so the beat frequency is clear of mainbeam clutter returns. Thus, the AM sideband frequency is displaced 14 Hz from the baseband carrier in a clutter-free region, except of course for amplitude and phase noise generated by the host radar's transmitter.

The power of the direct path signal, S_{DP}, at the bistatic receiver is proportional to $P_T G_T G_R/4\pi L^2$. The power of the bistatic path signal, S_B, at the bistatic receiver is proportional to $P_T G_T G_R \sigma_B/(4\pi)^2 R_o^2(L - R_o)^2$, assuming that the direct and bistatic path lengths are approximately equal in this geometry. Thus,

$$ \frac{S_B}{S_{DP}} = \frac{L^2\sigma_B}{4\pi R_o^2(L - R_o)^2} \tag{12.17} $$

For $R_o = 0.2L$ and $\sigma_B = 140$ m^2, $S_B/S_{DP} = 435/L^2$. When the missile is launched at $L = 30$ km, $S_B/S_{DP} = 5 \times 10^{-7}$. That is, the bistatic path signal power is 63 dB lower than the direct path signal power at the bistatic receiver. Thus, the degree of modulation, $M = 7 \times 10^{-4}$, such that the amplitude of the AM sideband is 0.0007 that of the detected carrier and is displaced from the carrier by 14 Hz in this example.

The principal constraint in detecting this AM sideband is the noise sideband energy imposed on the pulse transmitted by the noncooperative host radar. For example, in a high-PRF, airborne, pulse doppler radar, these sidebands are generated by STALO phase noise. Barton [154] reports that a high-quality X-band STALO, obtained by multiplication of a crystal source, will produce phase noise sidebands of -70 dB/Hz below the carrier at 10 Hz, uniformly decreasing to -80 dB at 20 Hz. Such a radar is typically designed for 80- to 90-dB clutter attenuation. Thus, when the host radar is designed for this type of performance, detection of the

AM missile launch alert sideband is possible with an approximate SNR of 10 dB. A cavity-stabilized klystron transmitter should give similar performance. Radars designed for less clutter attenuation may generate higher phase noise sidebands and may not be suitable host radars.

In any case, viability of the forward-scatter bistatic alert concept depends on the magnitude of the forward-scatter RCS, the missile launch range, and most importantly on the level of noise sidebands present in a specific host radar. Of course, the basic operational requirements of host radar transmitting and missiles in the host radar beam must also be satisfied.

12.2 FORWARD-SCATTER FENCES

As outlined in Chapter 8, the bistatic RCS of a target becomes large as the bistatic angle, β, becomes large, reaching the forward-scatter maximum, σ_F, at $\beta = 180°$, where

$$\sigma_F = 4\pi A^2/\lambda^2 \tag{12.18}$$

and A is the target's silhouette or shadow area, and λ is the wavelength. The target can be of simple or complex shape, and can be reflecting, absorbing, or a combination, as long as it "cuts a hole" in the transmitter beam. As a consequence, interest in forward-scatter bistatic fence operation has been revived [54, 101, 141, 142, 211, 212] as a counter to "stealth vehicles," which attempt to reduce their monostatic RCS by the use of shaping and radar absorbing materials (RAM).

References [1, 15, 16] show that operating exactly at $\beta = 180°$ yields information only that the target is somewhere between the transmitter and the receiver. Target range is indeterminate; target doppler shift is zero for all target velocity vectors; and both range and doppler resolution are lost. Implementation is also greatly complicated: The direct path signal and any forward-scattered clutter compete directly with the target signal, all at the same doppler frequency (zero doppler when the transmitter and receiver are stationary). Hence, doppler filtering cannot be used for clutter suppression. Of course, the receiver must also establish a LOS to the transmitting site. These problems are reduced as β moves away from 180°. In this case, the forward-scatter RCS diminishes at $\beta < 180°$, but a region of large β often exists where the bistatic RCS of all types of targets is enhanced, target doppler is separated from ground clutter, and the target echo is separated in time from the direct path signal, thus allowing a bistatic radar to detect targets and in some cases estimate the target's position, course, and speed in a forward-scatter fence configuration.

Although the forward-scatter RCS enhancement phenomenon was not known before or during World War II, targets were inadvertently detected on many occasions when in the vicinity of communication links, an emulation of the forward-scatter fence. These detections were, of course, the evidence needed to start development of radars in the 1930s. Chapter 2 provides historical details of these events. Almost 200 fences were deployed before and during World War II for air defense, and although implementation details are sketchy, they all appeared to use CW transmissions and the target aircraft direct path, beat frequency as the detection mechanism. (The fences operated between 25 and 80 MHz, which is near the resonant frequency of many aircraft; thus, the aircraft RCS was enhanced by this effect as well.) The fences were usually designed as a single trip wire, indicating only that an aircraft has crossed the transmitter-receiver direct path. The time of the crossing can be established by observing the occurrence of a zero beat frequency, either directly or by interpolating beat frequency measurements before and after a fence crossing. The French netted multiple fences together, which, possibly by measuring the time intervals between successive fence crossings and the time duration of detections within an elevation fan beam, provided an estimate of aircraft course, speed, and altitude.

In the 1980s a forward-scatter fence was developed and tested in the United States as an intrusion detector for military aircraft on the ground. It used pulse transmissions and a variation of the beat frequency detection method. The test bed is described in Section 2.3.4, and the basic technique is analyzed in Section 12.1.4 as a forward-scatter, missile launch alert, hitchhiking technique.

This section establishes the geometry for operating a single forward-scatter fence, and then develops equations for estimating a target velocity vector (course and speed) as it penetrates a net of fences, similar to that used in the French *maille en Z* system.

12.2.1 Single Forward-Scatter Fence

A forward-scatter fence is typically configured to (1) operate in regions of large β where the target forward-scatter RCS enhancement is significant and (2) ignore or excise regions of even larger β, near the direct path, where clutter and direct path signal levels are high, target doppler is low, and target range is indeterminate.

The forward-scatter RCS-enhanced region can start anywhere from $\beta = 110°$ to $\beta = 175°$, depending on frequency, target size, shape, and aspect angle. For example, a hypothetical target with rectangular (vertical and horizontal) shadow dimensions of 2 m $\times$ 5 m will generate a maximum forward-scatter RCS, $\sigma_F = 1260$ m^2 at $\lambda = 1$ m and $\beta = 180°$, from (12.18). To a first order, the 3-dB width of the forward-scattered lobe is 29° $\times$ 11°. Assuming the forward-scatter radiation pattern rolls off as $(\sin x)/x$, a system designed to detect 5-m^2 targets can operate out to about the fourth sidelobe, which is roughly 24 dB below σ_F and displaced about 40° from the center of the horizontal forward-scatter lobe. This displacement corresponds to $\beta = 140°$, as shown in Figure 12.8. These values are only approximate because as is shown in Figure 8.3, the patterns vary radically with geometry, and require a validated computer model [100, 102] to estimate the bistatic RCS, σ_B, at $\beta < 180°$. Under the simplifying assumptions in this example, however, target geometries with $\beta \gtrsim 140°$ set the minimum (constant) bistatic angle, β_{min}, that is feasible for forward-scatter operation. The contour for a constant bistatic angle is given by (3.11) and (3.12) and shown in Figure 3.3. The area within a contour of minimum β is the forward-scatter-enhanced region, i.e., where $\beta > \beta_{min}$, or $\approx 140°$ in this example.

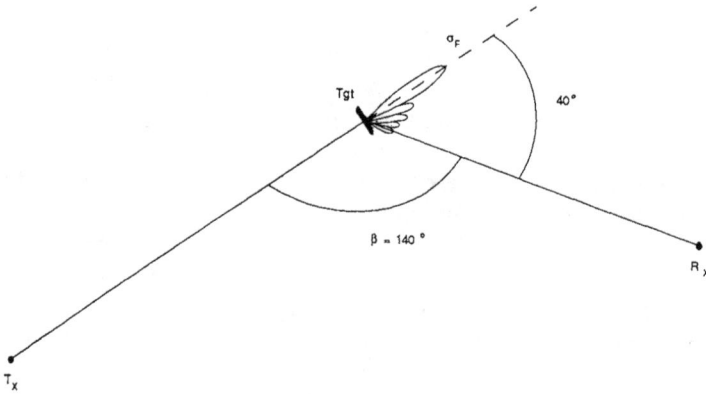

Figure 12.8 Geometry for establishing the approximate forward-scatter-enhanced region.

Three techniques can be used to excise strong clutter and direct path signals near the baseline: angle, range, and doppler excision. They can be used separately or together. Angle excision consists of generating low receiving (and possibly transmitting) sidelobes or a beam null along the baseline to attenuate the direct path signal. Terrain masking on the baseline can also be used when available. Range excision requires that the difference in propagation time, ΔT_n, between the transmitter-target-receiver path and the transmitter-receiver (direct) path, Figure 5.1, be greater than the uncompressed pulse width, τ_u, to prevent eclipsing of the target return. If some eclipsing loss, L_e, can be tolerated, then $\Delta T_n > \tau_u L_e$, and from (5.2),

$$(R_T + R_R) \geq L + c\tau_u L_e \qquad (12.19)$$

Equation (12.19) defines an ellipse, $(R_T + R_R) = L + c\tau_u L_e$ within which range measurements are proscribed.

Doppler excision requires that the target doppler return be resolved from the forward-scattered clutter, which is spread about zero doppler (baseband) due to internal clutter motion. Combining (7.5a) and (7.5b), solving for β, and assuming (1) $1/T$ is the one sided width of the forward-scattered clutter spread, (2) V_2 is at baseband ($= 0$), and (3) $V_1 \cos\delta_1$ is the projected target velocity vector on the bistatic bisector yields

$$\beta = 2\cos^{-1}(\lambda/2TV_1 \cos\delta_1) \tag{12.20}$$

When a minimum target $V_1 \cos\delta_1$ is specified, analogous to the monostatic minimum detectable velocity (MDV), (12.20) defines an ogive of maximum (constant) bistatic angle within which doppler measurements are proscribed. Again, Figure 3.3 is used to establish the max β contour, β_{max}.

Figure 12.9 shows the exclusion and enhanced RCS operating regions of a forward-scatter fence configuration for $\lambda = 1$ m, $\tau_u = 1$ μs, $L_e = 0.5$, $1/T = 10$ Hz, and $V_1 \cos\delta_1 = 50$ m/s, with the remaining parameters shown on the figure. In special cases such as when the receiving antenna is configured to measure the target angle of arrival, θ_R, in the angular region of $-80° < \theta_R < -60°$ in Figure 12.9, this measurement, combined with a range sum, $R_T + R_R$, measurement and a measurement of the baseline, L, can be used to estimate receiver-to-target range R_R by using (5.1). However, errors in the R_R estimate will be large because, in this geometry, ellipse eccentricity exceeds 0.9 and thus error slopes are large when $\theta_R \lesssim 60°$, as detailed in Section 5.2. Note that measurements can be made in two regions about the baseline; one when the target is in the upper region, and one in the lower.

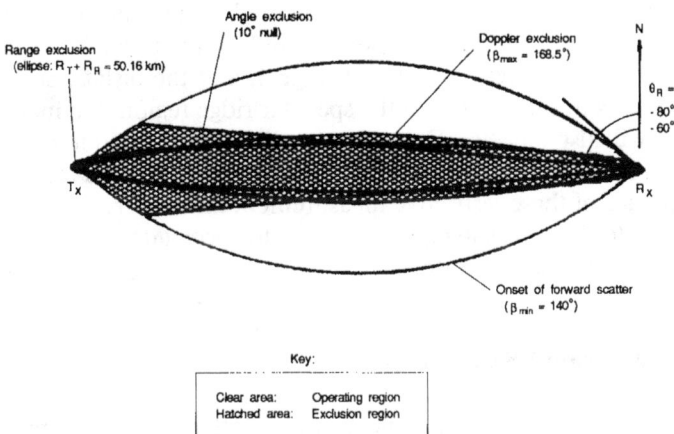

Figure 12.9 Forward-scatter fence configuration.

A combination of these measurements can provide an estimate of target course and speed, given a constant target velocity vector in these regions.

If the target forward-scatter RCS as a function of bistatic angle and target aspect angle is known or can be estimated, and the target flight paths can be predicted, then a forward-scatter detection contour can be generated by substituting the forward-scatter RCS values into the bistatic range constant, K, (4.3). The "variable" bistatic range constant is then used to generate a modified oval of Cassini, (4.2), which replaces the minimum β contour in Figure 12.9. In some cases an analytical expression can be generated to approximate the mean bistatic RCS of simple target shapes as a function of β, for a range of target aspect angles. In these cases the minimum β detection contour is readily calculated [142].

The enhanced RCS region may be truncated by the common coverage area, as described in Section 5.5.2 and shown in Figure 5.13 in particular. If so, forward-scatter fence operation is constrained to the intersection of the enhanced RCS region and the common coverage area, reduced by the excision region. Both transmitting and receiving antennas can be elevated to minimize truncation by the common coverage area for given target altitude, baseline, and terrain conditions. In the simplest case, assume a smooth 4/3 earth's radius, a minimum target altitude $h_t = 100$ m, and either R_T or R_R operating at maximum ranges comparable to the baseline, $L = 50$ km. Thus, $r_R = r_T = L$, and from (5.23) and (5.24), the transmitting and receiving antennas must be elevated to $h_R = h_T = 5$ m to prevent truncation by the common coverage area. If L were increased to 100 km, both antennas would have to be elevated to ≈ 190 m. Alternatively, with $h_R = h_T = 5$ m and $L = 100$ km, only targets at altitudes of $h_T \geq 490$ m are visible near either site. When the target is about halfway between the sites $r_R = r_T = L/2 = 50$ km, and the altitude of visible targets decreases to about 100 m.

When the forward-scatter fence is configured for low-altitude target coverage, bistatic clutter returns will also be present and compete with the target return. Because the clutter cell area increases at large β, and the bistatic scattering coefficient σ_B° increases in the vicinity of the specular ridge region, the magnitude of the clutter return will also increase. Doppler processing can be used to reject the clutter, but the task is made more difficult because the target doppler is reduced at large β. As a consequence of these restrictive measurement and coverage capabilities, a single forward-scatter fence is usually constrained to special deployments for trip wire alerting and cueing of other sensors.

12.2.2 Netted Forward-Scatter Fences

A grid of forward-scatter trip wire fences can be netted together to provide an estimate of a target's velocity vector (course and speed) through the fence and its point (and time) of fence penetration. The French *maille en Z* fence system used in

World War II was apparently configured this way. Available details are given in Section 2.1.4.

One possible grid of fences is shown in Figure 12.10, where two transmitters operating on separate frequencies, and two receivers, each receiving both frequencies, generate four fences in a "double-Z" configuration. A target is assumed to cross all four fences at separate times, and with a constant velocity vector, V, during the crossing time. The geometry of the double-Z configuration is assumed to be known *a priori*. In the example, the configuration is a rectangle with sides L (baseline) and S (baseline separation).

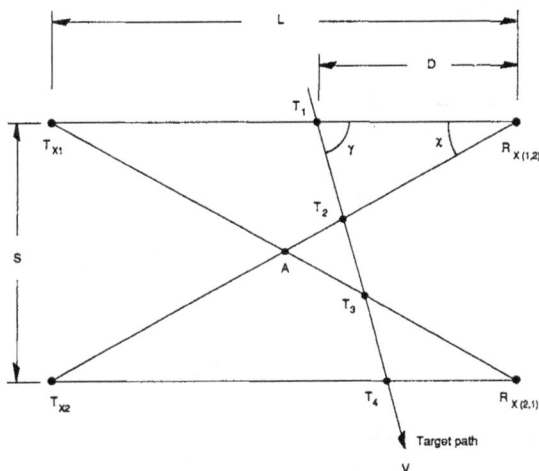

Figure 12.10 Geometry for double-Z grid of forward-scatter trip wire fences.

Four time-of-crossing measurements $T_1, \ldots, T_4$ are made, with $T_2 - T_1 = \Delta T_1$, $T_3 - T_2 = \Delta T_2$, $T_4 - T_3 = \Delta T_3$, and $\Delta T_1 + \Delta T_2 + \Delta T_3 = \Delta T$, the time between first and last fence crossing. Three target parameters are required: target velocity, V; angle between velocity vector and a horizontal baseline, γ; and point of penetrating the first horizontal baseline, D. Solving the geometry yields

$$\gamma = \tan^{-1}\left(\frac{\Delta T_2 \tan\chi}{\Delta T_1 - \Delta T_3}\right) \tag{12.21}$$

$$V = \frac{S \csc\gamma}{\Delta T_2(1 - \cot\gamma \tan\chi) + 2\Delta T_1} \tag{12.22}$$

$$D = \cos\gamma \left[S + V\Delta T_3(\tan\gamma \cot\chi - 1)\right] \tag{12.23}$$

where $\tan\chi = S/L$, as shown in Figure 12.10.

Consider an example using the following parameters and the geometry of Figure 12.10 (approximately to scale):

$$\left.\begin{array}{l} L = 40 \text{ km} \\ S = 20 \text{ km} \end{array}\right\} \quad \tan\chi = 0.5$$

$$\left.\begin{array}{l} T_1 = 0 \text{ s} \\ T_2 = 26.3 \text{ s} \end{array}\right\} \left.\begin{array}{l} \Delta T_1 = 26.3 \text{ s} \\ \\ \Delta T_2 = 17.7 \text{ s} \\ \Delta T_3 = 24.2 \text{ s} \end{array}\right.$$

$$\left.\begin{array}{l} T_3 = 44.0 \text{ s} \\ T_4 = 68.2 \text{ s} \end{array}\right\}$$

Substituting these parameters into (12.21) through (12.23) yields $\gamma = 76.2°$, $V = 0.3$ km/s, and $D = 17.2$ km. Note that if the target crosses the center of the fence, point A in Figure 12.10, $\Delta T_2 = 0$ and $\Delta T_1 = \Delta T_3$. Thus, the solution is indeterminate in the special case, except that the target's location at point A is known at time $T_2 = T_3$. The total measurement time, ΔT, can be reduced by reducing the baseline separation, S, but at a loss in accuracy of the estimates, especially when the target crosses near the center of the fence.

12.3 HYBRID RADARS

When a bistatic receiver is designed to operate directly with a cooperative monostatic radar, it is called a hybrid radar, and can sometimes improve the performance of the monostatic radar when compared to the performance of two netted monostatic radars. The principal example is countering three types of retrodirective ECM systems: retrojammers, augmented decoys, and active cancellation. This hybrid radar configuration exploits two bistatic features for performance enhancement: spatial diversity, or transmitting-receiving site separation, and covertness, because the second site must be quiet to counter the retrodirective ECM operation. In some hybrid configurations, the bistatic receiver can be a second monostatic radar that operates in the receive-only mode. When a hybrid radar is configured to counter a retrodirective ECM system, it is sometimes called a "counter-retro" system. (Of course, a bistatic radar by itself is also a counter, but these radars are seldom used alone.)

A retrodirective ECM system operates by measuring the AOA of a signal, usually a radar signal, and transmitting maximum or minimum power at that AOA. A retrodirective jammer, or retrojammer, is a typical example of a system transmitting maximum power. It might use a high-gain multibeam antenna when receiving and transmitting. A signal received on one beam identifies the AOA. The jamming signal is then retrodirected in the colinear transmit beam back to the signal source [192, 233]. The retrojammer operates as a self-screening escort or standoff jammer and can transmit noise or deception waveforms. The jammer's ERP is, of course, enhanced by the gain of the jammer's transmitting beam, which typically ranges from 10 to 30 dB, possibly more depending on aperture and frequency. These retrojamming systems are particularly useful for defending ships (with very large RCS) against aircraft and cruise missile attacks, defending bombers (with more modest RCS) against air defense systems, screening penetrating aircraft from air defense systems with standoff jammers, and jamming aircraft (and satellite) radars from ground-based sites. A second example of the maximum power retrodirective concept is the use of a passive retroreflector, such as a trihedral corner reflector or a Luneburg lens, by a decoy to augment its monostatic RCS, and thus match its RCS to that of attacking aircraft or missiles.

The crosseye deception ECM repeater is an example of a system transmitting minimum power to protect the platform—typically an aircraft—carrying the crosseye repeater [192]. In its basic configuration, the repeater uses a widely spaced, two-element receiving antenna, for example, one element on each wing tip of an aircraft. The radar signal received by one element is amplified and retransmitted by the second element. The radar signal received by the second element is phase shifted by 180°, amplified, and retransmitted by the first element. The composite repeated signal is a null positioned within the physical dimension monostatic of the antenna radar [105]. When the jamming-to-signal ratio is sufficiently high, the radar will angle track off the null, causing a significant bias in the radar's AOA estimate of the crosseye aircraft. A second example of the minimum power retrodirective concept is active cancellation [211], which is also called active loading [212]. Active cancellation generates a signal with an amplitude and phase that cancel the monostatic echo by destructive interference. The composite signal and echo again generate a null positioned in the direction of the radar. Crosseye typically operates at tracking radar frequencies and must generate a very narrow nullwidth, whereas active cancellation operates at much lower frequencies and can generate a broader nullwidth [212].

Many antenna configurations can be used for retrodirective operation, including the Rotman (and Mubis) lens, the Butler matrix, the aforementioned corner reflector and Luneburg lens, and phased arrays in multiple configurations, including digital beamforming arrays. The brute force approach of high-gain mechanically steered antennas for both receiving and transmitting can also be used in some cases.

Because the retrodirective system transmits maximum or minimum power back toward the source of the signal—the transmitter—a bistatic radar or a hybrid monostatic-bistatic radar, with its covert bistatic receiver sufficiently separated from its transmitter, will counter the retrodirective system. For the retrojammer, the bistatic receiver is positioned in the sidelobes of the retrojammer's mainbeam, thus improving the bistatic receiver's SJR by the jammers mainlobe-to-sidelobe ratio. An example of this improvement for a single, standoff, retrodirective noise jammer is given in Chapter 10. For the augmented decoy, the bistatic receiver is again positioned in the sidelobes of the passive retroreflector's mainbeam, thus reducing the decoy's SNR by the retroreflector's mainlobe-to-sidelobe ratio. For crosseye and active cancellation, the bistatic receiver is positioned outside the null, in which case crosseye and active cancellation operate as a simple repeater, enhancing the bistatic receiver's SNR. To counter a retrodirective system, the bistatic receiver must be covert and properly sited with respect to the monostatic radar. It should also generate coverage that enhances monostatic radar operation in a useful way.

While covert operation is inherent in the bistatic receiver, it also requires that all transmissions from the bistatic receiving site, including data links, be covert. This requirement can be satisfied by land lines, if the baseline is reasonably small, or special satellite links and possibly low probability of intercept (LPI) RF links otherwise. The knowledge that a counter-retro bistatic receiver is operating as part of the radar system is also an important element of the scenario. If the attacker is not aware of the configuration, a properly designed counter-retro system will be effective. If the attacker is aware of the configuration and the configuration is perceived to be effective, alternative penetration responses would probably be considered. The attacker may attempt to locate and disable the bistatic receiver—a difficult task if the receiver is indeed covert, or select attack directions outside of the counter-retro coverage areas, which, as will be detailed later, limit attack operations. Finally, the attacker may choose other methods to enhance penetration, which by definition means that the counter-retro configuration is successful. Often the other penetration methods are less effective. An example is the use of a broadcast jammer replacing the retrojammer for standoff jamming operations, as detailed in Section 10.1. In any case, the counter-retro configuration must be feasible in order to generate at least the perception of effectiveness. This requirement is considered next.

The bistatic receiver siting requirement is defined in terms of the bistatic angle, which is now measured with respect to the retrodirective system. Because the retrodirective system and the target may not be collocated, the retrodirective bistatic angle is defined as β_R to distinguish it from the target's bistatic angle, β. A conservative requirement for β_R is $\beta_R > \Delta\theta_J$, where $\Delta\theta_J$ is the retrodirective system's 3-dB transmitter beamwidth or nullwidth. For the retrojammer case, the jammer's first null is displaced from the peak of the jamming beam by approximately $\Delta\theta_J$. Figure 12.11 shows the geometry. When $\beta_R \approx \Delta\theta_J$ the bistatic receiver will be in the jammer's first null. For $\beta_R > \Delta\theta_J$ the bistatic receiver will walk through the jammer's sidelobes. Some performance improvement is obtained for $\Delta\theta_J > \beta_R > \Delta\theta_J/2$. For example, against the retrojammer when $\beta_R = \Delta\theta_J/2$ the bistatic receiver's SJR will be improved by 3 dB. Thus, while $\beta_R > \Delta\theta_J/2$ will yield improved SJRs, a more robust performance requirement is $\beta_R > \Delta\theta_J$. Figure 12.11 also applies to active cancellation and the augmented decoy, where $\Delta\theta_J$ represents the active canceller's nullwidth and the decoy retrodirective beamwidth.

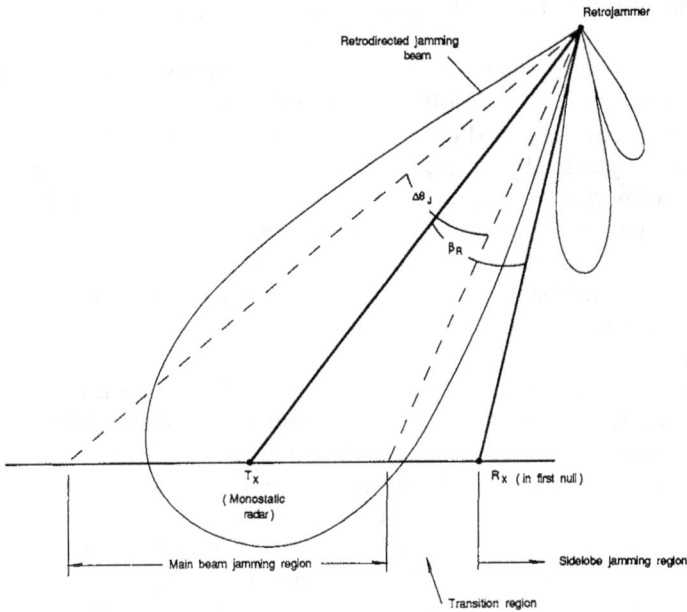

Figure 12.11 Hybrid radar geometry to counter a retrodirective jamming system.

These three retrodirective systems typically operate with $\Delta\theta_J$ ranging from 5° to 30°, depending on vehicle size, aspect angle, and frequency. However, for crosseye the required null width, $\Delta\theta_J$, is very small, <1°, with the null pointed within a fraction of a milliradian of the center of the radar antenna, when operating at X-band for example [105]. Thus, separate transmitting and receiving antennas located at the same site will generate an adequate β_R in nearly all scenarios of interest. As a consequence, the hybrid or bistatic radar configuration is not necessary to counter crosseye.

The region over which a hybrid radar will counter the retrodirective system is bounded by a contour of constant bistatic angle $\beta_R = \Delta\theta_J$. A family of constant bistatic angle contours is shown in Figure 3.3. As an example, if $\Delta\theta_J = 20°$ any retrodirective system located inside the $\beta_R = 20°$ contour will generate a bistatic angle greater than 20° and thus will be countered.

Figure 3.3 suggests bistatic geometries that are useful in countering a retrodirective system. For air defense operation, the bistatic receiver should be sited so that the baseline is roughly perpendicular to the expected direction of attack. Thus, in Figure 3.3 the attack direction should be from the north, or top of the page. In contrast, when the attack direction is from the east or west, or along the extended

228

baseline, the bistatic angle approaches zero and counter-retro operation is not possible.

The counter-retro coverage requirement is developed for an air defense scenario in which multiple, standoff, sidelobe retrojammers screen penetrating aircraft. The aircraft are assumed to be located within the monostatic radar's thermal noise-limited (unjammed) coverage area, and are further assumed to be screened by the retrojammers to an unacceptably small burn-through range from the monostatic radar, even after the use of sidelobe cancellation against each retrojammer. The objective of this counter-retro hybrid radar is to recover a significant fraction of the unjammed monostatic coverage area. This scenario generates the most stressing counter-retro requirement. Counter-retro operation against augmented decoys and active cancellation is treated as a subset of this scenario.

Monostatic and bistatic coverage is defined by maximum detection range contours, antenna scan limits, and LOS constraints. First consider an example of a monostatic radar with thermal noise-limited maximum detection range $\sqrt{\kappa}$ and a matched bistatic receiver with maximum range product κ (not always the case) as shown in Figure 12.12. The monostatic coverage is represented by a circle of radius $R_M = \sqrt{\kappa}$; the bistatic coverage is represented by an oval of Cassini, $R_T R_R = \kappa$, having a shape set by the length of the baseline L, (4.5). The counter-retro coverage is a circle of radius $r_\beta = L/(2 \sin\beta)$, (3.11), with a center located on the perpendicular bisector of the baseline and displaced from the baseline by a distance $d_\beta = L/(2 \tan\beta)$, (3.12).

These thermal noise-limited coverage areas are, of course, reduced by jamming. Expressions for calculating this reduction are developed in Chapter 10. Specifically, (10.5) gives the bistatic receiver's maximum range product under noise-jamming conditions, κ_J, as

$$\kappa_J = \kappa \left[\frac{kT_S}{kT_S + \sum_{i=1}^{n} (J_o)_i} \right]^{1/2} \tag{12.24}$$

where

κ = bistatic maximum range product (under thermal noise conditions),
k = Boltzmann's constant,
T_S = system noise temperature (under thermal noise conditions),
$(J_o)_i$ = jammer PSD of the ith jammer at the bistatic receiver, (10.1).

The equivalent monostatic range is $\sqrt{\kappa_J}$.

The term of interest in (10.1), which defines J_o, is f_J, the jammer's antenna pattern factor referenced to the receiver site. Figure 10.1(d) shows the geometry

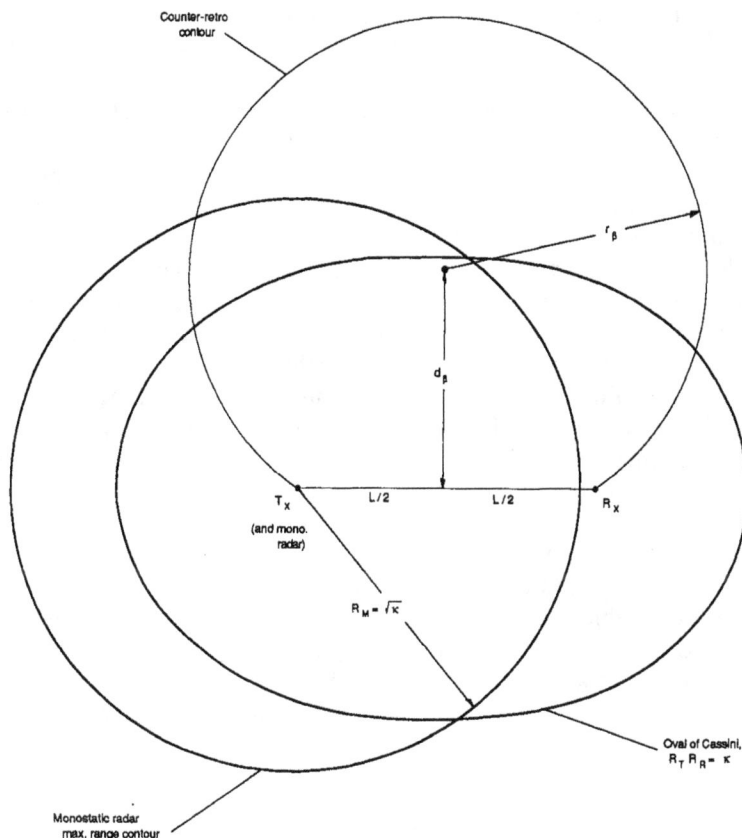

Figure 12.12 Generic thermal noise-limited coverage contours for hybrid radar, counter-retro operation.

defining f_J (and f_{RJ}, the radar receiving antenna pattern factor, referenced to the jammer site) for this scenario. Specifically, when a retrojammer operates within the bistatic receiver's counter-retro coverage area, then $f_J^2 = (G_J)_{SL}/G_J$, where $(G_J)_{SL}$ is the retrojammer's transmitting antenna (sidelobe) gain in the direction of the bistatic receiver and G_J is the retrojammer's transmitting antenna (peak) gain. Thus, (10.1) becomes

$$(J_o)_c = \frac{P_J(G_J)_{SL}G_R F'^2_J f^2_{RJ}\lambda^2}{(4\pi)^2 B_J R_J^2 L_J} \tag{12.25a}$$

$$= J_o[(G_J)_{SL}/G_J] \tag{12.25b}$$

where $(J_o)_c$ is the retrojammer's PSD at the bistatic receiving site when the retrojammer operates within the counter-retro coverage area. The remaining terms are defined by (10.1) but with $f_J = 1$ for the J_o term. This expression for $(J_o)_c$ is used in place of J_o in (12.24), and the bistatic maximum range product is now called $(\kappa_J)_c$ in this counter-retro scenario.

For the monostatic radar with maximum detection range under noise-jamming conditions $\sqrt{\kappa_J}$, the bistatic coverage improvement becomes a function of $G_J/(G_J)_{SL}$ and the baseline L. Thus, the counter-retro task is, for given retrojammer PSD levels and a required β_R, to select a baseline such that the joint coverage is optimized, where joint coverage is defined as the intersection of the counter-retro circle, the monostatic thermal noise-limited circle, the bistatic oval under jamming conditions, and the antenna scan sector. Note that when $L = 0$ the bistatic oval is a maximum from Figure 5.12, but the counter-retro circle vanishes. When L is large, the counter-retro circle becomes large, but the bistatic oval becomes small, and LOS requirements are difficult to satisfy. Consequently, L should be made as small as possible, yet it should provide acceptable joint coverage.

Consider an example of four standoff, retrodirective, spot noise jammers, each with a standoff range of $R_J = 1.5 \sqrt{\kappa}$, where $\sqrt{\kappa} = R_M$ is assumed to be 100 km. The bistatic receiver's thermal noise-limited maximum range product is $\kappa = R_T R_R$, as before. Table 12.3 lists the jammer and additional radar parameters to calculate performance under sidelobe noise-jamming conditions. Both monostatic and bistatic receivers are assumed to use 25-dB sidelobe cancellers against each retrojammer, the typical response discussed in Chapter 10. Each retrojammer operates in the spot noise-jamming mode, with $B_J = 2.5$ MHz assumed to cover the receiver's bandwidth.

Table 12.3

Parameters for the Counter-Retro Hybrid Radar Example under Sidelobe Noise-Jamming Conditions

$\kappa = 10^4$ km^2 (bistatic maximum range product)
$\sqrt{\kappa} = 100$ km (monostatic maximum detection range)
$G_R = 29$ dBi
$(G_R)_{SL} = -6$ dBi
$\lambda = 0.3$ m (L-band)
$G_J = 22$ dBi
$(G_J)_{SL} = -6$ dBi
$P_J = 1$ kW
$F_J' = 1$ (no multipath, diffraction, or refraction)
$f_{RJ}^2 = -60$ dB $[(G_R)_{SL}/G_R = -35$ dB plus -25-dB sidelobe cancellation of each jammer]
$f_J^2 = 1$ for monostatic radar $= (G_J)_{SL}/G_J = -28$ dB for bistatic counter-retro coverage
$B_J = 2.5$ MHz
$R_J = 1.5 \sqrt{\kappa} = 150$ km
$L_J = 3$ dB
$kT_s = 6 \times 10^{-21}$ W/Hz (≈ 2 dB noise figure)

Inserting the parameters of Table 12.3 into (10.1) yields $J_o = 6.4 \times 10^{-19}$ W/Hz for a single sidelobe retrojammer operating against the monostatic radar. Thus, from (12.24), $\sqrt{\kappa_J} = 0.22\sqrt{\kappa} = 22$ km for four sidelobe retrojammers operating against the monostatic radar. For the bistatic case, when each retrojammer is located in the counter-retro area, $(J_o)_c = 10^{-21}$ W/Hz, from (12.25b). Thus, $(\kappa_J)_c = 0.77\kappa = 7.7 \times 10^3$ km² for four sidelobe retrojammers operating against the bistatic receiver.

In this example, $G_J = 22$ dBi, which corresponds to a jammer beamwidth of $\Delta\theta_J = 13°$, assuming a circular antenna aperture. For counter-retro operation, $\beta_R > 13°$. Assume $\beta_R = 15°$ for the example. The remaining parameter is the baseline L, which should be as small as possible and yet optimize joint coverage. The selection of L is empirically determined. Figure 12.13 is the result of experiments, with $L = 50$ km, and is approximately to scale. Both monostatic and bistatic azimuth scan sectors are assumed to be 120°. The individual coverage areas are shown as dotted lines, with the joint coverage area shown as the hatched area under the solid line. The bistatic receiving site is located slightly forward of the line perpendicular to the attack direction, and its azimuth scan sector is tilted to the left to optimize joint coverage. Figure 12.13 shows that a significant fraction of the monostatic

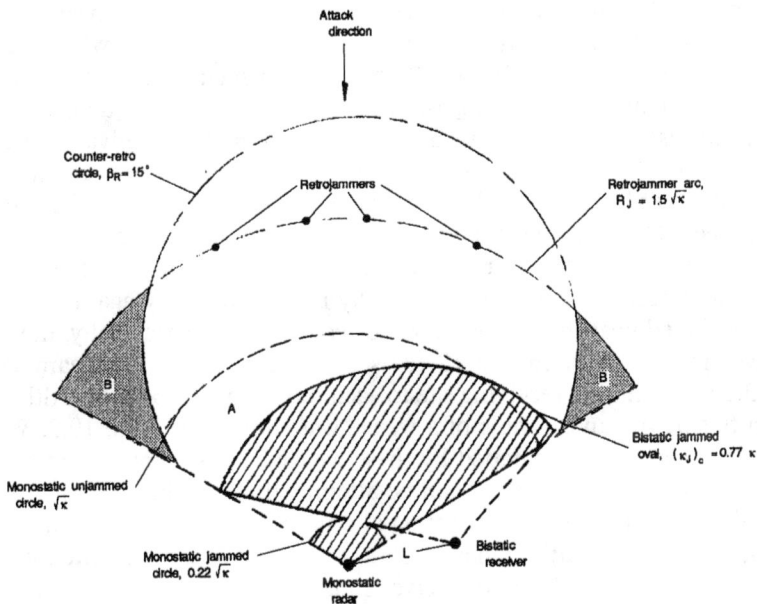

Figure 12.13 Counter-retro coverage for air defense retrojammer scenario, with $\beta_R = 15°$, $\sqrt{\kappa} = 100$ km, and $L = 50$ km.

radar's unjammed coverage area is recovered, with only the area labeled A on the left side of the scan sector beyond about 60 km from the monostatic radar left uncovered. A second bistatic receiving site located on the left side of the monostatic radar will, of course, recover this area but at an increase in cost. If two or more monostatic radars are deployed, a bistatic receiving site located between the two monostatic radars may be able to service both radars in time sequence. A data rate issue must be addressed in this case. Note that retrojammers located in the shaded areas labeled B near either edge of the monostatic radar scan sector are not countered, which is the area trade-off between the counter-retro circle and the bistatic oval. Again, a second bistatic site can recover some of the lost counter-retro coverage. When the terrain is relatively flat, LOS requirements can be satisfied by siting the bistatic receiver at roughly the same altitude as the monostatic radar.

To this point, only sidelobe jamming performance has been considered. When the aircraft is located near a retrojammer-to-monostatic radar LOS, retrojammer *mainlobe*-on-radar mainlobe jamming occurs, as is shown in Figure 10.1(a). In this case $f_{RJ}^2 = 1$ and J_o is increased by 10^6 to 6.4×10^{-13} W/Hz. Thus, $\sqrt{\kappa_J}$ is reduced to $0.01 \sqrt{\kappa} = 1$ km, which is the mainlobe jamming burn-through range. When the aircraft is located near a retrojammer-to-bistatic receiver LOS, retrojammer *sidelobe*-on-bistatic mainlobe jamming occurs, as is shown in Figure 10-1c. Again, $f_{RJ}^2 = 1$ and $(J_o)_c$ is increased by the same factor of 10^6 to 10^{-15} W/Hz. Thus, $(\kappa_J)_c$ is reduced to $2 \times 10^{-3} \kappa = 20$ km^2, and the bistatic coverage collapses to a small oval of Cassini centered on the receiving site, with radius $R_o \approx (\kappa_J)_c/L$, from (12.3). For $L = 50$ km, $R_o \approx 0.4$ km, which is smaller than the monostatic radar's mainlobe jamming burn-through range. Consequently, while the counter-retro bistatic receiver has an inherent $f_J^{-2} = 28$ dB advantage over the monostatic radar (the retrojammer sidelobe-on-bistatic mainlobe geometry), that advantage is offset by a very high mainlobe J_o and relatively large site separation. However, the f_J^{-2} advantage for retrojammer sidelobe-on-bistatic sidelobe jamming, Figure 10.1(d), remains a significant factor. Clearly, neither the monostatic nor the bistatic receiver has an operationally useful capability against penetrating aircraft positioned near the retrojammer-to-receiver LOS. Specifically, the counter-retro coverage shown in Figure 12.13 is reduced by four mainbeam jamming wedges, drawn from each receiving site to each retrojammer location. Although not shown in this figure, an example of a wedge was shown in Figure 10.3. When the coverage is unacceptably degraded by these wedges, the system response of triangulation and jammer engagement can be attempted, as outlined in Chapter 10. In fact, this type of response is a competing alternative to the hybrid radar response.

When the monostatic radar uses a high-gain scanning antenna—nearly always the case—and the bistatic receiver's maximum range product κ is matched to the monostatic radar's maximum detection range $\sqrt{\kappa}$, as was assumed for this example, the bistatic receiver must nearly always use an antenna of equal gain, G_R

= 29 dBi (and $\Delta\theta_R \approx 6°$) in the example. Thus, beam scan-on-scan problems arise in this counter-retro scenario. General solutions are outlined in Section 13.1 and are analyzed for a similar scenario in Section 12.1.1. Two solutions, multiple simultaneous receiving beams (and receivers) and pulse chasing (with a receiving antenna electronically scanning in at least the azimuth dimension), are again possible. Note that when the monostatic radar electronically scans in the azimuth dimension, a pair of these radars can be used in the hybrid configuration, one as the monostatic radar, and one as the pulse-chasing bistatic receiver.

For one receiving beam pulse chasing, the maximum pulse repetition frequency, PRF_1, must be selected such that only one pulse traverses the pulse-chasing scan sector at one time. For cosite operation, PRF_1 is given by (5.43), and in the example of Figure 12.13, PRF_1 is a minimum when the monostatic radar's beam is pointed near the right edge of its scan sector. In this case, $\theta_T = 70°$, $\theta_{R1} = 53°$, and $\theta_{R2} = -67°$, where θ_{R1} and θ_{R2} define the scan sector (Figure 5.16). Thus, PRF_1 = 2.8 kHz. When the monostatic radar operates with unambiguous range to $\sqrt{\kappa}$ = 100 km, $(PRF_M)_u$ = 1.5 kHz, from (5.37). Because $PRF_1 > (PRF_M)_u$, the requirement is satisfied. This geometry also establishes the maximum number, n, of simultaneous receiving beams, which in this case must cover the full receiving scan sector. In the example of Figure 12.12, $n = 120°/6° = 20$ receiving beams. Because a single transmitted pulse will traverse multiple receiving beams in the scan sector, a beam-pulse scheduling algorithm is required. This algorithm is similar to the pulse-chasing algorithm, but requires no beam scanning.

The geometry of Figure 12.13 can be used for counter-retro operation against augmented decoys and active cancellation. In these two cases, the retrodirective system is carried by a penetrating vehicle; thus, $\beta = \beta_R$. For active cancellation the monostatic jammed circle of 0.2 $\sqrt{\kappa}$ is analogous to the monostatic radar's detection range of the "loaded vehicle," and is equivalent to a monostatic RCS reduction of $(0.22)^{-4} = 28$ dB. Assuming the active cancellation null rolls off at 15°, the detection coverage would again be the hatched area of Figure 12.13. In this active cancellation geometry, the counter-retro circle can be reduced somewhat by reducing L. The result will be to increase the bistatic coverage area, which reduces area A; however, the counter-retro coverage area is decreased, which increases area B in Figure 12.13. Again the empirical method is used to maximize coverage.

For augmented decoys the monostatic and bistatic coverage areas are essentially reversed. When the augmented decoy's monostatic RCS is made equal to that of a penetrating aircraft, the monostatic radar's coverage extends out to the unjammed circle of radius $\sqrt{\kappa}$. In contrast, the bistatic coverage is intentionally reduced. When the decoy's retrodirected pattern rolls off 28 dB at $\beta = 15°$, the bistatic maximum range product is reduced from κ to 0.05κ, which at $L = 50$ km and $\kappa = 10^4$ km² generates an oval around the bistatic receiving site with radius R_o ≈ 10 km. Now, the fact that the monostatic radar detects the augmented decoy, and the bistatic receiver does not, becomes the discriminant. The baseline of 50 km is a reasonable choice for this case because the monostatic coverage is only slightly truncated by the counter-retro circle in Figure 12.13.

Scan-on-scan requirements remain unchanged from the retrojammer scenario, with one exception. If the hybrid radar is designed to counter only the augmented decoy, the scan-on-scan requirement can be finessed by cueing a single bistatic receiving beam to the decoy location by the monostatic radar. Unfortunately, cueing will work only in this scenario, because the monostatic radar must first detect the target in order to generate the cue. In the other scenarios, the monostatic radar cannot detect the target at useful cueing ranges.

12.4 CLUTTER TUNING

When both the bistatic transmitter and receiver are moving, for example, airborne or spaceborne, they generate a clutter doppler shift, (6.5), and a clutter doppler spread, (6.11). Because both platforms contribute to the shift and spread, this type of bistatic radar has an additional degree of freedom in controlling doppler from ground clutter by changing velocity vectors (V_T, δ_T, and V_R, δ_R) or geometry (θ_T, θ_R, R_T, R_R, and L). This concept is called *clutter tuning,* and applies to both SAR and MTI operation.

12.4.1 Synthetic Aperture Radar

The principal application of clutter tuning is forward-looking SAR. The transmitter flies with its velocity vector perpendicular to its look angle, and the receiver flies with its velocity vector colinear with its look angle, i.e., directly at the target area, as shown in Figure 12.14. In this case, $\delta_T = \theta_T \pm 90°$ and $\delta_R = \theta_R$. Thus, the bistatic SAR isorange resolution, τ_i, is from (7.17):

$$\tau_i = \frac{\lambda R_T}{V_T T \cos(\beta/2)} \tag{12.26}$$

where T is the array time.

Equation (12.26) shows that the resolution is independent of receiver range and velocity. Also the resolution is degraded (increased) by a factor $2/\cos(\beta/2)$ when compared to the cross-range resolution obtained if the transmitter were operating as a monostatic radar, (7.10).

The operational utility of forward-looking SAR is that the receiver can generate an SAR map directly on its velocity vector, which would not be possible if the receiver were operating as a monostatic radar. Also, as long as the transmitter

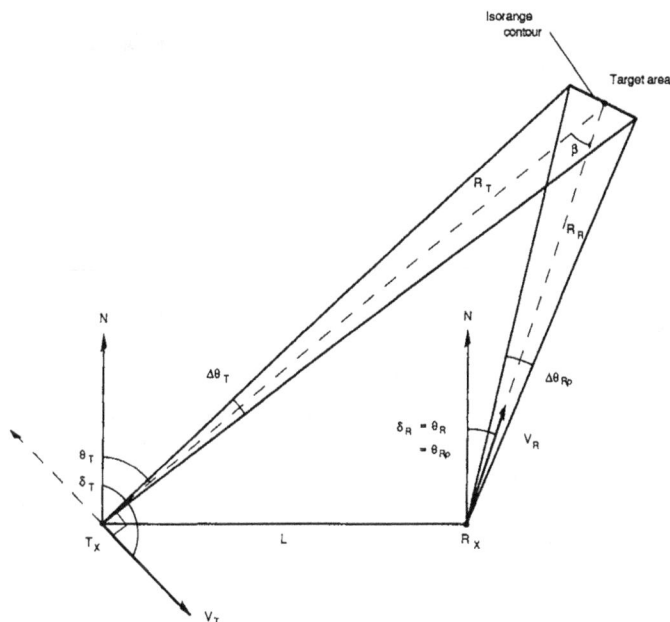

Figure 12.14 Geometry for forward-looking SAR.

is illuminating the target area the receiver can generate SAR maps continuously as it flies toward the target area. Commanded operation is also possible. In this case, the receiver requests transmitter illumination of a specific area at a specific time. Furthermore, the transmitter can stand back from the target area, in a sanctuary where it is less vulnerable to attack, while the receiver penetrates in RF silence. This bistatic radar configuration typically uses a receiver-centered operating region and a dedicated or cooperative transmitter as defined in Section 4.4. In all cases, receiver operation is constrained by the bistatic radar maximum range equation, (4.1).

However, both of these conditions, an SAR map directly on the receiver's velocity vector and a long transmitter standoff range, limit the bistatic SAR isorange resolution. Consider the following example, using the geometry of Figure 12.14 and the parameters listed in Table 12.4. From (12.26), $\tau_i = 20.7$ m.

When the bistatic transmitter also operates as a monostatic SAR, the monostatic SAR cross-range resolution, $\tau_c = 10$ m, from (7.10). Thus, the bistatic SAR isorange resolution is approximately twice as large as the monostatic cross-range resolution because the bistatic receiver does not generate bistatic doppler spread in this geometry. It would be exactly twice as large when $\beta = 0°$; i.e., when the bistatic receiver is positioned on the transmitter-to-target LOS.

Table 12.4
Example Parameters for Forward-Looking SAR Operation

$$V_T = V_R = 300 \text{ m/s}$$
$$R_T = 200 \text{ km}$$
$$\delta_T = \theta_T + 90°$$
$$\delta_R = \theta_R$$
$$\beta = 30°$$
$$\lambda = 0.03 \text{ m}$$
$$T = 1 \text{ s}$$

The bistatic receiver can improve its isorange resolution by increasing the array time T or flying a course offset from the target area, $(\delta_R - \theta_R) \neq 0$. As outlined in Section 13.5, uncancelled low-frequency phase noise in a dual STALO system, one in the transmitter and one in the receiver, constrains the array time required to achieve satisfactory integrated sidelobe ratios (ISLR) in the bistatic SAR image. For current STALOs (see Figure 13.6 below), $T = \approx1$ s for ISLR $= -30$ dB. If ISLR $= -25$ dB can be tolerated, T can be extended to ≈2 s. In this case, a factor of 2 improvement in bistatic isorange resolution is possible by using 2-s array times. In contrast, array times much greater than 2 s can be used in a monostatic SAR, yielding inherently better cross-range resolution. Alternatively, if the bistatic receiver can correlate low-frequency components of the phase noise by clutter-referenced autofocus techniques, phase noise in theory can approach the level achieved in a monostatic SAR, thus yielding $T > 2$ s. Analysis or tests of this bistatic autofocus approach apparently have not been reported.

Improvement in bistatic isorange resolution can also be achieved by initially choosing $(\delta_R - \theta_R)$ for a desired value of τ_i and then varying $(\delta_T - \theta_R)$ such that τ_i remains constant over the receiver (and transmitter) flight paths. (The receiver velocity can also be increased to improve τ_i, but this option is usually constrained by platform and operational requirements.) Solving (7.17) for $(\delta_R - \theta_R)$ yields

$$(\delta_R - \theta_R) = \sin^{-1}\left\{ (R_R/V_R) \left[\frac{\lambda}{\tau_i T \cos(\beta/2)} - \frac{V_T \sin|\delta_T - \theta_T|}{R_T} \right] \right\} \quad (12.27)$$

When the transmitter coordinates and velocity vector are provided to the receiver, it solves the time-varying bistatic triangle for R_T, R_R, θ_T, and β; selects a τ_i and maximum T permitted by the ISLR; and then calculates $(\delta_R - \theta_R)$ by (12.27). The sign of $(\delta_R - \theta_R)$ must equal the sign of $(\delta_T - \theta_T)$. That is, transmitter and receiver rotation about the target must be in the same direction, both clockwise or both counterclockwise. In this way, the clutter doppler spread from each platform adds.

Using the parameters of the previous example, assume that $R_R = 100$ km at the start of operation and that the transmitter flies a constant velocity, straight line course (which is in a counterclockwise direction about the target) as shown in Figure 12.15. Assume that $V_T = V_R = 300$ m/s throughout the operation, and that a constant isorange resolution $\tau_i = 10.4$ m is desired. Figure 12.15 shows the spiral,

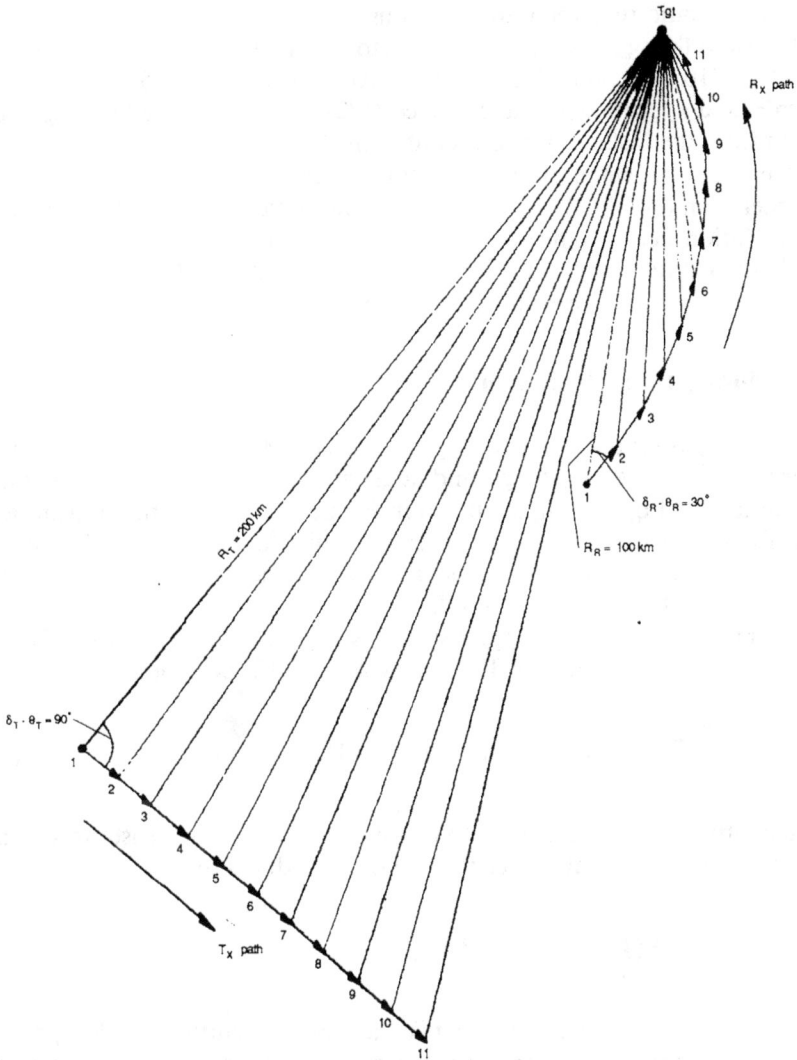

Figure 12.15 Bistatic receiver offset (spiral) flight path to maintain constant isorange resolution, $\tau_i = 10.4$m.

counterclockwise receiver flight path to achieve $\tau_i = 10.4$ m. The figure is approximately to scale.

The paired transmitter and the receiver positions in Figure 12.15, (1,1), (2,2), et cetera, are spaced 10 km apart (elapsed time of 33.3 s). For example, at the initial position (1,1), $R_T = 200$ km, $R_R = 100$ km, $(\delta_T - \theta_T) = -90°$, $(\delta_R - \theta_R) = 30°$, and $\beta = 30°$. At end game (11,11), $R_T = 223$ km, $R_R = 6$ km, $(\delta_T - \theta_T) = -116°$, $(\delta_R - \theta_R) = -2.3°$, and $\beta = 53°$.

If the bistatic receiver were operating as a monostatic SAR with the same array time and flying the same spiral trajectory of Figure 12.15, its cross-range resolution, $\tau_c = 10$ m at point 1, from (7.10). At point 11, $\tau_c = 7.5$ m. The difference between τ_i and τ_c at these points is the $\cos(\beta/2)$ term and the reduced spread contribution from the bistatic transmitter at point 11.

Consequently, when either array times greater than ≈ 2 s or receiver flight paths other than $(\delta_R = \theta_R)$ are required, a monostatic SAR will yield better cross-range resolution performance than a dual STALO bistatic SAR. In any case the survivability feature of a sanctuary transmitter and an RF silent penetration receiver remain a useful operational capability for the bistatic SAR.

12.4.2 Moving Target Indication

A second application of clutter tuning is broadside MTI. Now the transmitter and receiver counter-rotate about a ground target area in an attempt to cancel the clutter spread for MTI operation, as shown in Figure 12.16. Both the transmitter and receiver fly with their velocity vectors perpendicular to their look angles, one rotating clockwise and the other counterclockwise about the target, or vice versa. In this case, $\delta_T = \theta_T \pm 90°$ and $\delta_R = \theta_R \mp 90°$.

The mainbeam clutter doppler spread is given by (6.11) or (6.12). When the approximations of (6.17) and (6.18) are invoked, (6.12) becomes

$$\Delta f_{TR} \approx \frac{R_T \Delta\theta_T}{\lambda}[(V_T/R_T)\sin(\delta_T - \theta_T) + (V_R/R_R)\sin(\delta_R - \theta_R)] \tag{12.28}$$

The transmitting beam is assumed to establish the length of the isorange contour. For the given counter-rotation geometry, (12.28) reduces to

$$\Delta f_{TR} \approx \frac{R_T \Delta\theta_T}{\lambda}(V_T/R_T - V_R/R_R) \tag{12.29}$$

When $V_T/R_T = V_R/R_R$, $\Delta f_{TR} \approx 0$. That is, the receiver's clutter doppler spread cancels that of the transmitter when their respective angular rates about the target, V_T/R_T and V_R/R_R, are equal and opposite.

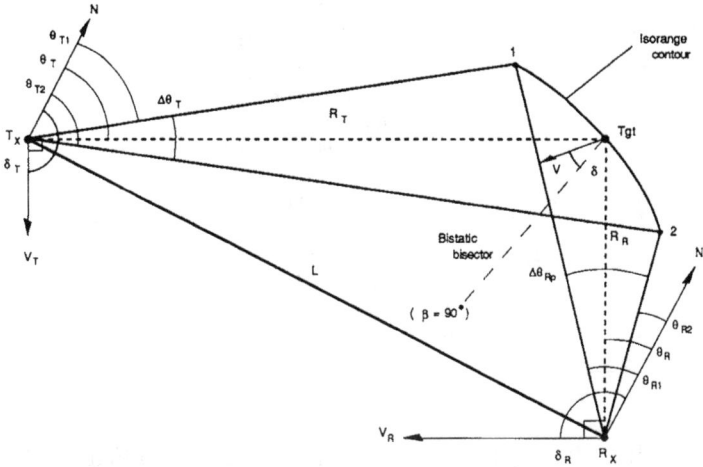

Figure 12.16 Geometry and kinematics for broadside MTI, with a counter-rotating transmitter and receiver, when $\beta = 90°$.

The operational utility of broadside MTI is that when the approximations of (6.17) and (6.18) are satisfied, the ground clutter spectrum collapses to a line at zero doppler. Thus, slowly moving ground targets with aspect angles of $\delta \neq \pm 90°$, (6.4b), will be separated in doppler from ground clutter and, given an adequate SNR, can be detected. [As developed in Section 6.1, when $\delta = \pm 90°$, the target's aspect angle is normal to the bistatic bisector, and from (6.4b) the target bistatic doppler is also zero.] The transmitter and receiver can be anywhere on their respective counter-rotating circles—at all bistatic angles—and the circles can be of any radius, as long as the bistatic radar maximum range equation, (4.1), is satisfied and $V_T/R_T = V_R/R_R$.

Two complications arise in the use of (12.29) for broadside MTI: (1) unless the transmitting (or receiving) beamwidth is small, less than a few degrees, the approximations of (6.17) and (6.18) will not accurately reflect the clutter doppler spread across an isorange contour, resulting in imperfect cancellation; and (2) when perfect cancellation is not obtained, the clutter doppler spread becomes one-sided at most bistatic angles; that is, the bistatic doppler shift from both ends of the isorange contour will be either positive or negative, depending on the direction of transmitter and receiver rotation.

Consider the following example, using the geometry of Figure 12.15 and the parameters listed in Table 12.5. Because $V_T/R_T = V_R/R_R = 0.003$ rad/s, the approximate solution for Δf_{TR} is 0 from (12.29). An exact solution for Δf_{TR} is developed using the additional parameters listed in Table 12.6. The bistatic doppler

Table 12.5
Example Parameters for Broadside MTI Operation

$$R_T = 100 \text{ km}$$
$$R_R = 50 \text{ km}$$
$$\beta = 90°$$
$$V_T = 300 \text{ m/s}$$
$$V_R = 150 \text{ m/s}$$
$$\Delta\theta_T = 5°$$
$$\Delta\theta_R = 10°$$
$$\lambda = 0.03 \text{ m}$$
$$(\delta_T - \theta_T) = +90°$$
$$(\delta_R - \theta_R) = -90°$$

shifts, f_{TR1} at point 1 $(\theta_{T1}, \theta_{R1})$ and f_{TR2} at point 2 $(\theta_{T2}, \theta_{R2})$, in Figure 12.16 are -35 and -34 Hz, respectively, from (6.5). The doppler shift from the target position (θ_T, θ_R) is exactly zero. Thus, the clutter doppler spread, Δf_{TR}, is from (6.8): 0 Hz $-$ $(-35$ Hz$) = +35$ Hz. Note that if (6.11) had been used, $\Delta f_{TR} = 1$ Hz, which, ignoring the sign, is the difference in clutter doppler shift from each edge of the isorange contour. This value is clearly not correct for this special counter-rotation case.

For this example, the mainbeam clutter doppler is spread between 0 and -35 Hz. Targets with bistatic doppler returns outside this region will be detected in a thermal noise background—ignoring any spread caused by internal clutter motion and contributions from sidelobe clutter. The minimum target velocity

Table 12.6
Additional Parameters for Calculating Exact Values of Clutter Doppler Spread in the Example of Figure 12.16 and Table 12.4

Parameter	Source
$\beta = 90°$	Figure 12.16 (given)
$L = 111.8$ km	(3.22)
$e = 0.745$	(3.3a)
$\theta_R = -26.6°$	(3.15)
$\delta_R = -116.6°$	$\delta_R = -90° + \theta_R$
$\theta_T = 63.4°$	(3.17)
$\delta_T = 153.4°$	$\delta_T = 90° + \theta_T$
$\theta_{T1} = 60.9°$	$\theta_{T1} = \theta_T - \Delta\theta_T/2$
$\theta_{T2} = 65.9°$	$\theta_{T2} = \theta_T + \Delta\theta_T/2$
$\theta_{R1} = -31.2°$	(6.13)
$\theta_{R2} = -21.2°$	(6.14)

required for thermal noise-limited detection is called the minimum detectable velocity (MDV), defined using (6.4b) as

$$\frac{\lambda(f_{TR})_{min}}{2\cos(\beta/2)} > V\cos\delta > \frac{\lambda(f_{TR})_{max}}{2\cos(\beta/2)} \qquad (12.30)$$

where

$V\cos\delta$ = MDV, the target velocity vector projected onto the bistatic bisector,

$(f_{TR})_{max}$ = maximum clutter doppler shift,

$(f_{TR})_{min}$ = minimum clutter doppler shift.

For the geometry of Figure 12.16, $(f_{TR})_{max} = 0$ Hz, $(f_{TR})_{min} = -35$ Hz, $\lambda = 0.03$ m, and $\beta = 90°$. Hence, when -0.74 m/s $> V\cos\delta > 0$ is satisfied, the target and clutter doppler returns will be separated and detection of the target will be thermal noise-limited. In this geometry all moving targets with $-90° < \delta < +90°$, or approximately closing the baseline, will have a positive, or up, doppler as defined in Section 6.1, and thus will be thermal noise-limited. Alternatively, when $V = 1$ m/s, detection will be thermal noise-limited only when $+137.7° < \delta < -137.7°$, so that -0.74 m/s $> V\cos\delta$.

If the direction of rotation of the transmitter and receiver were reversed in Figure 12.16, $(f_{TR})_{max} = +35$ Hz and $(f_{TR})_{min} = 0$ Hz. Thus, $0 > V\cos\delta > +0.74$ m/s, and all targets with $+90° < \delta < -90°$, or approximately opening the baseline, will be thermal noise-limited. When $V = 1$ m/s and $-42.3° < \delta < +42.3°$, detection will be thermal noise-limited.

As the transmitter and receiver continue their counter-rotation about the target in Figure 12.16, such that their velocity vectors remain perpendicular to their look angles, $\delta_T = \theta_T + 90°$ and $\delta_R = \theta_R - 90°$, respectively, the bistatic angle will take on all values, $0° \leq \beta \leq 180°$. For each value of β, a unique value of Δf_{TR} exists and is calculated using the equations shown in Table 12.6.

Results are plotted in Figure 12.17 for each end of the isorange contour, f_{TR1}, or point 1 $(\theta_{T1}, \theta_{R1})$, and f_{TR2}, or point 2 $(\theta_{T2}, \theta_{R2})$. In this counter-rotation geometry, f_{TR1} is always negative; f_{TR2} is also negative, except for very small and very large β. Thus, for most β, $(f_{TR})_{max} = 0$, and f_{TR2} usually sets the value for $(f_{TR})_{min}$. Figure 12.17 shows small clutter doppler spread for $\beta \lesssim 120°$, where the isorange contour eccentricity, $e \lesssim 0.9$. For $\beta > 120°$ and $e \gtrsim 0.9$, the spread increases significantly, exceeding 1000 Hz in regions near $\beta = 170°$. The peak value of f_{TR2} at $\beta \approx 172°$ corresponds to $e = 0.9978$, when $R_T = 100$ km and $R_R = 50$ km, and from (3.17), $\theta_T = 87.3°$. Thus, as outlined in Appendix E, the peak in f_{TR2} is caused by the θ_{Rp}

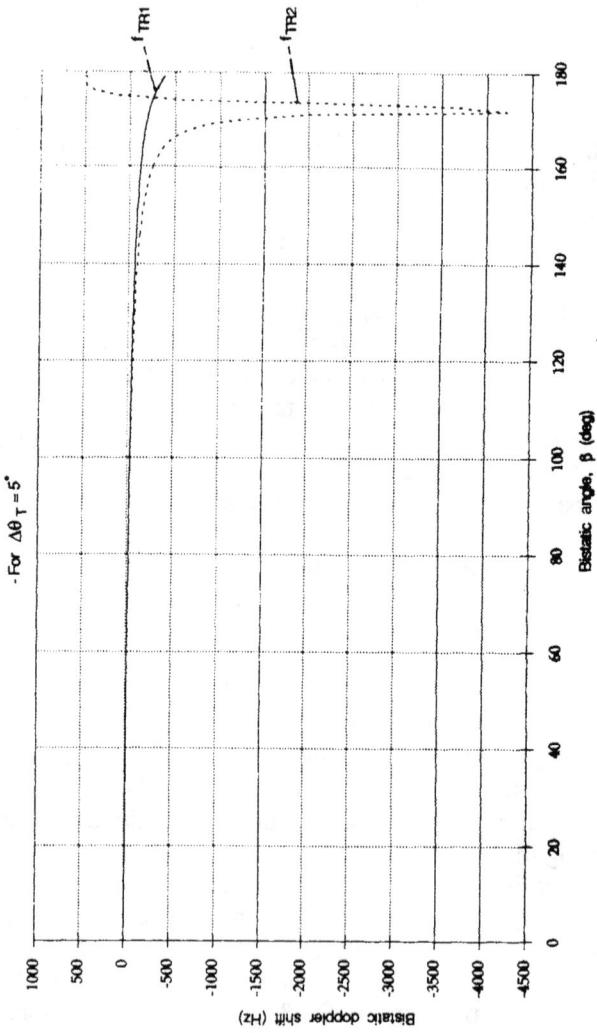

Figure 12.17 Bistatic clutter doppler spread for the counter-rotating transmitter-receiver case for $\Delta\theta_T = 5°$.

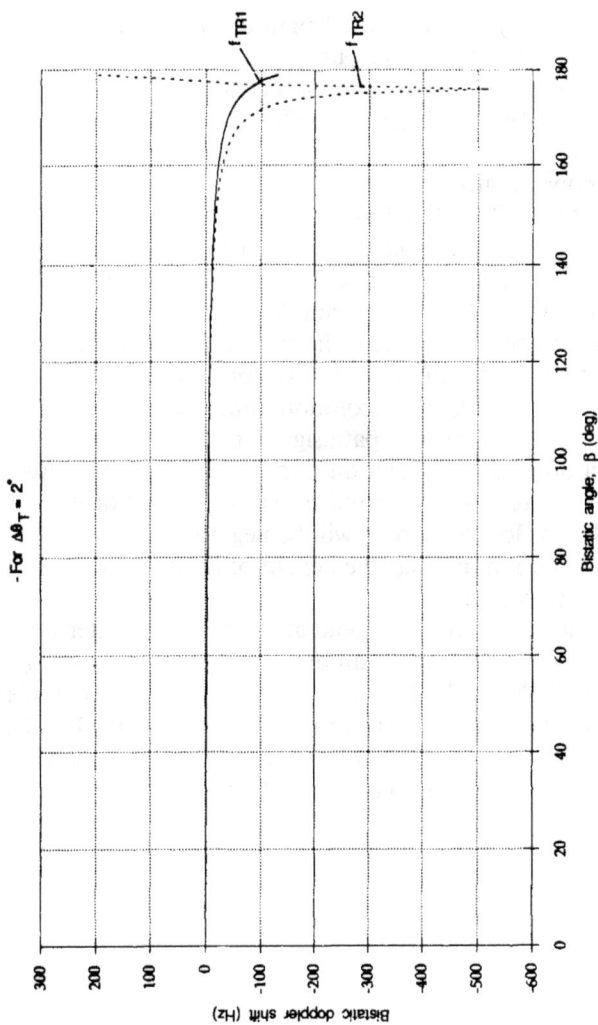

Figure 12.18 Bistatic clutter doppler spread for the counter-rotating transmitter-receiver case for $\Delta\theta_T = 2°$.

excursion. No solution exists for Δf_{TR} at $\beta = 180°$, the forward-scatter case, because $e = 1$. In this case, the isorange contour collapses to the baseline, and the expressions become indeterminate.

The preceding example assumed that $\Delta \theta_T = 5°$, $\Delta \theta_R = 10°$, $R_T = 100$ km, and $R_R = 50$ km. By using the approximation of (6.18), which is $\Delta \theta_{Rp} \approx (R_T/R_R)\Delta \theta_T$, then $\Delta \theta_R \approx \Delta \theta_{Rp}$. In this case the length of the isorange contour is established by either $\Delta \theta_T$ or $\Delta \theta_R$. If $\Delta \theta_R$ were used, the results shown in Figure 12.17 would be almost unchanged, with only a slight increase in f_{TR2} at β near 170°. Thus, the example applies for ($\Delta \theta_T = 5°$, $\Delta \theta_R \geq 10°$) and the converse ($\Delta \theta_R = 10°$, $\Delta \theta_T \geq 5°$).

Figure 12.18 shows results for ($\Delta \theta_T = 2°$, $\Delta \theta_R \geq 4°$), or ($\Delta \theta_R = 4°$, $\Delta \theta_T \geq 2°$) with all other parameters in Table 12.5 unchanged. As expected, the clutter doppler spread decreases with a peak near 500 Hz at $\beta = 176°$. In general, the smaller the beamwidth for either the transmitter or the receiver, the smaller the clutter doppler spread, and the closer to $\beta = 180°$ the bistatic MTI can operate.

In these and nearly all other cases of interest, the bistatic MTI configuration would attempt to avoid operation at $\beta \gtrsim 170°$ for three reasons: (1) the clutter doppler spread becomes large; (2) resolution (and accuracy) are significantly degraded, Chapter 7; and (3) the direct path signal from the transmitter is large and must be excised, Section 12.2. One solution is to fly arcs about the target area, with both the transmitter and receiver reversing course at the end of their arc. On one arc, the one-sided clutter doppler spread will be negative; on the other, the spread will be positive. This solution also has the benefit of limiting, or even eliminating, operation over hostile territory.

Note that a monostatic radar will generate significant clutter doppler spread in a broadside MTI mode. This spread can be greatly reduced by using a displaced phase center aperture (DPCA) [1], but at an increase in system complexity. Thus a trade-off can be made between the single-platform monostatic DPCA complexity and the dual-platform bistatic complexity. The monostatic configuration is usually preferred, unless the bistatic transmitter can be placed in sanctuary, where it is less vulnerable to attack.

Chapter 13
SPECIAL PROBLEMS AND REQUIREMENTS

Bistatic and multistatic radars are subject to problems and special requirements that are either not encountered or encountered in less serious form by monostatic radars. They include beam scan-on-scan coverage losses, with pulse chasing as one remedy to recover these losses, increased sidelobe clutter levels, precise time and phase synchronization, and adequate phase stability between transmitter and receiver. Each of these topics is considered in this chapter.

13.1 BEAM SCAN-ON-SCAN

If high-gain narrow-beam scanning antennas are used by both the transmitter and receiver in a bistatic surveillance radar, inefficient use is made of the radar energy because only the volume common to both beams can be observed by the receiver at any given time. Targets illuminated by the transmitting beam outside of this volume are lost to the receiver, as are targets outside of this volume but still in the receiving beam. Figure 13.1 shows the geometry. Four remedies can be considered to mitigate the beam scan-on-scan problem: step scan, floodlight beams, multiple simultaneous beams-receivers-signal processors, and time-multiplexed multi-beams-receivers-signal processors, which in the limit is called pulse chasing. Pulse-chasing requirements are detailed in Section 13.2.

The step scan remedy consists of fixing the transmitting beam position and scanning the receiving beam across the transmitting beam. The transmitting beam is then stepped one beamwidth and the receiving beam scan is repeated, and so forth until the transmitting beam has stepped across the surveillance sector. This remedy increases the surveillance frame time by approximately the number of required transmitting beam steps, depending on the geometry, and is usually not acceptable for area surveillance. It can be considered for limited surveillance regions, for example, in an over-the-shoulder geometry, where surveillance is only required near the extended baseline. In this case the transmitting and receiving

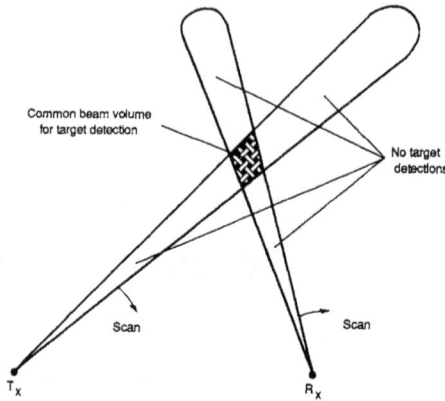

Figure 13.1 Beam scan-on-scan coverage problem shown in two dimensions (bistatic plane).

beams are nearly aligned, which requires only a few receiving beam positions to scan the transmitting beam.

The floodlight and multiple simultaneous beam (multibeam) remedies can be evaluated by using a variant of the monostatic surveillance radar range equation [1], where a penalty in the bistatic maximum range product, κ, is incurred for a particular scan-on-scan remedy. This penalty is expressed as the ratios G_{Tv}/G_T and G_{Rv}/G_R, where G_{Tv} and G_{Rv} are variable transmitting and receiving antenna power gains, respectively, required for the remedy, under the constraints $G_{Tv} \leq G_T$ and $G_{Rv} \leq G_R$. The bistatic radar maximum range equation, (4.1b), can now be written as

$$(R_T R_R)_{\max} = \kappa \left(\frac{G_{Tv} G_{Rv}}{G_T G_R} \right)^{1/2} \tag{13.1}$$

where κ is the bistatic maximum range product, and the term in parentheses (≤ 1) is the reciprocal of the beam scan-on-scan loss. (Following standard convention, loss is defined as a term ≥ 1.)

The first step is to express G_T and G_R in terms of surveillance radar parameters. The surveillance frame time, t_s, is defined as [1]:

$$t_s = \frac{t_o \Omega}{\Omega_o} \tag{13.2}$$

where

t_o = required time on target = n/PRF,
n = number of pulses integrated,

PRF = pulse repetition frequency,
 Ω = required angular region to be searched $\approx \theta_A \, \theta_E$
 θ_A = total azimuth coverage,
 θ_E = total elevation coverage,
 Ω_o = solid angular beamwidth $\approx \theta_a \, \theta_e$
 θ_a = azimuth beamwidth
 θ_e = elevation beamwidth.

To a first order, G_T and G_R can be defined as follows [1]:

$$G_T \approx \frac{4\pi}{\Omega_{oT}} \tag{13.3}$$

$$G_R \approx \frac{4\pi}{\Omega_{oR}} \tag{13.4}$$

where Ω_{oT} and Ω_{oR} are the solid angular beamwidths of the transmitting and receiving beams, respectively. In the monostatic and special bistatic case, when $G_T = G_R$,

$$G_T = G_R = \frac{4\pi}{\Omega}\left(\frac{t_s}{t_o}\right) \tag{13.5}$$

from (13.2) through (13.4). Thus, (13.1) can be written as

$$(R_T \, R_R)_{max} = \kappa (G_{Tv} \, G_{Rv})^{1/2} \left(\frac{\Omega}{4\pi}\right)\left(\frac{t_o}{t_s}\right) \tag{13.6}$$

Equation (13.6) represents the base case, bistatic surveillance range equation.
 Consider the following example, with

t_s = 10 s
t_o = 0.1 s
Ω = 120° x 30° = 1.1 rad²

From (13.5), $G_T = G_R = 30.6$ dBi. Thus, a monostatic radar with transmitting and receiving antenna gains of 30.6 dBi can search a 1.1 rad² region in 10 s, dwelling in each beam position for 0.1 s, out to maximum range $\sqrt{\kappa}$. A bistatic radar with the same antenna gains and no beam scan-on-scan losses, such that $G_{Tv} = G_T$ and $G_{Rv} = G_R$, can search the same region in the same time, out to a maximum range product κ, from (13.6).
 When floodlight and multibeam remedies are considered, bistatic range performance will equal, or be less than, this base case performance depending on the

required beam configuration, which in turn, defines $(G_{Tv}, G_{Rv})^{1/2}$. Five transmitting-receiving beam combinations are of interest, and are shown in Table 13.1, along with the base case.

Table 13.1
Possible Transmitting-Receiving Beam Configurations to Remedy Beam Scan-on-Scan Problems

		Transmitting Beam Configuration	
		One Beam	Flood Beam
Receiving Beam Configuration	One beam	Base case	Case 3 Case 3A*
	Multiple simultaneous beams	Case 1	Case 4
	Flood beam	Case 2	Case 5

*Tailored flood beam.

Case 1: One Beam T_x, Multibeam R_x

For this case, the single transmitting beam gain, G_{Tv}, is assumed to satisfy (13.5). When t_s/t_o multiple, identical, receiving beams are designed to cover the search region Ω, each receiving beam will have the gain of the single transmitting beam, and thus will also satisfy (13.5). Consequently, $(R_T R_R)_{max} = \kappa$, from (13.6). The penalty for this configuration is the cost of implementing t_s/t_o beams, receivers, and signal processors, where in the example $t_s/t_o = 100$. Because each receiving beam need process only those range cells illuminated by the transmitting beam at any time, the processing load can be reduced by identifying these cells, given the geometry and transmitting beam position, and time multiplexing the signal processor to operate in these range cells. Cost can be further reduced by time multiplexing a cluster of receiving beams (at RF), receiver (at IF or video), or signal processors (after A/D conversion) to cover only the volume illuminated by the transmitting beam at any time. These schemes are similar in concept to pulse chasing (Section 13.2), but do not require such fast scanning or multiplexing rates. Pulse chasing can be considered an upper bound to these multiplexing schemes, where only one or two beams, receivers, or processors are used. In any case these range and angle multiplexing schemes require a range cell or beam prediction algorithm similar to that of pulse chasing.

Case 2: One Beam T_x, Flood Beam R_x

Again, the single transmitting beam gain is assumed to satisfy (13.5). When the flood receiving beam covers the search region, $\Omega_o = \Omega$, and from (13.4), $G_{Rv} =$

$4\pi/\Omega$. Consequently, $(R_T R_R)_{max} = \kappa \sqrt{t_o/t_s}$. In the example, $(R_T R_R)_{max}$ is thus reduced by a factor of 10, and the equivalent monostatic range, $\sqrt{\kappa}$, is reduced by a factor of $\sqrt{10}$. In addition to this range penalty, mainbeam clutter levels are significantly increased and angle measurement accuracy and target resolution are decreased. As a result this remedy is not usually considered for surveillance.

Case 3: Flood Beam T_x, One Beam R_x

The flood transmitting beam is assumed to cover the search region, Ω, and the single receiving beam is assumed to satisfy (13.5), which is the reverse of case 2. Thus, again $(R_T R_R)_{max} = \kappa \sqrt{t_o/t_s}$. In contrast to case 2, however, mainbeam clutter levels are reduced by approximately Ω/Ω_{oR} and angle measurement accuracy and target resolution are restored. However, sidelobe clutter levels remain high, as outlined in Section 13.3. The Sanctuary bistatic radar test program used this beam configuration, actually with four receiving beams, and is detailed in Section 2.3.1. In special bistatic geometries, the flood transmitting beam can be tailored to illuminate only the region covered by the receiving beam when it is at a given pointing angle. Furthermore, if a minimum signal-to-noise power ratio required for detection, $(S/N)_{min}$, is acceptable at all bistatic ranges, the flood transmitting beam can also be tailored to satisfy this requirement. Thus, the term $\sqrt{t_o/t_s}$ can be eliminated. This special case is treated as case 3A, tailored flood beam T_x, one beam R_x. This case is analyzed at the end of this section because it is different from cases 1 to 5.

Case 4: Flood Beam T_x, Multibeam R_x

The flood transmitting beam is assumed to cover the search region, Ω, and t_s/t_o multiple receiving beams are designed to cover Ω, with each beam having a gain that satisfies (13.5). When the returns in each beam are integrated for t_o seconds, $(R_T R_R)_{max} = \kappa \sqrt{t_o/t_s}$. In this beam configuration, however, the integration time can, in concept, be extended to the surveillance frame time, t_s. Thus, much of the $\sqrt{t_o/t_s}$ range loss can be recovered. With these long integration times, targets can migrate through multiple range, doppler, and angle cells, which requires complex target association algorithms in the signal processor. Furthermore, integration efficiency is usually reduced because target returns decorrelate, which, in turn, requires noncoherent integration. This latter effect can be represented by the ratio E_{t_s}/E_{t_o}, where E_{ts} is the integration efficiency over time t_s and E_{t_o} is the integration efficiency over time t_o. Consequently, $(R_T R_R)_{max} = \kappa \sqrt{E_{t_s}/E_{t_o}}$. Finally, cost represents the major issue in this configuration because t_s/t_o beams, receivers, and signal processors (100 in the example) must be implemented, with no recourse to time multiplexing, as was possible for case 1.

Case 5: Flood Beam T_x, Flood Beam R_x

Both the transmitting and receiving beams are assumed to cover the search region Ω. When the returns in each beam are integrated for t_o seconds, $(R_T R_R)_{\max} = \kappa(t_o/t_s)$. When the returns in each beam are integrated for t_s seconds, as in case 4, $(R_T R_R)_{\max} = \kappa \sqrt{(t_o/t_s)(E_{t_s}/E_{t_o})}$. In this beam configuration, the multibeam receiving cost penalty is avoided, but at the expense of significantly reduced range performance, increased mainbeam clutter, and decreased angle measurement accuracy and target resolution. It is seldom considered for surveillance.

Case 3A: Tailored Flood Beam T_x, One Beam R_x [229]

The flood transmitting beam is assumed to illuminate only the region covered by the receiving beam at a given look angle; furthermore, the transmitting beam is assumed to generate a beam gain G_{Tv} such that the minimum signal-to-noise power ratio, $(S/N)_{\min}$, is constant at all bistatic ranges within that receiving beam. Equation (13.1) is used for this development, where the single receiving beam power gain is assumed to remain constant over all receiving look angles; thus, $G_{Rv} = G_R$ and (13.1) becomes

$$(R_T R_R)_{\max} = \kappa (G_{Tv}/G_T)^{1/2} \tag{13.7}$$

Substituting expressions for R_T and R_R, (3.18) and (3.19), into (13.7), recalling that $\beta = \theta_T - \theta_R$, and solving for G_{Tv}, yields

$$G_{Tv} = G_T \left(\frac{L^4}{\kappa^2} \right) \left(\frac{\cos^2\theta_T \cos^2\theta_R}{\sin^4(\theta_T - \theta_R)} \right) \tag{13.8}$$

When κ, L, and G_T are specified and a particular θ_R is selected, G_{Tv} will vary from a maximum of G_T if the product of the two terms in parentheses of (13.8) equals 1, to a vanishingly small value as $\theta_T \to 90°$. [Equations (3.18), (3.19), and (13.8) yield no solutions when θ_T or $\theta_R = \pm 90°$ because the bistatic triangle collapses to a line at these points. This problem can be finessed by specifying a minimum value of R_R, which for $\theta_R < 90°$ requires that $\theta_T < 90°$.]

An example geometry for the tailored transmitting flood beam approach is shown in Figure 13.2, where the bistatic radar operates in the cosite region (Section 4.4). In this example, $\kappa = (100 \text{ km})^2$, $L = 50$ km, $G_T = 40$ dBi, and the receiving beam is pointed at $\theta_R = -40°$. The transmitting beam pattern is superimposed on the figure. At $\theta_T = -14.5°$, $G_{Tv} = G_T = 40$ dBi, where the intersection of $\theta_R = -40°$ and $\theta_T = -14.5°$ is a point on the specified maximum range oval of Cassini, $(R_T R_R)_{\max} = \kappa = (100 \text{ km})^2$. The transmitting beam roll-off pattern is calculated by

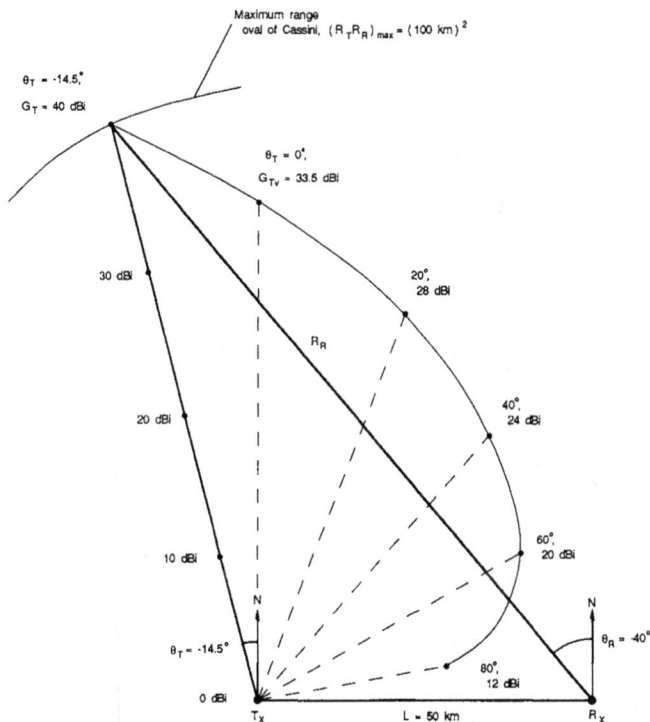

Figure 13.2 Example for tailored transmitting flood beam approach, case 3A, with tailored beam pattern superimposed on geometry for $\theta_R = -40°$ and $L_t = 1$; resulting S/N along R_R is a constant $(S/N)_{min}$.

using (13.8) and is shown on the figure in 20° increments. Because G_{Tv} satisfies (13.8) at all points along the receiving beam, except, of course, at $\theta_T = 90°$, $(S/N)_{min}$, which is a constant term in κ, will be generated at all points along the receiving beam. Note that this case is analogous to a monostatic air surveillance radar using a cosecant-squared antenna pattern, where the echo signal is independent of monostatic range for a constant altitude target [1].

The general case of a tailored transmitting flood beam is shown in Figure 13.3, where the required value of G_{Tv}, normalized with respect to $G_T L^4/\kappa^2$, is plotted with receiving look angle, θ_R, as a parameter. In the previous example, $G_{Tv} = G_T$ is represented as a horizontal line at $\kappa^2/L^4 = 12$ dB. The intersection of this line and the $\theta_R = -40°$ curve occurs at $\theta_T = -14.5°$, with the transmitting beam roll-off pattern continuing down the $\theta_R = -40°$ curve. This roll-off is identical to that shown in Figure 13.2.

Because the transmitting beam pattern varies for each θ_R, a transmitting phased array antenna with control of the element taper for each receiving beam

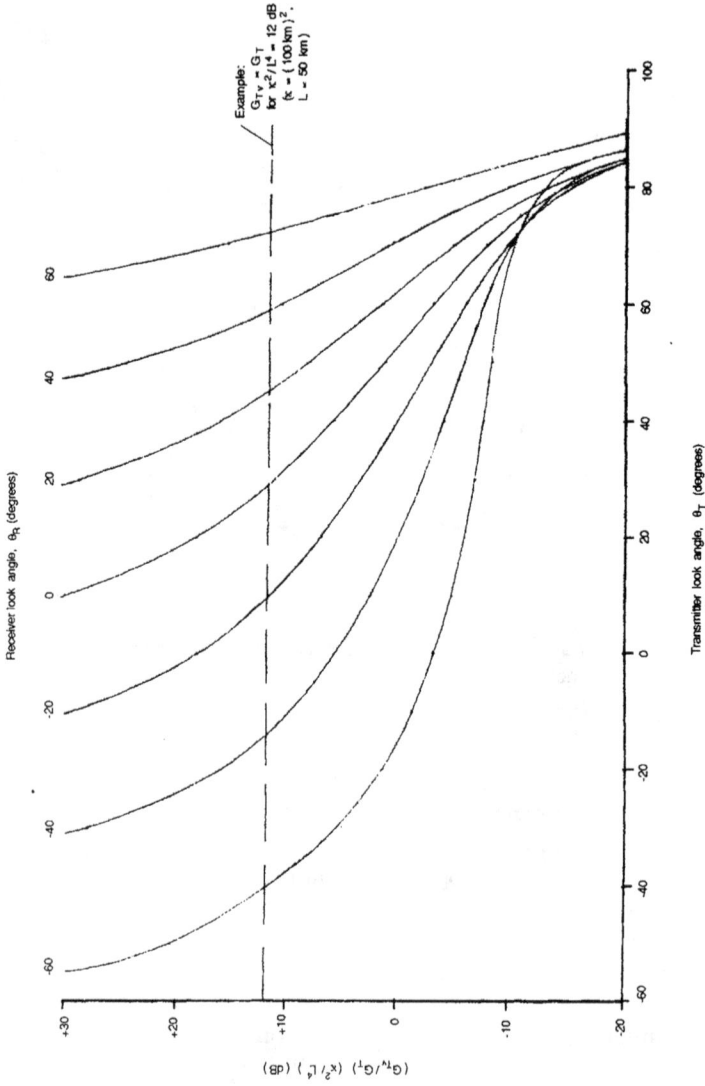

Figure 13.3 General case for tailored transmitting flood beam approach, case 3A, with variable θ_R and $L_t = 1$. (Courtesy of R. G. Martin, Westinghouse.)

dwell is a typical requirement for this scheme. This antenna taper will cause a modest reduction in peak transmitting gain, G_T, particularly when θ_R scans toward $-60°$. As shown in Figure 13.3 for the region near $\theta_R = -60°$, the transmitting beam roll-off on the right side of the beam peak is less than that of a typical $(\sin x)/x$ pattern; thus, the transmitting beam sidelobes must be increased and the nulls filled at the expense of peak gain. This taper loss, $L_t > 1$, will reduce κ by $\sqrt{L_t}$. Note that a similar scheme can be used to tailor a flood receiving beam, case 2; but, as with case 2, the mainbeam clutter levels are increased, and angle measurement accuracy and resolution are decreased. Thus, this tailored receiving beam scheme is not usually considered for surveillance.

Table 13.2 summarizes results for these beam scan-on-scan remedies. In short, no simple and inexpensive remedies are available for the beam scan-on-scan problem, without suffering a penalty in surveillance performance.

Table 13.2
Results for Beam Scan-on-Scan Remedies, Referenced to the Bistatic Maximum Range Product κ

Case	T_x	R_x	Performance	Issues
Base (Monostatic equivalent)	One beam (over Ω)	One beam (over Ω)	κ	Beam scan-on-scan, requiring pulse chasing
1	One beam	t_s/t_o beams (over Ω)	κ	Cost, unless time multiplexing is invoked, which requires cell-beam prediction algorithms
2	One beam	Flood (over Ω)	$\kappa\sqrt{t_o/t_s}$	Reduced range Increased clutter levels Decreased angle accuracy and resolution
3	Flood (over Ω)	One beam	$\kappa\sqrt{t_o/t_s}$	Reduced range
3A	Tailored flood	one beam	$\kappa\sqrt{L_t}$	Phased array transmitting antenna Constant $(S/N)_{min}$ at all bistatic ranges.
4	Flood	t_s/t_o beams, integration for t_s seconds	$\kappa\sqrt{E_{t_s}/E_{t_o}}$	Cost Target cell migration Modestly reduced range
5	Flood	Flood and integration for t_s seconds	$\kappa\sqrt{t_o/t_s}\cdot\sqrt{E_{t_s}/E_{t_o}}$	Significantly reduced range Increased clutter levels Decreased angle accuracy and resolution Target cell migration

13.2 PULSE CHASING

The concept of pulse chasing [49, 73, 126, 127, 129] has been proposed as a means to reduce the complexity and cost of multibeam bistatic receivers, which are one solution to the beam scan-on-scan problem, Section 13.1. The fundamental requirements for pulse chasing have been derived by Jackson [73]. This section is a summary and extension of Jackson's analysis. The simplest pulse-chasing concept replaces the multibeam receiving system (n beams, receivers, and signal processors) with a single beam, receiver, and signal processor. Figure 13.4 shows the geometry

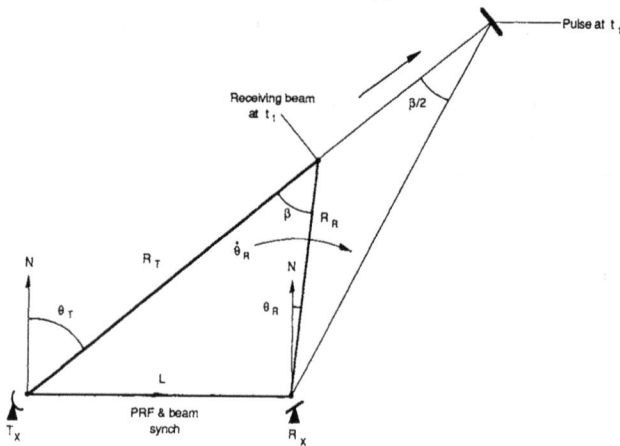

Figure 13.4 Pulse chasing for the single-beam continuous scan case. (Courtesy of IEE [73].)

for two dimensions on the bistatic plane. The single receiving beam rapidly scans the volume (or area in two dimensions) covered by the transmitting beam, essentially chasing the pulse as it propagates from the transmitter, hence, the term "pulse chasing." In addition to the usual requirements for solving the bistatic triangle, pulse chasing requires knowledge of θ_T and pulse transmission time [126], which can be provided to the receiving site by a data link. Alternatively, if the transmitting beam scan rate and PRF are uniform, the receiver can estimate these parameters as the transmitting beam passes by the receiving site [127]. Because the single beam chases one pulse at a time, a maximum allowable transmitting pulse repetition frequenty, PRF_1, exists such that only one pulse traverses the bistatic coverage area at one time. Calculation of PRF_1 is given in Section 5.7.

The receiving beam scanning rate must be at the transmitter's pulse propagation rate, modified by the usual geometric conditions. This rate, $\dot{\theta}_R$, is given by [73] and subsequently verified by Moyer and Morgan: *

$$\dot{\theta}_R = c \tan(\beta/2)/R_R \tag{13.9}$$

For operation in the cosite region (Section 4.4), $\dot{\theta}_R$ can vary from $1°/\mu s$ near the baseline to $0.01°/\mu s$ when $R_T + R_R > L$. (Typical contours of constant $\dot{\theta}_R$ are shown on Figure 14 of Jackson [73].) These rates and rate changes require an inertialess antenna such as a phased-array antenna and fast switching phase shifters. The phased-array receiving antenna must nearly always electronically scan in azimuth. When the transmitting beam is narrow in both azimuth and elevation, pulse chasing is required in azimuth and elevation in order to match the receiving beam to the transmitting beam volume, called the common beam volume, or common beam area in two dimensions. Normally a phased-array antenna used for surveillance is programmed to scan in increments of a beamwidth. Fractional shifts of a beamwidth can be achieved by changing the phase of a few (symmetric) pairs of phase shifters in the array. In this way, a pseudocontinuous beam scan can be generated, with the required rates and rate changes [128].

Because of pulse propagation delays from the target to the receiver, the pointing angle of the receiving beam, θ_R, must lag the actual pulse position, as shown in Figure 13.4. For an instantaneous pulse position that generates a bistatic angle $\beta/2$, $\theta_R = \theta_T - \beta$. In terms of the bistatic triangle, the required receiving beam pointing angle is [73]:

$$\theta_R = \theta_T - 2 \tan^{-1}\left(\frac{L \cos\theta_T}{R_T + R_R - L \sin\theta_T}\right) \tag{13.10a}$$

$$= \theta_T - 2 \tan^{-1}\left(\frac{\cos\theta_T}{e^{-1} - \sin\theta_T}\right) \tag{13.10b}$$

where e is isorange contour (ellipse) eccentricity.

The minimum receiving beamwidth, $(\Delta\theta_R)_{\min}$, required to capture all returns from a range cell intersecting the common beam area is approximated by [73]:

$$(\Delta\theta_R)_{\min} \approx [c\tau_u \tan(\beta/2) + \Delta\theta_T R_T]/R_R \tag{13.11}$$

where τ_u is the uncompressed pulse width. The approximation assumes that (1) the range cell can be approximated as $c\tau_u/2 \cos(\beta/2)$ and (2) respective rays from the transmitting and receiving beams are parallel at the common beam area. As discussed in Section 7.1, these assumptions are reasonable when the target is at long range from both the transmitter and receiver, $R_T + R_R \gg L$, and when the uncompressed range cell is small compared to the baseline, $L \gg c\tau_u$.

*Lee R. Moyer, *IEEE Trans. on Aerospace and Electronic Systems*, Vol. 38, No. 1, Jan. 2002, correspondence, p. 300.

Equation (13.11) shows that the required minimum receiving beamwidth changes as the receiving beam scans out the transmitting beam. Phased-array receiving antennas operating with a digital beamformer [218, 219] can accommodate this change. For other arrays, changing the receiving beamwidth "on the fly" becomes complex and costly. One compromise is to select a constant receiving beamwidth, $(\Delta\theta_R)_c$, that satisfies (13.11) for one geometry, and accept a loss in radar performance for other geometries. This loss is analogous to a beam scan-on-scan loss (Section 13.1), and can be defined with an example.

Assume that the bistatic radar uses a relatively short, uncompressed pulsewidth, so that $c\tau_u \tan(\beta/2)/R_R$ is small compared to $\Delta\theta_T R_T/R_R$ in (13.11). This situation is typical unless the target is close to the baseline, where $\beta \rightarrow 180°$. Then, (13.11) simplifies to $(\Delta\theta_R)_{min} \approx \Delta\theta_T R_T/R_R$. In this case, the compromise is to calculate the maximum value of R_T/R_R, i.e., $(R_T/R_R)_{max}$, required for anticipated target geometries and set the constant receiving beamwidth, $(\Delta\theta_R)_c = \Delta\theta_T(R_T/R_R)_{max}$. The loss in radar performance can be expressed as a beam mismatch loss (>1), which reduces the bistatic maximum range product κ by a factor of $\sqrt{(\Delta\theta_R)_c/\Delta\theta_R}$. For example, when $(R_T/R_R)_{max} = 2$ and $\Delta\theta_R = \Delta\theta_T$, κ is reduced by $\sqrt{2}$. (Expressions for defining a contour of constant bistatic range ratio are given in [73], where Figure 4 plots contours for $R_T/R_R = 1$, 1.2, 1.5, and 2.0. Incidentally, these contours are orthogonal to contours of constant bistatic angle.)

Other pulse-chasing concepts are possible. In one concept, the n (fixed) beam receiving antenna is retained with $\Delta\theta_R = \Delta\theta_T$ for each of the n receiving beams. Two receiver-signal processors (RSPs) are then time-multiplexed across the n beams. One RSP steps across the even-numbered beams and the other RSP steps across the odd-numbered beams, so that returns in beam pairs are processed simultaneously: (1,2), (2,3), (3,4), *etcetera*. This leapfrog sequence is required to capture all returns in the common beam area.

A second concept uses two beams and two RSPs step scanning over the required volume. It uses an identical leapfrog sequence. Both concepts relax the fractional beam scan requirements by either sampling or stepping the beams in units of a beamwidth. Because both dwell in each beam for twice the pulse propagation time across the beam, the beam dwell time, T_b, is approximately $2(\Delta\theta_R)_{min}R_R/c$ and the stepping rate is T_b^{-1}. The approximation assumes negligible phase shift delays and settling times. Either a dual manifold phased array or a digital beam-forming array [18, 19] can be used to generate the two leapfrog beams, with the former a less complex and costly implementation.

A third concept uses one RSP and one relatively broad beam, step scanning in increments of a half beamwidth over the required volume. The beamwidth must be $\geq 2(\Delta\theta_R)_{min}$ in order to capture all returns. Alternatively, the RSP can be time multiplexed across multiple interlaced beams, where the peak of one beam is positioned at the edge (3-dB point) of the adjacent beam, *etcetera*. Again, each beamwidth must be $\geq 2(\Delta\theta_R)_{min}$ in order to capture all returns.

In principle, pulse chasing operation is compatible with moving target indication (MTI) operation, for example, by sending successively chased pulses through a delay-line canceller. Phase synchronization between transmitter and receiver (Section 13.5), must, of course, be established for this operation. Particular attention must be paid to the pointing precision of the receiving beam for pulse chasing MTI operation. Specifically, the receiving beam must precisely retrace its scan pattern on successive pulse chasing sweeps to capture the same clutter samples over the MTI processing time. Otherwise, the range-limited clutter cell area (see Figure 5.15) will shift from sweep to sweep, which generates different samples of clutter, causing it to decorrelate. This situation is analogous to monostatic antenna scanning modulation, which limits the achievable MTI improvement factor [1]. In the bistatic case, both transmitting antenna scanning and receiving antenna scan retracing errors degrade the improvement factor.

Pulse chasing was identified in the 1960s [16], but was not tested until the 1970s, in the Coordinated Scan Bistatic Radar (CSBR) research and development test program [143]. CSBR used a multiple-beam Rotman receiving antenna and one RSP multiplexed across the multiple beams. The multiplexer was driven by a beam steering computer synchronized to the transmitter PRF. Tests were successful, but no long-range bistatic surveillance radars with multiple receiving beams were available for either development or retrofit at the time, and the CSBR program ended. A second implementation is planned for the Royal Signals and Radar Establishment Bistatic Radar Programme [144]. It was briefly described in the 1990 Military Microwaves Conference [233], but without pulse chasing test results. Reports of pulse chasing MTI operation are not available.

13.3 SIDELOBE CLUTTER

As with a monostatic radar, a bistatic radar must contend with sidelobe clutter. When both transmitter and receiver are ground based and separated by a baseline range L, only ground clutter from regions having an adequate LOS to both the transmitter and the receiver will enter the receiver antenna sidelobes. That region is defined for a smooth earth as the common coverage area, A_C. From (5.23), (5.24), and (5.26), when the height of target clutter, h_t, equals 0 and $L \geq (r_R + r_T) = 130(\sqrt{h_R} + \sqrt{h_T})$, A_C for ground clutter is zero, and no sidelobe (or mainlobe) clutter enters the receiving antenna. Targets with adequate LOS to both transmitter and receiver can be detected in a thermal noise-limited background. This situation is analogous to a monostatic radar detecting targets at ranges greater than r_T.

This development applies to ocean scenarios, but seldom is valid for land scenarios. In land scenarios variable terrain can decrease clutter levels by masking a clutter LOS when $L < r_R + r_T$, or increase clutter levels by generating a clutter LOS when $L > r_R + r_T$.

When the transmitter or the receiver is elevated or airborne, LOS restrictions are greatly reduced. However, two clutter problems unique to bistatic radars are encountered in this situation. The first occurs when a floodlight transmitting beam is used. To a first order, sidelobe clutter levels are reduced only by the one-way receiving antenna sidelobes, in contrast to two-way sidelobe clutter reduction for a monostatic radar.

The second problem occurs when the transmitter or receiver is moving, for example, when airborne. Now the bistatic clutter doppler returns skew and spread depending on the geometry for each clutter patch and the kinematics of the transmitting and receiving platforms. Doppler skew is defined in terms of isodoppler contours, or isodops, given by (6.5) for two dimensions (on the bistatic plane). The skew is range- and angle-dependent. The range-dependent skewing effect is not present in an airborne monostatic radar. Clutter spread in a particular sidelobe range cell is centered on the doppler skew present in the range cell, and is given by (6.12), again for two dimensions.

These skewing and spreading effects, along with increased clutter levels, can greatly complicate the ability of a bistatic radar to detect targets in clutter. Remedies include conventional doppler filtering and high time-bandwidth waveforms; the judicious use of masking when available; control of the geometry, especially when a dedicated or cooperative transmitter is available; design of very low receiving (and transmitting when possible) antenna sidelobe levels; sidelobe blanking of discrete clutter returns; range or range-doppler averaging in the constant false alarm rate (CFAR) unit for homogeneous clutter; and spatial excision of clutter returns. One implementation of this last technique relies on knowledge of the geometry and kinematics to predict the clutter doppler skew and spread in a given area. Then a filter or gate is set to excise mainbeam clutter returns in that area. The amount of range-doppler space excised by this procedure can be as high as 8% [45].

13.4 TIME SYNCHRONIZATION

Time synchronization is required between the bistatic transmitter and receiver for range measurement. Timing accuracy on the order of a fraction of the transmitter's (compressed) pulse width is also a typical requirement. Time synchronization can be accomplished directly by receiving a signal from the transmitter, demodulating the signal, if necessary, and using the demodulated signal to synchronize a clock, or local oscillator, in the receiver. The transmitting signal can be sent via land line, communication link, or directly at the transmitter's RF if an adequate LOS exists between transmitter and receiver. If an adequate LOS is not available, it can be sent via a satellite communication link or via a scatter path, if the scatterer has adequate LOS to both the transmitter and receiver [49]. In this case the scatterer must lie in the common coverage area, as defined by (5.26). Transmission via tropospheric scatter can also be used in special cases [50]. In all of these direct time synchronization schemes, implementation is straightforward, much like the initial synchronization process in communication systems. They can also be used for any type of transmitting PRI modulation: stable, staggered, jittered, and random. With time synchronization established, target range is calculated via (5.1) or similar methods.

For stable PRIs, time synchronization can be accomplished indirectly by using identical stable clocks at the transmitting and receiving sites. The clocks can be synchronized (i.e., matched) periodically. Direct time synchronization methods can be used for this task, for example, whenever the transmitter and receiver are within LOS or located together, if one or both are mobile. Alternatively, the clocks can be slaved to a second source, such as the Navstar global positioning system (GPS) or Loran C [50, 82, 130]. Indirect time synchronization can also be used with a dedicated or cooperative transmitter using random PRIs if a random code sequence is established *a priori* and is known by the receiving site.

For direct time synchronization, the required clock stability between updates is, to a first order, $\Delta\tau/T_u$, where $\Delta\tau$ is the required timing accuracy and T_u is the clock update interval. The update interval typically ranges from a minimum of the transmitter's interpulse period to a maximum of the transmitter's antenna scan period. The former usually requires a dedicated link between transmitter and receiver; the latter can be implemented whenever the transmitting beam scans past the receiving site, given an adequate LOS, and is sometimes called "direct breakthrough" [130].

For example, when the required timing accuracy, $\Delta\tau$, is specified as 0.1 μs and the clock update interval, T_u, is the transmitter's antenna scan period of 10 s, required lock stability is 10^{-8}, or one part in 10^8 over 10 s. Figure 13.5 plots typical stability data for commercially available (*circa* 1980–1990), high performance clocks. Fractional frequency stability is defined as the square root of the two sample, or Allan variance, $\sigma_y^2(\tau)$, where τ is the averaging time, or duration of each sample average. The curves are drawn to show general trends in stability, and do not represent the capability of a single device. Because quartz oscillators are an integral part of rubidium and cesium clocks, the rubidium and cesium curves converge to the quartz curve for $\tau \lesssim 10$ s, where the quartz oscillator establishes stability performance for all clocks. For the example, a quartz clock will satisfy the 10^{-8} stability requirement for $\tau = 10$ s.

When direct breakthrough time synchronization is used, multipath and other propagation anomalies, as well as radio-frequency interference (RFI), will degrade the accuracy of updating. Errors of ± 1 μs have been measured when a direct LOS is available [131]. They increase to ± 5 μs over a tropospheric propagation path [130]. Thus, in the above example, propagation errors dominate any error contribution from the clock.

Because two clocks are used in indirect time synchronization, clock stability is, to a first order, $\Delta\tau/2T_u$. For example, when $\Delta\tau = 0.1$ μs and $T_u = 5$ h, required clock stability is 2.8×10^{-12}, or about 4 parts in 10^{11} over 5 h (1.8×10^4 s). From

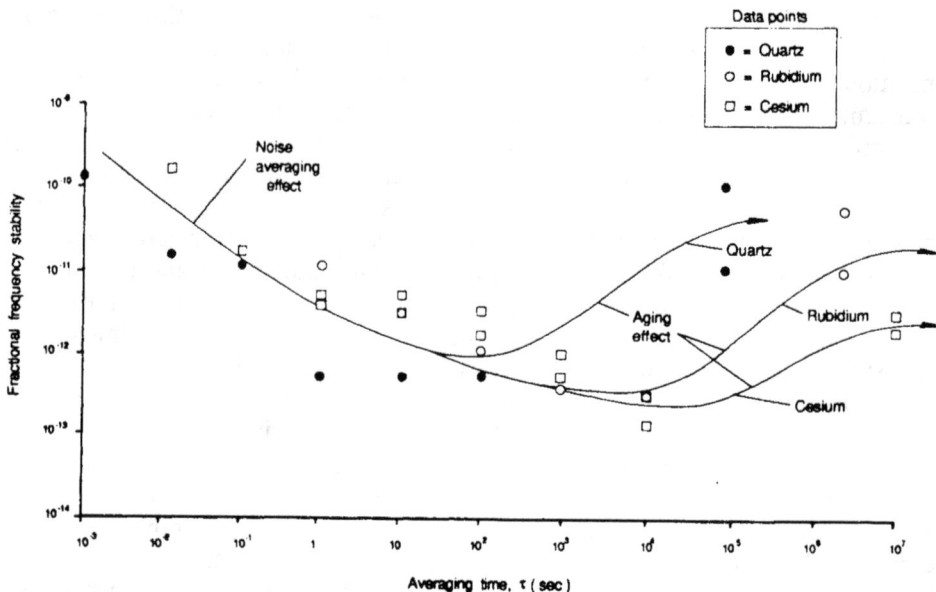

Figure 13.5 Local oscillator stability data and trends for commercially available (1980-1990) standards.

Figure 13.5, a rubidium or cesium clock is required to satisfy this long-term requirement. If the stable clocks are slaved to a second source, estimated timing accuracies of 0.5 μs for Loran-C, and < 0.1 μs for Navstar GPS are reported [82].

13.5 PHASE SYNCHRONIZATION AND STABILITY

As with monostatic radars, doppler or MTI processing can be used by the bistatic receiver to reject clutter or chaff, given that phase coherence, or synchronization, can be established between clocks, or local oscillators, in the transmitter and receiver. If noncoherent MTI is acceptable for clutter rejection, the bistatic receiver can use a clutter reference to establish phase synchronization, exactly as a monostatic radar would, whenever a clutter patch is visible to both the transmitter and receiver.

In one bistatic noncoherent MTI implementation, called phase priming, an oscillator at the receiver was phase-synchronized at the PRF rate by using a small sample of close-in clutter returns [132]. Phase coherence was obtained within about 10 μs and extended over 1 ms. The process was found insensitive to the clutter signal level, but quite sensitive to pulse-to-pulse fluctuations in the clutter signal.

If coherent processing is required, phase synchronization can be established with methods similar to those used for time synchronization: directly by phase locking the receiver to the transmitting signal, or indirectly by using matched local oscillators in the transmitter and receiver, which are again synchronized (i.e., matched) periodically to remove any long-term frequency drift. Phase accuracy, or stability, requirements are the same as those for coherent processing by a mono-static radar. They can range from less than one degree to many tens of degrees of RF phase over a coherent processing interval, depending on the type and duration of coherent processing.

Direct phase synchronization can be implemented as in direct time synchro-nization: via land line, communication link, or directly at RF. If a direct RF link is used, adequate transmitter-to-receiver LOS is again required, but it is subject to multipath. It is also subject to phase reversals if coherent operation is required across transmitter antenna sidelobes. However, this latter problem can be over-come by a Costas loop for phase reversals near 180° [134]. An extension of direct path phase locking is the use of the direct path signal as a reference signal in a correlation processor [135].

For direct phase synchronization on a pulse-to-pulse basis [133], oscillator stability is $\Delta\phi/2\pi f \Delta T_{rt,}$ where $\Delta\phi$ is the allowable rms phase error (in radians), f is the transmitter frequency (in hertz), and $\Delta T_{rt,}$ is the difference in propagation time (in seconds) between the transmitter-target-receiver path and transmitter-receiver (direct) path; see Figure 5.1(b). For example, in a bistatic synthetic aperture radar, which is sensitive to high frequency or "sinusoidal" phase errors over $\Delta T_{rt,}$, $\Delta\phi \approx$ 4°= 0.07 rad [133]; thus, for f = 10 GHz and ΔT_{rt} = 1 ms, required oscillator stability is approximately 10^{-9}, or 1 part in 10^9 over 1 ms. From Figure 13.5, a quartz oscillator will satisfy this stability requirement.

For indirect phase synchronization using matched local oscillators in the transmitter and receiver, phase stability is required over the coherent integration time, T. Thus, oscillator stability is $\Delta\phi/2\pi f T$. In the bistatic SAR example, pro-cessing is now sensitive to lower frequency or "quadratic" phase errors over T, and $\Delta\phi$ can be relaxed to 90° [133]. Thus, for T = 1 s and f = 10 GHz, required stability is 2.5 x 10^{-11}, or 4 parts in 10^{10} over 1 s, which again can be satisfied by a quartz oscillator.

In most types of SAR images, the Integrated Sidelobe Ratio (ISLR) is an important criterion for image quality. It is a measure of the energy from a particular target that appears at image locations other than that corresponding to the target. Typically a -30 to -40 dB ISLR allocation for local oscillator phase noise is desired [53]. When a single oscillator is used, as in the monostatic case, these levels can be achieved for reasonably long coherent integration terms (T < 10 s) because the high level of phase noise near the carrier frequency, i.e, the low-frequency components of the phase noise power spectrum, are partially cancelled in the demodulation process. Thus, in this monostatic case, ISLR levels are typically set by the high-frequency "floor" of the phase noise power spectrum. However, for the bistatic case, both phase synchronization techniques use two oscillators with uncorrelated phase noise power spectra; thus, these low-frequency components do not cancel, which, in turn, set the ISLR levels.

Specifically, for both monostatic and bistatic SARs, the ISLR contribution due to phase noise, in decibels, is given as [53]:

$$\text{ISLR} = 10 \log \int_{1/T}^{\infty} S(f) df \tag{13.12}$$

where $S(f)$ is the power spectrum of the phase noise. For a monostatic radar, $S(f)$ is

$$S_M(f) = 2M^2 (2\sin\pi f \tau_M)^2 L(f) \tag{13.13a}$$

where M is the ratio of the RF to the local oscillator frequency, τ_M is round-trip propagation time, and $L(f)$ is the single-sideband phase noise of the local oscillator. The sin $(\cdot)$ term in (13.13a) occurs because one local oscillator is used in the demodulation process, and results in partial cancellation of low-frequency components of the phase noise spectrum.

For bistatic radar, $S(f)$ is

$$S_B(f) = 4M^2 L(f) \tag{13.13b}$$

The extra factor of 2 arises from the addition of two uncorrelated but identical power spectra; the sin $(\cdot)$ term is absent because the spectra are uncorrelated. Thus, for the bistatic case, (13.12) becomes

$$(\text{ISLR})_B = 10 \log \int_{1/T}^{\infty} 4M^2 L(f) df \tag{13.14}$$

For high-quality quartz oscillators, $L(f) \approx -160$ dBc/Hz (with respect to the carrier in a 1 Hz bandwidth) at $f \geq 1$ kHz. For $f \leq 10$ Hz, the region of interest for SAR operation, $L(f)$ increases 30 dB/decade of frequency decrease [53]. In this region, $L(f)$ is modeled as

$$L(f) = L_1 f^{-3} \tag{13.15}$$

where L_1 is the value of $L(f)$ at $f = 1$ Hz for a specific oscillator. Thus

$$(\text{ISLR})_B = 10 \log (2M^2 L_1) + 20 \log T \tag{13.16}$$

Figure 13.6 plots (13.16) for typical, high-quality, 10 MHz quartz oscillators with $L_1 = -90$ and -100 dBc/Hz, respectively. The RF is assumed to be 9.3 GHz, so $M = 930$. When a -30 dB threshold is set for acceptable ISLR levels, the coherent integration time T cannot exceed 2.4 s for $L_1 = -100$ dBc/Hz.

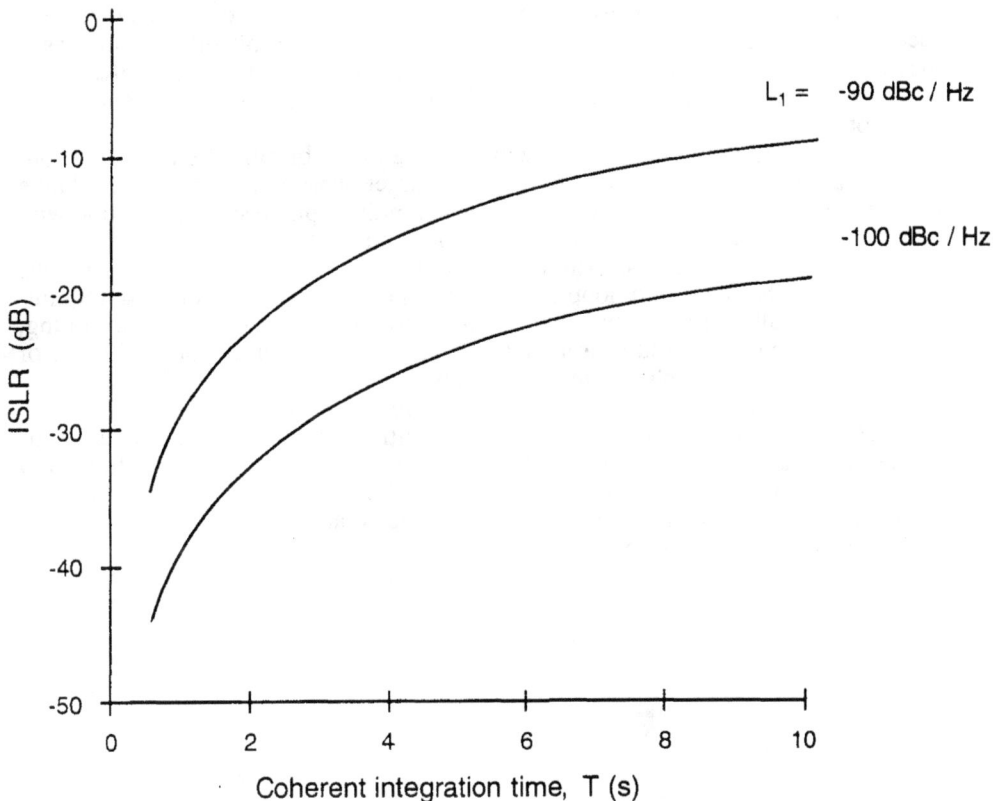

Figure 13.6 Bistatic SAR Integrated Sidelobe Ratio (ISLR) using matched quartz local oscillators with uncorrelated power spectra and $M = 930$.

The foregoing analysis assumed that the phase noise power spectra of the two bistatic local oscillators were uncorrelated. Because the low-frequency components of the spectrum generate the high ISLR for long T [the missing sin($\cdot$) term in (13.13b)], methods analogous to monostatic SAR autofocus* might be used to correlate these low frequency components. For bistatic operation, any phase noise in the transmitter's local oscillator will modulate clutter returns in each bistatic range cell (or, alternatively, in each subaperture). The low-frequency phase component of these

---

*Autofocus is a technique that extracts residual phase errors from the partially processed SAR data in each range-doppler cell or in each subaperture, and then removes them prior to the final SAR image formation. Autofocus is typically required for fine resolution monostatic SAR images with $T > 10$ s [232].

returns can, in concept, be measured by the receiver and then used to synchronize the receiver's local oscillator on a cell-by-cell or subaperture-by-subaperture basis, yielding partially correlated phase noise power spectra for the two local oscillators. In this case, T can be extended. Reports of this bistatic autofocus concept apparently have not been published.

When direct phase synchronization is used by a bistatic SAR, any motion-compensation phase shift needed to track the target phase must correct for relative motion between transmitter and receiver. When indirect phase synchronization with matched local oscillators is used, this correction is not required [133].

Either time or phase errors can dominate synchronization requirements, depending on the range and doppler accuracies needed. While all of these requirements can usually be met, implementation is more complicated, time consuming, and costly when compared to a monostatic system, which uses one oscillator, or clock, for both time and phase synchronization.

Finally, the monostatic single oscillator configuration allows reasonably long ($T < 10$ s) SAR coherent integration times without the use of autofocus techniques, whereas the bistatic dual oscillator configuration is currently constrained to relatively short ($T < 3$ s) coherent integration times. Perhaps this constraint can be relaxed through the use of the bistatic autofocus, and the results will be tested during the next bistatic resurgence.

Appendix A
EARLY PUBLICATIONS OF BISTATIC
RADAR PHENOMENOLOGY

A.1 INTRODUCTION

The earliest documentation of bistatic radar phenomenology appears to have been by British Post Office (herein denoted as GPO) engineers in 1932 [169]. Similar documentation was published in the United States by Bell Telephone Laboratory (BTL) engineers in 1933 [156], followed by a U.S. patent in 1934 issued to Naval Research Laboratory engineers [21]. Up until 1934 the NRL work had been classified, but when the BTL results were published, NRL quickly declassified their work and applied for a patent. This appendix reviews these three publications to place them in the context of later bistatic radar developments, and to round out the early history of bistatic radars. They are reviewed in reverse order of publication, simply because more information is contained in the later ones.

A.2 1934 NRL PATENT

Six figures were included in the NRL patent [21] by Taylor, Young, and Hyland, all of which show a bistatic radar configuration. They are reproduced as Figures A.1 through A.6. Each figure shows the bistatic geometry and an associated signal amplitude versus time plot, but no ranges or other units are specified, except for "max" and "min" signal levels on Figures A.1 and A.2. A cw transmit waveform was specified for all configurations, but with no mention of frequency. Figure A.1 shows a bistatic transmitter emitting both sky and ground (direct path) waves, with no target present. The receiver detects the constant amplitude ground wave, as shown as a straight line signal level versus time plot. Figure A.2 shows the same configuration, but now with a target present. A sky wave is "reradiated" from the target, arrives at the receiving site, and generates interference with the ground wave. This effect was not characterized in the patent, but as early as 1947 Fink [161]

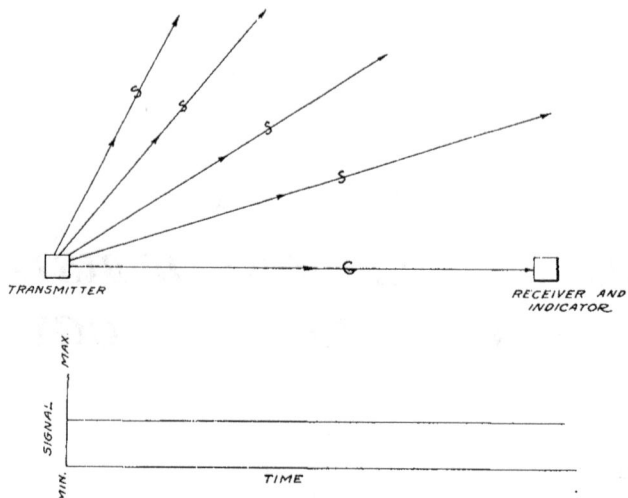

Figure A.1 Figure 1 of NRL patent.

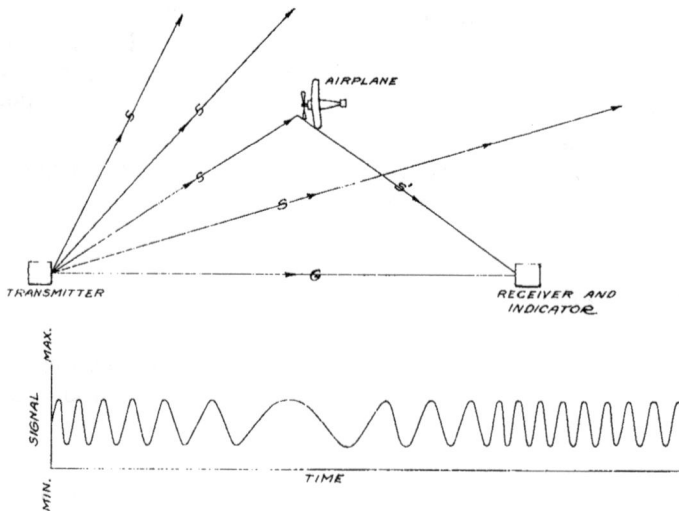

Figure A.2 Figure 2 of NRL patent.

ascribed "wave interference" to Taylor and Young in their 1922 and 1930 experiments. Later, Skolnik [1] reports the effect as "cw wave-interference."

The interference effect shown in the signal amplitude versus time plot of Figure A.2 is the beat frequency, or beat note, between the constant frequency, ground wave signal, and the doppler-shifted signal "reradiated" from the moving target. It can be calculated directly from the bistatic target doppler equation, (6.4b) of Chapter 6:

$$f_{Tgt} = (2V/\lambda) \cos\delta \cos(\beta/2) \qquad (A.1)$$

where

f_{Tgt} = target bistatic doppler at the receiving site,
$V \cos\delta$ = target velocity vector projected onto the bisector of the bistatic angle,
V = target velocity,
δ = angle between target velocity vector and bistatic bisector,
λ = wavelength.

For example, assume $\lambda = 5$ m ($f = 60$ MHz), $V = 120$ mph $= 54$ m/s, with the airplane flying parallel to the baseline of length L and offset from the baseline by $0.32\,L$ as is the approximate geometry shown in Figure A.2. When the airplane is some distance to the right of the receiver, both β and δ are small angles, and $f_{Tgt} \approx 2V/\lambda \approx 22$ Hz. As the airplane passes the receiver, $\delta = 54°$, $\beta = 72°$, and $f_{Tgt} = 10$ Hz. When the airplane is midway between the transmitter and receiver, its position on Figure A.2, $\delta = 90°$ and $f_{Tgt} = 0$. The target doppler becomes negative as the airplane continues toward the transmitter because $\delta > 90°$. When the airplane passes the transmitter, $f_{Tgt} = -10$ Hz, and when it is some distance to the left of the transmitter, $f_{Tgt} \approx -22$ Hz. That is, the absolute value of the target doppler is symmetric about the perpendicular bisector of the baseline for this geometry. Although not quantified in Figure A.2, the signal frequency approximates these doppler characteristics: maximum doppler at or beyond either the transmitter or the receiver, and minimum (near zero) doppler at the midpoint. A negative doppler is not shown because the probable receiving configuration was a homodyne receiver using the external ground wave signal as a local oscillator. A homodyne receiver is sometimes characterized a superheterodyne receiver with zero IF. In this configuration the doppler-shifted target signal is heterodyned in the detector, or mixer, with the ground wave signal to produce the doppler beat note. The sign of the doppler beat note is lost in the heterodyning process [1].

Figure A.3 shows a more complex interference pattern, which is characterized in the patent as multiple "reradiations" from the target. NRL ascribes this effect to

268

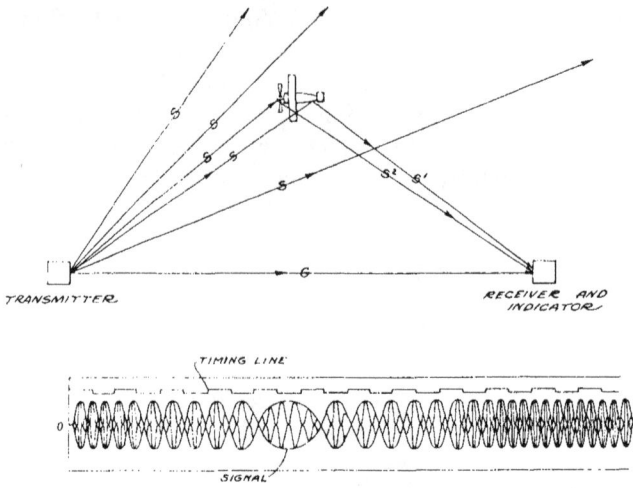

Figure A.3. Figure 3 of NRL patent.

a change in polarization as the aircraft's propeller rotates. They also suggest that this effect can be used for "identification of intervening or adjacent objects," for example, aircraft versus automobiles, tanks, or ships.

Figure A.4 shows a modification of the first configuration where the ground wave is attenuated, leaving the reradiated sky wave as the dominant signal. Now

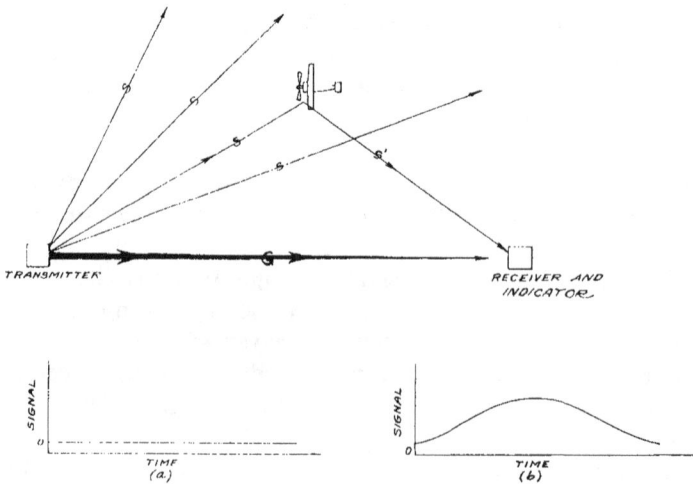

Figure A.4 Figure 4 of NRL patent.

the interference pattern is suppressed, and a smooth change in signal amplitude occurs as the target flies parallel to the baseline and past the two sites. The shape of the curve was not discussed in the patent. However a similar shaped curve can be produced by taking a straight line cut across contours of constant signal-to-noise ratios, or ovals of Cassini, shown in Figure 4.2 of Chapter 4. Although ovals of Cassini cannot be reconstructed for the NRL configuration, because the bistatic radar constant K is not known, the change in signal amplitude can be characterized as follows.

When the cut is parallel to the baseline and displaced some distance from the baseline, which is approximately the geometry shown in Figure A.4, the signal amplitude is low when the target is beyond the receiver. It rises monotonically to a relatively flat peak when the target is midway between transmitter and receiver, i.e., when it crosses the perpendicular bisector of the baseline. It then falls monotonically as the target flies past the transmitter. Because the ovals are symmetric about the perpendicular bisector of the baseline, the signal amplitude curve is also symmetric about its peak, as shown in Figure A.4. Thus, while the relationship between changes in signal amplitude and target position was not discussed in the patent, the phenomenon, which was likely based on observations, is clearly in evidence, as shown in Figure A.4.

Two different bistatic configurations are shown in Figures A.5 and A.6. The first is a shoreline system with an apparent directional transmitting antenna for detecting ships at small bistatic angles. The second is a cross channel (or a ship-to-ship) system also for detecting ships, but now at very large bistatic angles. For the

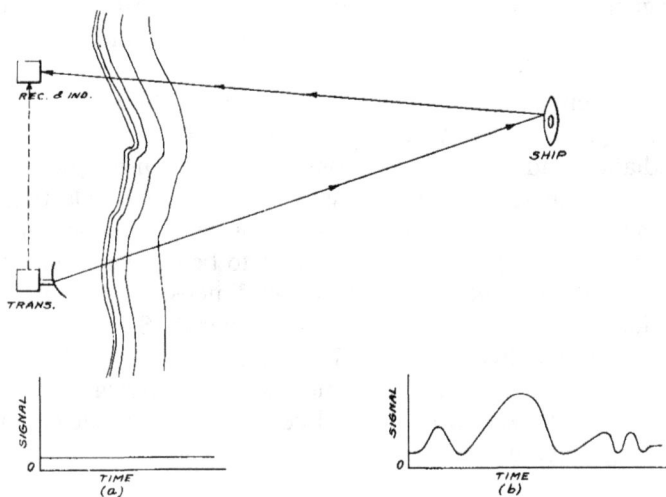

Figure A.5 Figure 5 of NRL patent.

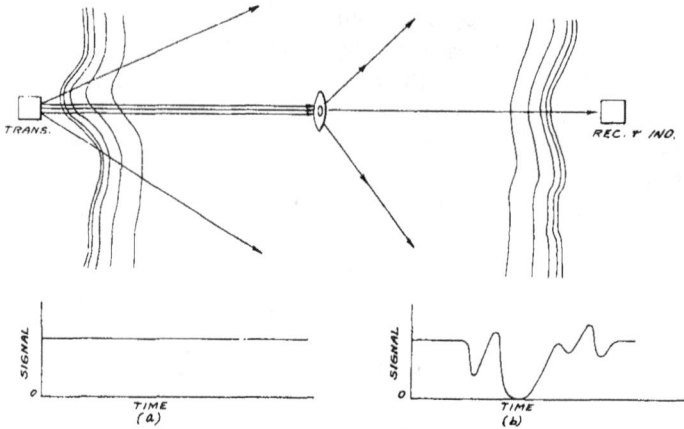

Figure A.6 Figure 6 of NRL patent.

second configuration, the transmitting antenna can be either directional or nondi-rectional. NRL stated that "a violent fluctuation takes place in the received signal," when a ship comes between the transmitter and receiver. "This fluctuation is dis-tinguished by an absence of signal for certain positions of the ship." Although not known at the time, this effect appears to be the forward-scatter RCS enhancement phenomena with its characteristic $4\pi A^2/\lambda^2$ peak, and deep nulls off this peak. An example of the forward-scatter pattern is shown in Figure 8.3 of Chapter 8. The absence of signal could also be caused by the ship physically blocking the trans-mitter-to-receiver LOS, as discussed by Page in his review of the early Taylor and Young experiments [162].

The NRL patent also claimed that their systems exploited target "reradiation" of the incident signal. Taylor, Young, and Hyland took great pains to distinguish between reradiation and reflection RCS phenomena because a previous patent for an aircraft radio altimeter described ground scattering as a reflection. They may have been concerned that their patent would not be granted without such a dis-tinction. This reradiation phenomenon appears to be similar to target resonance, where the wavelength is of the order of the target dimension. Of course, both effects can occur simultaneously, along with forward-scatter RCS enhancement, depend-ing on target size, shape, frequency, and bistatic angle.

Finally, the NRL patent briefly mentioned an electromagnetic impulse wave-form to replace the CW waveform. It could be used for both "detection and prox-imity" (i.e., range measurement). This statement appears to be a precursor of pulsed radars.

In summary, the NRL patent defined a bistatic radar, using a CW transmitter and both directive and broad beam antennas, to detect moving targets, such as aircraft and ships. The patent showed bistatic geometries with both small and large bistatic angles for multiple uses; the characteristic (forward-scatter) signal fluctuations at large bistatic angles; and the characteristic signal frequency (doppler) and amplitude changes as a function of time and, by implication, bistatic geometry. It did not, however, analytically characterize these changes. In fact, the patent disclosure appears to have been based entirely on experimental work. The theoretical basis for bistatic—and of course monostatic—radar would have to await further experiments and a pondering of the data.

A.3 1933 BTL PAPER

The BTL paper [156] reported results of propagation experiments at 64 to 81 MHz (λ = 4.7 to 3.7 m) over distances up to 125 miles. The effects of terrain reflection and diffraction were measured over a variety of propagation paths. The transmitter was a broadcast station ("Station W2XM") with a fixed location at the BTL Holmdel Laboratory in New Jersey. The receiver was a double detection receiver installed in an automobile (or airplane) and driven (flown) up to 125 miles from the transmitting site.

In the process of reporting and analyzing their propagation measurements, BTL included a short, almost parenthetical, section on their accidental observations of the effects of an airplane, a "moving, conducting body" on the measured "radio field." This section clearly captured NRL's full attention. It is reproduced as follows:

FIELD FLUCTUATIONS FROM MOVING BODIES

It is well known that the motion of conducting bodies, such as human beings, in the neighborhood of ultra-short-wave receivers produces readily observable variations in the radio field. This phenomenon extends to unsuspected distances at times. Thus, while surveying the field pattern in the field described above, we observed that an airplane flying about 1500 feet (458 meters) overhead and roughly along the line joining us with the transmitter, produced a very noticeable flutter, of about four cycles per second, in the low-frequency detector meter. We then made a trip to the near-by Red Bank, NJ, airport, distant about 5½ miles (8.8 kilometers) and observed even more striking reradiation phenomena. Near-by planes gave field variations up to two decibels in amplitude, and an airplane flying over the Holmdel Laboratory and towards this landing field was detected just as the Holmdel operator announced "airplane overhead." These were all fabric wing planes. If the reradiation field to which such an airplane is exposed is of inverse distance amplitude type while the directly received ground fields are of more nearly inverse distance square type, . . . it is easy to see that at five miles an overhead airplane is exposed to a field intensity about ten times (20 decibels) that existing at the ground, and for ordinary airplane heights a high energy transformation loss in the reradiation process can occur and still give marked indications in the receiver meter. This airplane reradiation was noticed at various subsequent times, sometimes when the airplane itself was invisible. A set of theoretical beat frequency versus distance curves are given in Fig. 7.

The inverse distance and inverse-distance-squared propagation loss cited by BTL refers to measured fields in terms of millivolts per meter. These measurements are equivalent to the more commonly used R^{-2} and R^{-4} propagation losses in terms of power, where R^{-2} is the one-way free-space loss and R^{-4} is the one-way space loss in the presence of diffraction.

The curves shown on BTL's Figure 7, reproduced here as Figure A.7, can be replicated almost exactly by again using the bistatic target doppler equation, (A.1),

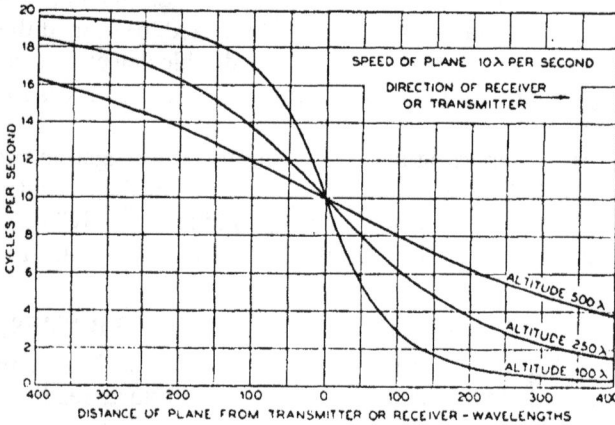

Figure A.7 —Beat frequencies produced by reflection from a moving airplane. (© 1933 IRE [156].)

and the geometry approximation shown in Figure A.8. When the baseline range L is much greater than the airplane altitude, as implied in Figure A.7, the baseline and airplane flight path are nearly colinear. Thus, $\delta \approx 180° - \beta/2$, and

$$f_{\text{Tgt}} = \frac{2V}{\lambda} \cos^2(\beta/2) \tag{A.2}$$

Because $V = 10 \lambda$ per second, as specified in Figure A.7, (A.2) becomes (ignoring the sign of the doppler):

$$|f_{\text{Tgt}}| \approx 20 \cos^2(\beta/2) \tag{A.3}$$

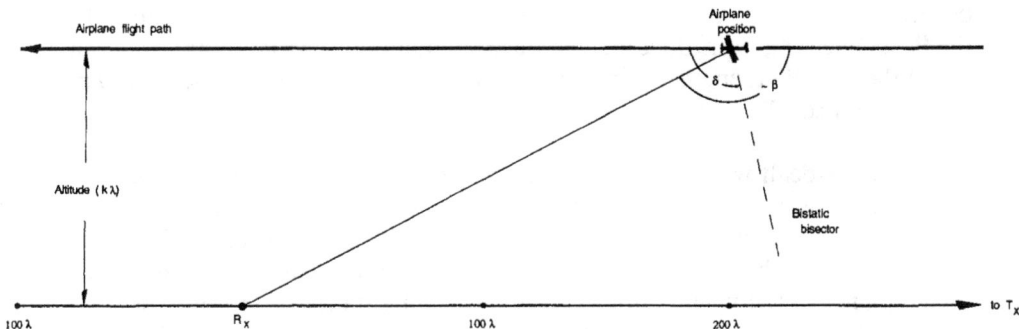

Figure A.8 Geometry (vertical profile) approximation for BTL observations [156] and calculations of Figure A.7.

For example, when the airplane is at 100λ altitude and 200λ ground range from the receiver, which is the geometry of Figure A.8, $\beta/2 \approx 76.7°$ and $|f_{Tgt}| \approx 1.1$ Hz. When the airplane is directly over the receiver, at "zero" ground range, $\beta/2 \approx 45°$ and $|f_{Tgt}| \approx 10$ Hz. In fact, for all airplane altitudes $\ll L$, $|f_{Tgt}| \approx 10$ Hz at this point. In the limits of this geometric approximation, $|f_{Tgt}| \to 0$ and 20 Hz because $\beta \to 180°$ (to the right) and $0°$ (to the left), respectively, of the receiver for all altitudes. These points match those on Figure A.7, which shows that the beat frequency phenomenon is a manifestation of the bistatic doppler effect.

Two cautionary notes should be made about the beat frequency curves of Figure A.7. First, the curves are approximate because the bistatic angle is an approximation. (It should be referenced to the target-to-transmitter LOS rather than the airplane flight path.) Second, the curves are shown approaching $f_{Tgt} = 0$ Hz as the airplane approaches the transmitting site. However, when the exact geometry is used, $f_{Tgt} = 0$ Hz midway between the transmitting and receiving sites, and then repeats a mirror image of the curves shown in Figure A.7 as the airplane passes the transmit site—with appropriate corrections for the bistatic angle approximation. Thus, the curves are a reasonable representation of half the beat frequency profile, but only when the aircraft altitude is a small fraction of the baseline.

A.4 1932 GPO PAPER

Although the British GPO paper [169] is not readily available, both Swords [159] and Watson-Watt [170] have reported its contents. Swords reports that observations of the beat frequency effect, which was also called the flutter effect, were made in 1931 by Post Office engineers on their 60-MHz experimental communication link between Colney Heath and Dollis Hill. The flutter in the received signal occurred when aircraft from Hatfield aerodrome flew in the vicinity. In June 1932 the Post Office engineers documented their serendipitous observations [169], much as did the U.S. BTL engineers in 1933 [156]. Relevant parts are extracted and analyzed by Watson-Watt, as follows [170]:

> ... Part V dealt with "Interference from Aeroplanes", and reported that the five-metre wave-length signals from a transmitter 14 miles distant from a receiving station at Colney Heath, and partly screened off by an earth hill on the direct line between the stations, were interfered with by the passage of aeroplanes in the neighbourhood of the receiving station. "The only feasible explanation of the phenomena seems to be that interference is set up between the directly received waves and those reradiated from the aeorplane. . . . The beats [surges of intensity] usually recurred at a rate between five and fifteen per second. An aeroplane travelling at 88 feet per second directly towards the antenna would [on the theory adopted to explain the effects] give beats at a rate of 11 per second—It should be noted that this type of interference which has been noted on many occasions . . . [has also] been experienced from distant aeroplanes." No maximum distances were reported; the specimen log reproduced in the report shows distances up to 2½ miles maximum, and flying heights up to 2,000 feet. The trend of the report does not go beyond treating these effects as being an interesting (and quantitatively explicable) nuisance.

The geometry of the British Post Office observations and the U.S. BTL observations appears to be identical. Using the approximate geometry of Figure A.8 with the corresponding equation (A.2) to calculate the beat, or doppler frequency $|f_{Tgt}|$, and assuming $V = 88$ ft/s $= 26.8$ m/s, yields $|f_{Tgt}| = 10.7 \cos^2(\beta/2)$ Hz. The maximum beat frequency occurs when the aeroplane approaches the receiver on the extended baseline at either low altitude or at long range so that $\beta \rightarrow 0°$, the pseudomonostatic case. Thus, $|f_{Tgt}| \approx 10.7$ Hz, which closely matches the Post Office 11-Hz calculation.

When the aeroplane is directly over the receiver, $\beta = 45°$ and $|f_{Tgt}| = 5.35$ Hz, which approximates the minimum observed value. If observations were made when the aeroplane flew between the receiver and transmitter, near the baseline, β would grow large and $|f_{Tgt}| < 5.35$ Hz. Because these observations were not reported, either that flight path was not taken or the equipment was not capable of measurement below about 5 Hz. The only way a 15-Hz beat frequency could be observed is if an aeroplane flew faster. For example, when $V = 123$ ft/s $= 37.5$ m/s, $|f_{Tgt}| = 15$ Hz in the pseudomonostatic case.

Thus, again the beat frequency phenomenon appears to be a manifestation of the bistatic doppler effect. In fact all of the terms, BTL's and GPO's "beat frequencies," NRL's "sky wave interference with the ground wave," Skolnik's "cw wave-interference" [1], Swords's "flutter effect," and Fink's "wave interference" [161] appear to be a manifestation of the bistatic doppler effect.*

Author's Note: For about six months after reaching this finding, I had indulged in an exercise of self-congratulation over what George Polya in his seminal book, HOW TO SOLVE IT, called an "act of sagacity." (Dr. Polya was once my professor of mathematics at Stanford University in the early 1950s, although he surely didn't remember.) Then while I was searching Guelrac's radar history [25] for bistatic data, I came across an account of radar-type tests using a klystron developed by Hansen and the Varian brothers at Stanford University in 1937. Guerlac continues [25]:

> Late in 1937, a few months after they had successfully operated the first klystron, Hansen and the Varian brothers began work on an aircraft detection system, using a 10-cm klystron that gave about a watt of cw power and provided a detection range of possibly three or four miles. The earliest antenna reflector was a 16-ft parabolic cylinder mounted on a Sperry 60-in. searchlight frame; this provided a very narrow, high beam, about 15°–20° high and a degree or so wide. The system detected moving objects by the beat method, or, as they preferred to say, by the Doppler frequency of the moving object.

So the act of sagacity belongs to Hansen and the Varian brothers.

Incidentally, it was not clear from Guerlac's account whether these tests used a bistatic or monostatic configuration. In any case detection ranges on automobiles and trains were small, about a quarter mile, and this radar-related work apparently was not pursued. The klystron, of course, found considerable application in radars of all types.

Appendix B
WIDTH OF A BISTATIC RANGE CELL

B.1 INTRODUCTION

The width of a bistatic range cell, ΔR_B, is usually given as [73]

$$\Delta R_B \approx \frac{c\tau}{2 \cos(\beta/2)} \tag{B.1}$$

where

  c = speed of light,
  τ = compressed pulsewidth,
  β = bistatic angle.

As shown in Figure B.1, ΔR_B is measured along the bisector of the bistatic angle, line $\overline{AB}$, and the bistatic angle is measured at the intersection of the receiver-to-target LOS and the inner edge of the bistatic range cell, or the smaller of the two confocal ellipses (isorange contours) defining the range cell.

This width can also be defined as

$$\Delta R_B' \approx \frac{c\tau}{2 \cos^2(\beta/2)} \tag{B.2}$$

where $\Delta R_B'$ is measured along the receiver-to-target LOS, line $\overline{AC}$ in Figure B.1. When $\overline{ABC}$ is approximated as a right triangle, $\Delta R_B = \Delta R_B' \cos(\beta/2)$, so that (B.1) and (B.2) are equivalent. Note that $c\tau/2 = \Delta R_M$, the width of a monostatic range cell.

An intuitive explanation of (B.1) and (B.2) is that the bistatic range cell broadens as β increases from the pseudomonostatic geometry on the extended baseline ($\beta = 0$, $\Delta R_B = \Delta R_B' = \Delta R_M$). This broadening occurs because the separation

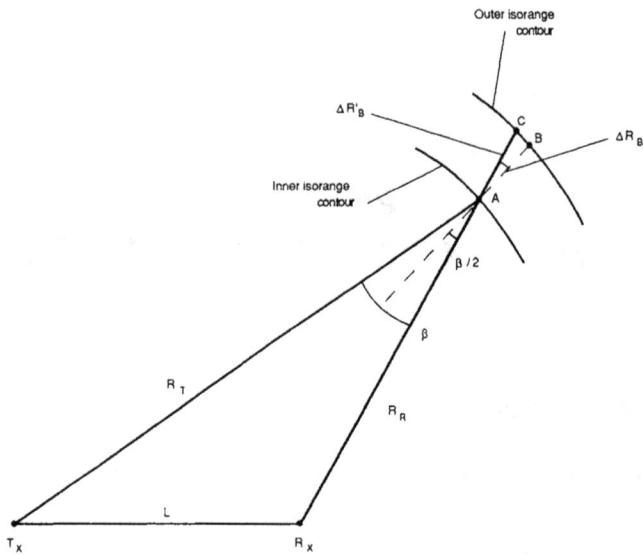

Figure B.1 Geometry for approximation to the width of a bistatic range cell.

between two confocal ellipses defining the bistatic range cell increases from a minimum on the extended baseline to a maximum at the point where the perpendicular bisector of the baseline intersects the two ellipses, as shown in Figure B.2 for (B.1).

As the ellipses become more eccentric, i.e., flatter, β_{max} increases. In the limit when the ellipse eccentricity, $e \rightarrow 1$, $\beta_{max} \rightarrow 180°$, and from (B.1) and (B.2),

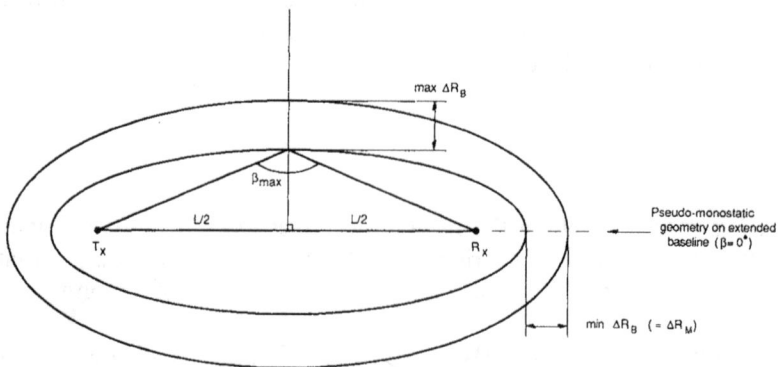

Figure B.2 Maximum and minimum bistatic range cells.

$\Delta R_B \to \infty$. Clearly, the separation between confocal ellipses cannot approach infinity anywhere, and thus (B.1) and (B.2) are approximations to the exact expression for ellipse separation. Apparently neither the approximate nor the exact expressions have been published. This appendix first develops the approximate expression for $\Delta R_B'$, (B.2), and then develops an exact, but implicit, expression for ΔR_B and compares the approximate expression of (B.1) with the exact expression.

B.2 DEVELOPMENT OF THE APPROXIMATE EXPRESSION FOR THE WIDTH OF A BISTATIC RANGE CELL, $\Delta R_B'$

The approximate expression for $\Delta R_B'$, (B.2), is developed following the method of Moyer [90]:

$$\Delta R_B' = \frac{\partial R_R}{\partial(R_T + R_R)} \Delta(R_T + R_R) \tag{B.3}$$

where R_T, R_R, and $\Delta R_B'$ are defined in Figure B.1. Now $\partial R_R / \partial(R_T + R_R)$ is given by (5.8b) as

$$\frac{\partial R_R}{\partial(R_T + R_R)} = \frac{1 + e^2 + 2e \sin\theta_R}{2(1 + e \sin\theta_R)^2} \tag{B.4}$$

where $e = L/(R_T + R_R)$ is the eccentricity of the inner ellipse. Also from (3.13b) and (3.14b),

$$R_R = \frac{L(1 - e^2)}{2e(1 + e \sin\theta_R)} \tag{B.5}$$

$$R_T = \frac{L(e^2 + 1 + 2e \sin\theta_R)}{2e(1 + e \sin\theta_R)} \tag{B.6}$$

Combining (B.4) through (B.6) yields

$$\frac{\partial R_R}{\partial(R_T + R_R)} = \frac{2e^2 R_T R_R}{L^2(1 - e^2)} \tag{B.7a}$$

$$= \frac{2R_T R_R}{(R_T + R_R)^2 - L^2} \tag{B.7b}$$

From the law of cosines,

$$\cos\beta = \frac{R_T^2 + R_R^2 - L^2}{2R_TR_R} \tag{B.8a}$$

$$= \frac{(R_T + R_R)^2 - L^2}{2R_TR_R} - 1 \tag{B.8b}$$

Thus,

$$\frac{\partial R_R}{\partial(R_T + R_R)} = (\cos\beta + 1)^{-1} \tag{B.9a}$$

$$= [2\cos^2(\beta/2)]^{-1} \tag{B.9b}$$

Combining (B.9b) with (B.3) yields

$$\Delta R_B' = \frac{\Delta(R_T + R_R)}{2\cos^2(\beta/2)} \tag{B.10}$$

When $\Delta(R_T + R_R) = c\tau$,

$$\Delta R_B' = \frac{c\tau}{2\cos^2(\beta/2)} \tag{B.11}$$

which is (B.2). For the pseudomonostatic case on the extended baseline, $\beta = 0$ and $\Delta R_B' = c\tau/2 = \Delta R_M$, the width of the monostatic range cell.

B.3   DEVELOPMENT OF AN EXACT EXPRESSION FOR THE WIDTH OF A BISTATIC RADAR CELL, ΔR_B

B.3.1   Geometry and Definitions

Figure B.3 establishes the geometry for the width of a bistatic range cell, ΔR_B. Consider two confocal ellipses with foci at T_x and R_x, separated by L, the baseline. The inner confocal ellipse is defined by the semimajor axis, a:

$$R_T + R_R = 2a \tag{B.12}$$

The outer confocal ellipse is defined by its corresponding semimajor axis, a', where $a' > a$:

$$R_T' + R_R' = 2a' \tag{B.13}$$

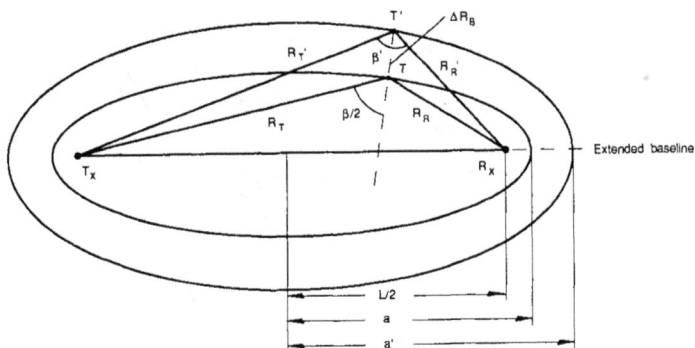

Figure B.3 Geometry for bistatic range cell, ΔR_B.

The ellipses are completely defined by the parameters a, a', and L. The difference $(a' - a)$ is the separation between ellipses on the extended baseline, and is defined as the width of a pseudomonostatic range cell, ΔR_M:

$$\Delta R_M \equiv (a' - a) \tag{B.14}$$

An arbitrary point on the inner ellipse, T, is defined by R_T, R_R, and a corresponding β, which is given by the law of cosines as

$$\beta = \cos^{-1}\left(\frac{R_T^2 + R_R^2 - L^2}{2R_TR_R}\right) \tag{B.15}$$

A point, T', is located at the intersection of the bisector of β and the outer ellipse. Associated with this point are the ranges R_T', R_R', and a corresponding β'. Note that $(R_T' + R_R') > (R_T + R_R)$ and $\beta > \beta'$, but the bisector of β *does not* bisect β' except when it lies on the baseline or the perpendicular bisector of the baseline. The distance $\overline{TT'}$ is defined as the width of bistatic range cell, ΔR_B.

B.3.2 Development

Unfortunately, a convenient analytical method is not available for developing an explicit expression for ΔR_B, in terms of the ellipse parameters a, a', and L for any given point on the inner ellipse defined by β. An implicit expression is developed using the geometry shown in Figure B.4, as follows:

$$R_T' = [\overline{T_xB}^2 + (\overline{TB} + \Delta R_B)^2]^{1/2} \tag{B.16}$$
$$= [R_T^2 \sin^2(\beta/2) + (R_T \cos(\beta/2) + \Delta R_B)^2]^{1/2}$$

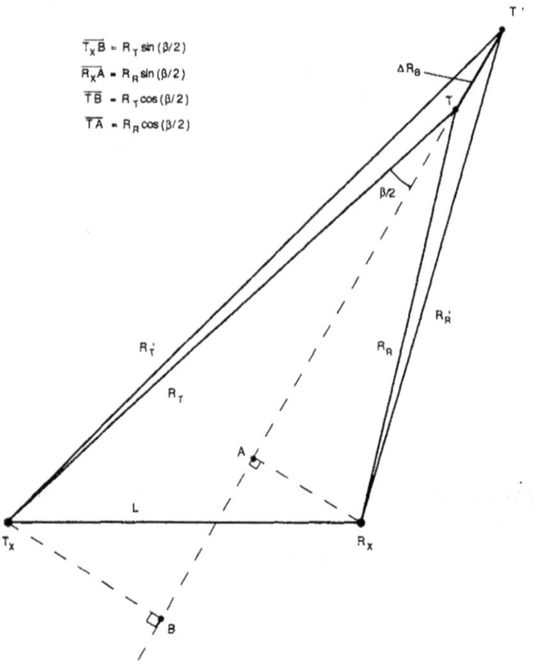

Figure B.4 Geometry for an implicit expression of the bistatic range cell, ΔR_B.

and

$$R'_R = [\overline{R_x A}^2 + (\overline{TA} + \Delta R_B)^2]^{1/2} \tag{B.17}$$
$$= [R_R^2 \sin^2(\beta/2) + (R_R \cos(\beta/2) + \Delta R_B)^2]^{1/2}$$

Substituting (B.16) and (B.17) into (B.13) yields

$$[R_T^2 \sin^2(\beta/2) + (R_T \cos(\beta/2) + \Delta R_B)^2]^{1/2} \tag{B.18}$$
$$+ [R_R^2 \sin^2(\beta/2) + (R_R \cos(\beta/2) + \Delta R_B)^2]^{1/2} = 2a'$$

Solving (B.12) and (B.15) for R_T and R_R yields

$$R_T = a + \left(\frac{L^2/2 - 2a^2}{1 + \cos\beta} + a^2\right)^{1/2} \tag{B.19}$$

$$R_R = a - \left(\frac{L^2/2 - 2a^2}{1 + \cos\beta} + a^2\right)^{1/2} \tag{B.20}$$

Substituting (B.19) and (B.20) into (B.18) yields an implicit expression for ΔR_B in terms of the ellipse parameters a, a', and L as a function of any point in the inner ellipse defined by β.

For the sanity check, when $\beta = 0°$, $\Delta R_B = a' - a$. When β is a maximum on the bisector of the baseline, ΔR_B is maximum, where

$$\beta_{max} = 2\cos^{-1}(1 - L^2/4a^2)^{1/2} \tag{B.21}$$

and

$$(\Delta R_B)_{max} = (a'^2 - L^2/4)^{1/2} - (a^2 - L^2/4)^{1/2} \tag{B.22}$$

As an aside, because the semiminor axes of the two ellipses, b and b', are defined as

$$b = (a^2 - L^2/4)^{1/2} \tag{B.23}$$
$$b' = (a'^2 - L^2/4)^{1/2} \tag{B.24}$$

(B.21) and (B.22) become

$$\beta_{max} = 2\cos^{-1}(b/a) \tag{B.21a}$$

and

$$(\Delta R_B)_{max} = (b' - b) \tag{B.22a}$$

B.3.3 Discussion

Equation (B.18) is the exact expression for the width of a bistatic range cell, ΔR_B, defined as the separation of two confocal ellipses, $(R_T + R_R) = 2a$ and $(R'_T + R'_R) = 2a'$, $a' > a$, and common baseline L, where the separation is measured along the bisector of the bistatic angle β defined on the inner ellipse, $(R_T + R_R) = 2a$.

This definition of the bistatic range cell, ΔR_B, is not altogether arbitrary. Separation measured along the bistatic bisector is consistent with the assertion of Chapter 3 that equivalent monostatic operation is obtained by locating the monostatic radar on the bisector of the bistatic angle.

The bistatic angle could be chosen differently, for example, by locating it on the outer ellipse. An expression similar to (B.18) would be obtained. Averaging the outer and inner bistatic angles could be considered but that poses problems because no ellipse represents $(\beta + \beta')/2$ for all (β, β'). That is, an ellipse drawn through a given $(\beta + \beta')/2$ will not pass through any other values of $(\beta + \beta')/2$. Thus, the preceding definition, while somewhat arbitrary, is also somewhat pragmatic.

B.4 ERROR ANALYSIS AND EXAMPLES

The exact width of the bistatic range cell, ΔR_B, given implicitly in (B.18), and normalized with respect to the baseline L, is plotted on Figure B.5 for the case $a = 0.55L$ and $a' = 0.65L$. The ellipses are highly eccentric: $e = L/2a = 0.91$ and e'

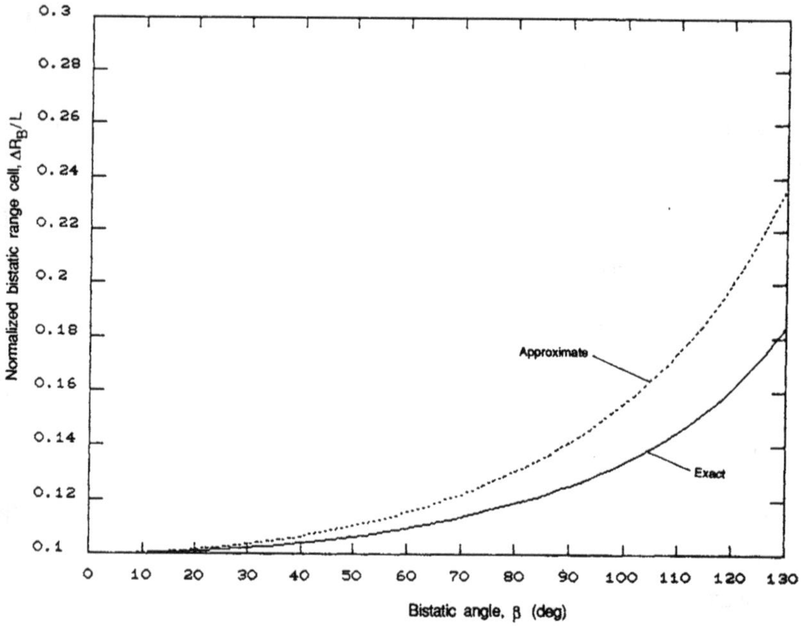

Figure B.5. Exact and approximate values of bistatic range cell, ΔR_B, for $a = 0.55L$ and $a' = 0.65L$.

$= L/2a' = 0.77$. The normalized pseudomonostatic range cell, $\Delta R_M/L$, is $(a' - a)L = 0.1$ and is shown in Figure B.5 for $\beta = 0°$. The maximum value for ΔR_B at β_{max} is given by (B.22). For this case $\beta_{max} = 130.8°$ and $(\Delta R_B)_{max}/L = 0.186$, as shown in Figure B.5.

Equation (B.1), the approximation to ΔR_B, is also plotted in Figure B.5. At $\beta = 0$ the approximation is exact: $\Delta R_B/L = 0.1$. However, as β increases, (B.1) increases faster than (B.18), so that at β_{max} the approximation gives $\Delta R_B/L = 0.24$. The relative error, ϵ, is

$$\epsilon = \frac{(\Delta R_B)_A - (\Delta R_B)_E}{(\Delta R_B)_E} \tag{B.25}$$

where the subscripts A and E denote approximate and exact values, respectively. The percent error is shown in Figure B.6 for this example. The maximum error occurs at β_{max}, where the approximate value is 29% higher than the exact value.

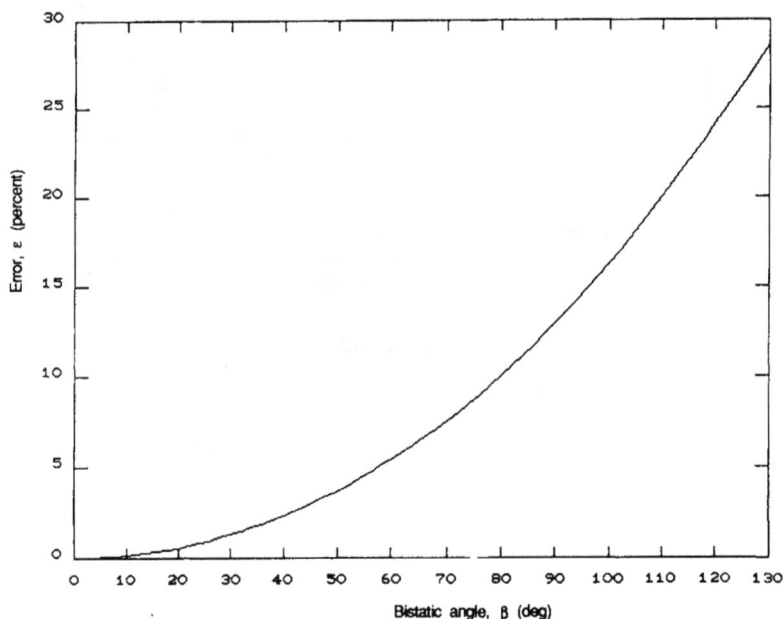

Figure B.6 Error between exact and approximate values of bistatic range cell, ΔR_B, for $a = 0.55L$ and $a' = 0.65L$.

As the separation between ellipses on the extended baseline decreases, the error decreases. Figures B.7 and B.8 show the case for $a = 0.55L$ (as before), but now $a' = 0.56L$. Now the maximum error is 4.1%. Also, as the ellipses become less eccentric, the error decreases. Figures B.9 and B.10 show that for $a = 1.1L$ and $a' = 1.2L$ (with the same pseudomonostatic range cell of $0.1L$ as in Figure B.5, but with eccentricities of only $e = 0.45$ and $e' = 0.42$) the maximum error is now 1.1%. The error rapidly diminishes as the ellipse separation or the eccentricities decrease.

The maximum error, ϵ_{max}, is given by substituting (B.14) and (B.21) into (B.1); then (B.1) into (B.25); and (B.22) into (B.25). Thus,

$$\epsilon_{max} = \frac{a(a' - a)}{[(a^2 - L^2/4)(a'^2 - L^2/4)]^{1/2} - (a^2 - L^2/4)} - 1 \tag{B.26}$$

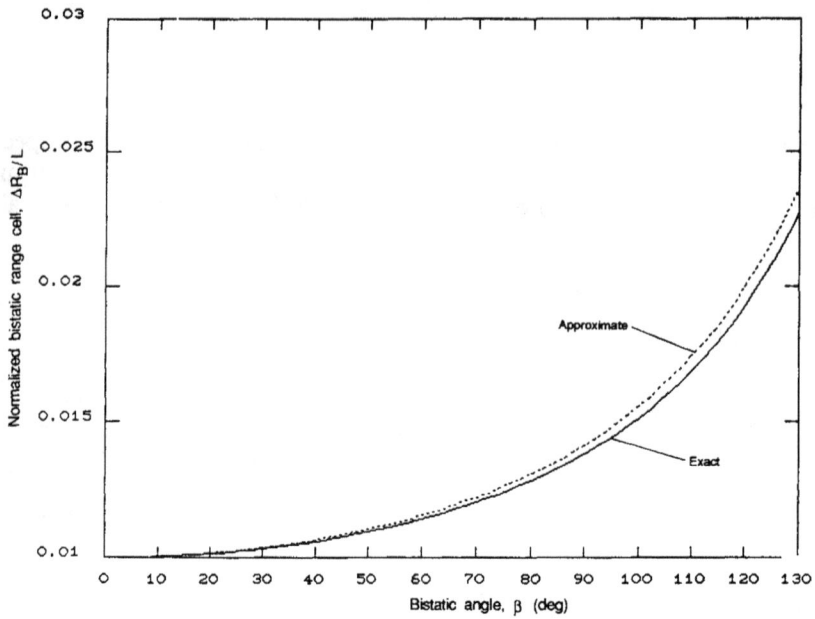

Figure B.7 Exact and approximate values of bistatic range cell, ΔR_B, for $a = 0.55L$ and $a' = 0.56L$.

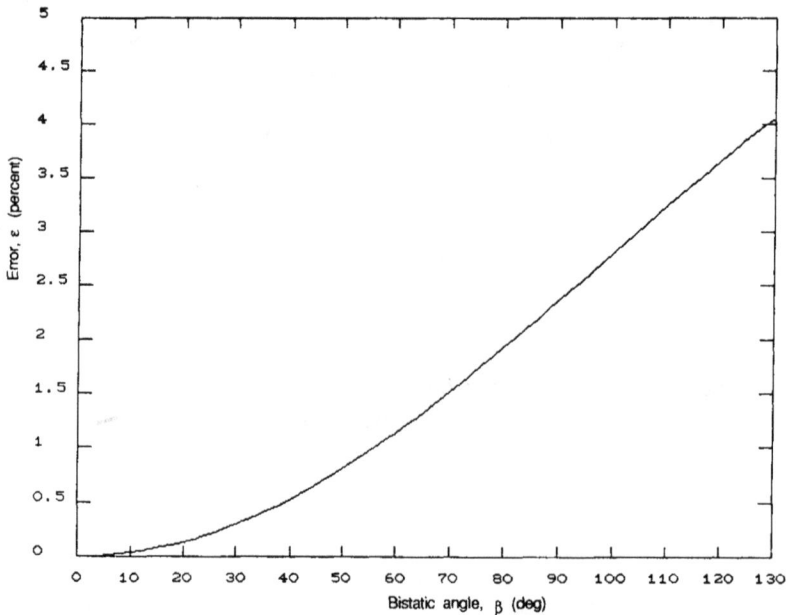

Figure B.8 Error between exact and approximate values of bistatic range cell, ΔR_B, for $a = 0.55L$ and $a' = 0.56L$.

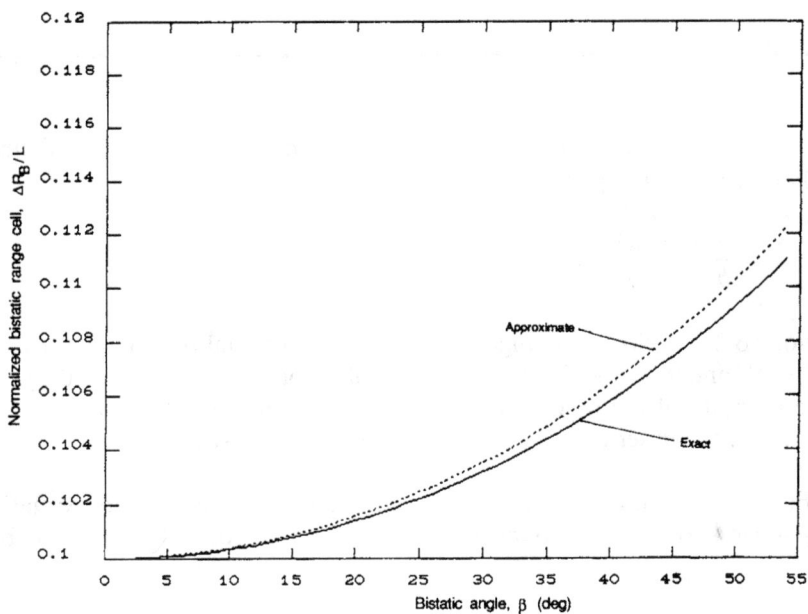

Figure B.9 Exact and approximate values of bistatic range cell, ΔR_B, for $a = 1.1L$ and $a' = 1.2L$.

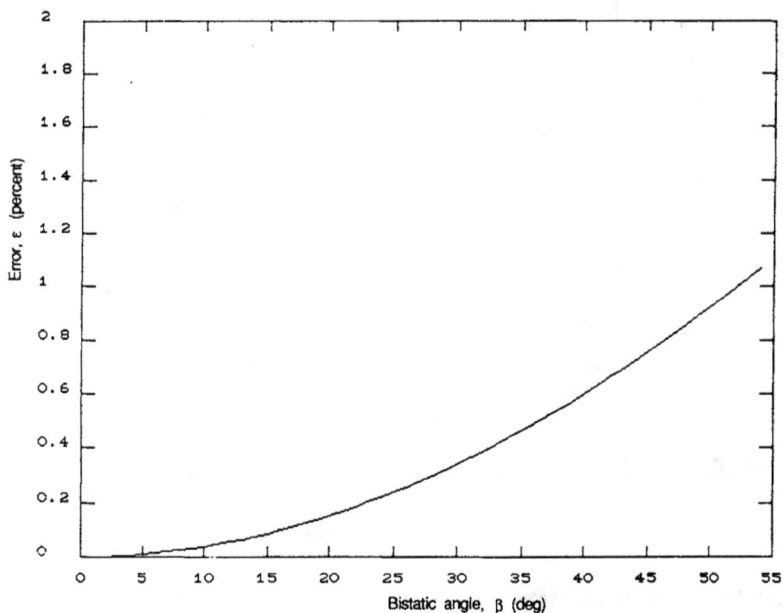

Figure B.10 Error between exact and approximate values of bistatic range cell, ΔR_B, for $a = 1.1L$ and $a' = 1.2L$.

In terms of semimajor axes, (a, a'), and semiminor axes (b, b'), as defined by (B.23) and (B.24), (B.26) can also be written as

$$\epsilon_{max} = \frac{a(a' - a)}{b(b' - b)} - 1 \tag{B.27}$$

Equation (B.26) is plotted on Figure B.11 for three normalized values of $(a' - a)/L$: 0.1, 0.01, and 0.001, and for normalized values of a/L ranging from 0.51 to 100. The eccentricity of the inner ellipse is also shown on the ordinate. When $\epsilon \rightarrow 1$, $2a \rightarrow L$, i.e., the inner ellipse collapses to the baseline. At this point the first term of (B.26) $\rightarrow \infty$ and $\epsilon_{max} \rightarrow \infty$.

Figure B.11 shows that the maximum error in using the approximation to ΔR_B given by (B.1) is always positive. In fact, values of ΔR_B calculated by the

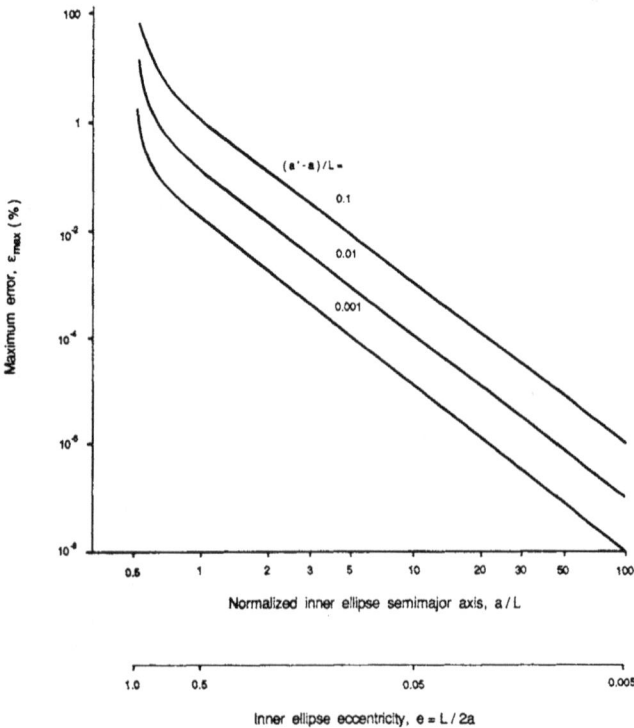

Figure B.11 Maximum error between exact and approximate values of bistatic range cell, ϵ_{max}.

approximate method, (B.1), are always larger than those calculated by the exact method, (B.18). Unless, however, the inner ellipse is highly eccentric, $\epsilon > 0.5$, *and* the pseudomonostatic range cell $\Delta R_M = (a' - a)$ is a significant fraction of the baseline L, $\Delta R_M > 0.1L$, the maximum error in using the approximate method is less than a few percent.

Consider the following example. The bistatic transmitter and receiver are separated by $L = 10$ km. A target is on the inner range ellipse $R_T + R_R = 2a = 30$ km. The compressed pulse width τ is 6.7 μs. Thus, the pseudomonostatic range cell $\Delta R_M = c\tau/2 = 1.0$ km, and a' for the outer ellipse is 16 km. The maximum bistatic angle occurring on the bisector of the baseline is, from (B.21), 38.9°. The approximation to ΔR_B at this point is, from (B.1), 1.59 km. The maximum error is, from (B.26) or Figure B.11, 0.39%. For $\beta < 38.9°$, $\epsilon < 0.39\%$.

Appendix C
APPROXIMATION TO THE LOCATION EQUATION

C.1 INTRODUCTION

One solution for the receiver-to-target range, R_R, is

$$R_R = \frac{(R_T + R_R)^2 - L^2}{2(R_T + R_R + L\sin\theta_R)} \tag{C.1}$$

where

R_T = transmitter-to-target range,
L = baseline range,
θ_R = receiver look angle (in North coordinates).

This appendix develops a simplified approximation to (C.1) for the special case of a bistatic radar (1) using the direct range sum estimation method, Section 5.1, and (2) operating in a short-range over-the-shoulder geometry, where the range sum $(R_T + R_R)$ is slightly greater than the baseline L. Errors associated with this approximation are also developed. Specifically, the approximation generates a 10% error in the estimate of R_R for $0° \leq \theta_R \leq 180°$ and $L \geq 0.82(R_T + R_R)$. Because the isorange contour eccentricity, $e = L/(R_T + R_R)$, $e > 0.82$ for this special case.

C.2 DEVELOPMENT

Assume

$$(R_T + R_R) = L + \Delta L \tag{C.2}$$

where $L \gg \Delta L$. Thus,

$$L \gg (R_T + R_R) - L \tag{C.3}$$

In the direct range sum estimation method, the time interval ΔT_{rt} between reception of the transmitted pulse and reception of the target echo is measured. The range sum $(R_T + R_R)$ is then calculated by (5.2) as

$$(R_T + R_R) = c\Delta T_{rt} + L \tag{C.4}$$

or

$$c\Delta T_{rt} = (R_T + R_R) - L \tag{C.5}$$

Thus,

$$L \gg c\Delta T_{rt} \tag{C.6}$$

Combining (C.1) and (C.4) yields

$$R_R = \frac{(c\Delta T_{rt} + L)^2 - L^2}{2(c\Delta T_{rt} + L + L \sin\theta_R)} \tag{C.7}$$

Expanding and simplifying yields

$$R_R = \frac{c\Delta T_{rt}(1 + c\Delta T_{rt}/2L)}{c\Delta T_{rt}/L + 1 + \sin\theta_R} \tag{C.8}$$

From (C.6), $c\Delta T_{rt}/L \ll 1$ and (C.8) becomes

$$R_R \approx \frac{c\Delta T_{rt}}{1 + \sin\theta_R} \tag{C.9}$$

In this case an estimate of the baseline length L is not needed. (An estimate of the angular location of the transmitter from the receiver is still required because that sets the baseline from which θ_R is measured.)

As an aside, when θ_R is near $+90°$, in a near-over-the-shoulder operation where the target is close to the extended baseline, (C.9) reduces to

$$R_R \approx \frac{c\Delta T_{rt}}{2} \tag{C.10}$$

Equation (C.10) represents the pseudomonostatic operating point. It can be derived exactly by combining (C.1) and (C.4) and assuming $\theta_R = 90°$.

Equation (C.10) implies that neither an estimate of the baseline length nor an estimate of the transmitter's angular location is required for estimating R_R in a near-over-the-shoulder operation. This implication, however, is deceptive because the definition of over-the-shoulder operation requires that the transmitter's angular location be known, so that the receiver can "turn its back on the transmitter," so to speak. Once that angle is known, θ_R can be referenced to the baseline and (C.9) can be used.

C.3 ERROR ANALYSIS, DISCUSSION, AND EXAMPLES

The relative error between the approximate and exact expressions for R_R, (C.9) and (C.1), respectively, is given as

$$\epsilon = \frac{(R_R)_A - (R_R)_E}{(R_R)_E} \tag{C.11}$$

where the subscripts A and E denote approximate and exact values, respectively. Substituting (C.1) and (C.9) into (C.11) yields:

$$\epsilon = \frac{c\Delta T_{rt}(1 - \sin\theta_R)}{(c\Delta T_{rt} + 2L)(1 + \sin\theta_R)} \tag{C.12}$$

Because $c\Delta T_{rt} = (R_T + R_R) - L$ and eccentricity, $e = L/(R_T + R_R)$, (C.12) can be expressed as:

$$\epsilon = \frac{e^{-1} - 1}{e^{-1} + 1} \cdot \frac{1 - \sin\theta_R}{1 + \sin\theta_R} \tag{C.12a}$$

Also, $L/c\Delta T_{rt}$ can be expressed as $(e^{-1} - 1)^{-1}$.

Equations (C.12) and (C.12a) are plotted on Figure C.1 as a function of the receiver look angle θ_R, for values of $L/c\Delta T_{rt}$ ranging from 3 to 300, or e ranging from 0.75 to 0.997. Errors are plotted for θ_R in the northern hemisphere, $-90° \le \theta_R \le 90°$. They are identical for θ_R in the southern hemisphere, i.e., $90° \le \theta_R \le 270°$. As θ_R approaches 90°, the error for all values of $L/c\Delta T_{rt}$ approaches zero. This geometry is for the target on the extended baseline beyond the receiver, i.e., exact over-the-shoulder operation, where θ_R is not required for an estimate of R_R. That is, the approximate expression is exact. As θ_R approaches $-90°$, the error for all values of $L/c\Delta T_{rt}$ approaches infinity. This geometry is for the target on the baseline to the left of receiver, where the approximate expression approaches infinity. Note that the exact expression remains valid for target locations on the extended

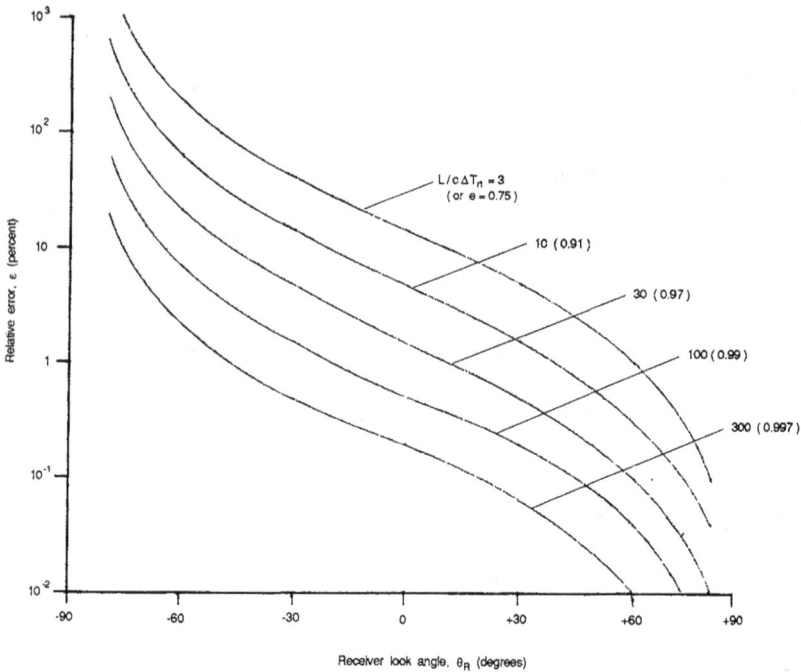

Figure C.1 Relative error between approximate and exact expressions for receiver-to-target range, R_R.

baseline beyond the transmitter. No solution exists, of course, for the target on the baseline, between transmitter and receiver for either expression.

Equations (C.12) and (C.12a) and Figure C.1 show that the relative error is always positive; that is, the approximation gives values greater than or equal to that of the exact expression, where equality occurs at $\theta_R = 90°$. The error is less than 1% for $L \geq 3c\Delta T_{rt}$ (or $e > 0.75$) and $60° \leq \theta_R \leq 120°$. Because $e = L/(R_T + R_R)$, an equivalent statement is that the baseline L must be at least three quarters of the range sum $(R_T + R_R)$ to generate an error of 1% or less in the estimate of R_R, when the target lies within $\pm 30°$ of the extended baseline. For example, if $L = 200$ km and $\theta_R = 60°$, $(R_T + R_R) = 266.7$ km for a 1% error in the estimate of R_R. From (C.1), $R_R = 35.36$ km, the exact value. From (C.5) and (C.9), $R_R = 35.73$ km, the approximate value. From (C.11), $\epsilon \approx 0.01$.

If errors in the estimate of R_R of 10% can be tolerated, the receiver look angle, θ_R, can be extended to 0° and 180°, i.e., $\pm 90°$ from the extended baseline, with $e \geq 0.82$, i.e., $L \geq 0.82(R_T + R_R)$. Using the previous example of $L = 200$ km, but with $\theta_R = 0°$, $(R_T + R_R) = 244$ km. From (C.1), $R_R = 40$ km, the exact value. From (C.9), $R_R = 44$ km, the approximate value. From (C.11), $\epsilon = 0.1$. The relative error grows very quickly as θ_R looks back toward the transmitter, i.e., $\theta_R < 0°$.

Appendix D
AREA WITHIN A MAXIMUM RANGE OVAL
OF CASSINI

The area within a maximum range oval of Cassini is developed in this appendix. The development is in two parts: (1) the area within a single oval, where the baseline, $L < 2\sqrt{\kappa}$ and κ = bistatic maximum range constant, and (2) the area within two identical ovals, surrounding the transmitter and receiver, where $L \geq 2\sqrt{\kappa}$. The development uses the polar coordinate system shown in Figure 4.1.

The area within a closed curve, A_B, defined in polar coordinates (r,θ) is

$$A_B = \frac{1}{2} \int_0^{2\pi} [f(\theta)]^2 \, d\theta \tag{D.1}$$

The equation for an oval of Cassini in polar coordinates is given by (4.5) as

$$R_T^2 R_R^2 = (r^2 + L^2/4)^2 - r^2 L^2 \cos^2\theta \tag{D.2}$$

When the oval is a maximum range oval of Cassini, (4.1a) applies:

$$(R_T R_R)_{max} = \kappa \tag{D.3}$$

Substituting (D.3) into (D.2) and solving for $r = f(\theta)$ yields:

$$r = (L/2)[\cos2\theta \pm (16\kappa^2/L^4 - \sin^2 2\theta)^{1/2}]^{1/2} \tag{D.4}$$

Substituting (D.4) into (D.1), manipulating terms, and noting that the oval (or ovals) are symmetric in each quadrant, and also that

$$\int_0^{\pi/2} \cos2\theta \, d\theta = 0 \tag{D.5}$$

---

yields:

$$A_B = 2\kappa \int_0^{\pi/2} \left(1 - \frac{L^4}{16\kappa^2} \sin^2 2\theta\right)^{1/2} d\theta \tag{D.6}$$

where only the positive value of the radical is evaluated.

Equation (D.6) can be integrated directly when $L^4/16\kappa^2 < 1$, i.e., $L < 2\sqrt{\kappa}$, the case where a single oval surrounds both transmitter and receiver. In this case the area within the single oval, A_{B1}, is [152]:

$$A_{B1} = \pi\kappa \left[1 - \left(\frac{1}{2}\right)^2 \left(\frac{L^4}{16\kappa^2}\right)^1 \left(\frac{1}{1}\right) - \left(\frac{1 \cdot 3}{2 \cdot 4}\right)^2 \left(\frac{L^4}{16\kappa^2}\right)^2 \left(\frac{1}{3}\right) - \cdots \right] \tag{D.7}$$

$$\approx \pi\kappa[1 - (1/64)(L^4/\kappa^2) - (3/16384)(L^8/\kappa^4)] \tag{D.7a}$$

For the two-oval case where $L \geq 2\sqrt{\kappa}$, (D.6) must be modified by a substitution of variables:

$$\sin^2 \xi = \frac{L^4}{16\kappa^2} \sin^2 2\theta \tag{D.8}$$

Thus, the area within two ovals, A_{B2}, becomes

$$A_{B2} = \frac{8\kappa^2}{L^2} \int_0^{\pi/2} \left[\cos^2 \xi \left(1 - \frac{16\kappa^2 \sin^2 \xi}{L^4}\right)^{-1/2}\right] d\xi \tag{D.9}$$

With this substitution of variables, ξ can assume all values from 0 to $\pi/2$, whereas θ is constrained to values of

$$\theta \leq \tfrac{1}{2} \sin^{-1}(4\kappa/L^2) \tag{D.10}$$

which is the angular region where the two small ovals exist.

Performing the integration yields [152]:

$$A_{B2} = \frac{2\pi\kappa^2}{L^2} \left[1 + \left(\frac{1}{2^3}\right)\left(\frac{16\kappa^2}{L^4}\right) + \left(\frac{3^2}{2^4 \cdot 3! \cdot 2!}\right)\left(\frac{16\kappa^2}{L^4}\right)^2 \right.$$
$$\left. + \left(\frac{3^2 \cdot 5^2}{2^6 \cdot 4! \cdot 3!}\right)\left(\frac{16\kappa^2}{L^4}\right)^3 + \cdots \right] \tag{D.11}$$

$$\approx \frac{2\pi\kappa^2}{L^2}(1 + 2\kappa^2/L^4 + 12\kappa^4/L^8 + 100\kappa^6/L^{12}) \tag{D.11a}$$

For the monostatic case, $L = 0$ and (D.7) becomes

$$A_M = \pi\kappa \tag{D.12}$$
$$= \pi(R_M)^2_{\text{max}} \tag{D.12a}$$

as expected because the oval becomes a circle of radius $(R_M)_{\text{max}}$ and area A_M. Note that $A_M > A_{B1}$, for the case of a bistatic radar with maximum range product, κ, and a monostatic radar with equivalent monostatic maximum range $\sqrt{\kappa}$.

Appendix E
RELATIONSHIPS BETWEEN PARAMETERS IN TARGET LOCATION AND CLUTTER DOPPLER SPREAD EQUATIONS

E.1 INTRODUCTION

This appendix develops exact and approximate relationships between parameters in target location and clutter doppler spread equations. The parameters are shown in Figure E.1, using the North-referenced coordinate system of Figure 3.1. Because exact relationships are required, the isorange contour must be an ellipse, rather than the tangent approximation (perpendicular to the bistatic angle bisector). Initially, this requirement would seem to be intimidating, but the results are surprisingly tractable when ellipse eccentricity, e, is used in the development.

The isorange contour of interest is defined by an ellipse as

$$R_T + R_R = 2a \tag{E.1}$$

where $2a$ is the major axis of the ellipse. Eccentricity of the ellipse, e, is

$$e = L/2a \tag{E.2a}$$
$$= L/(R_T + R_R) \tag{E.2b}$$

In all expressions involving e, when the target lies on the baseline, $e = 1$ and the parameters become indeterminate.

E.2 TARGET LOCATION

When L, $(R_T + R_R)$, and θ_R are measured, R_R and R_T are calculated as follows. From the law of cosines:

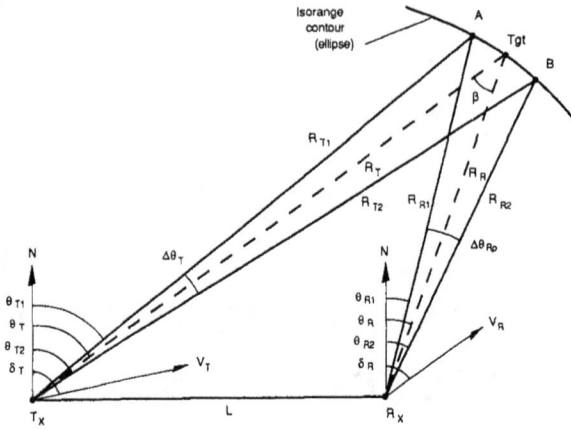

Figure E.1 Geometry for target location and clutter doppler spread.

$$R_T^2 = R_R^2 + L^2 - 2R_R L \cos(90 + \theta_R) \tag{E.3}$$

$$R_R^2 + 2R_R L \sin\theta_R = R_T^2 - L^2 \tag{E.4}$$

$$2R_R^2 + 2R_R L \sin\theta_R + 2R_T R_R = (R_T + R_R)^2 - L^2 \tag{E.5}$$

$$R_R = \frac{(R_T + R_R)^2 - L^2}{2(R_T + R_R + L \sin\theta_R)} \tag{E.6}$$

and

$$R_T = (R_R^2 + L^2 + 2R_R L \sin\theta_R)^{1/2} \tag{E.7}$$

Combining (E.2b) and (E.6) yields

$$R_R = \frac{L(1 - e^2)}{2e(1 + e \sin\theta_R)} \tag{E.8}$$

Combining (E.2b) and (E.7) yields

$$R_T = \frac{L(e^2 + 1 + 2e \sin\theta_R)}{2e(1 + e \sin\theta_R)} \tag{E.9}$$

When L, $(R_T + R_R)$, and θ_T are measured, R_R and R_T are calculated in a similar manner. From the law of cosines:

$$R_R^2 = R_T^2 + L^2 - 2R_T L \cos(90 - \theta_T) \tag{E.10}$$

$$R_T = \frac{(R_T + R_R)^2 - L^2}{2(R_T + R_R - L \sin\theta_T)} \tag{E.11}$$

and

$$R_R = (R_T^2 + L^2 - 2R_T L \sin\theta_T)^{1/2} \tag{E.12}$$

Combining (E.2b) and (E.11) yields

$$R_T = \frac{L(1 - e^2)}{2e(1 - e \sin\theta_T)} \tag{E.13}$$

Combining (E.2b) and (E.12) yields

$$R_R = \frac{L(e^2 + 1 - 2e \sin\theta_T)}{2e(1 - e \sin\theta_T)} \tag{E.14}$$

E.3 PARAMETER RATIOS

From the law of sines:

$$\frac{R_R}{\sin(90 - \theta_T)} = \frac{R_T}{\sin(90 + \theta_R)} = \frac{L}{\sin\beta} \tag{E.15}$$

Thus,

$$\frac{R_R}{R_T} = \frac{\cos\theta_T}{\cos\theta_R} \tag{E.16a}$$

$$\frac{R_R}{L} = \frac{\cos\theta_T}{\sin\beta} \tag{E.16b}$$

and

$$\frac{R_T}{L} = \frac{\cos\theta_R}{\sin\beta} \tag{E.16c}$$

Combining (E.8), (E.13), and (E.16a) yields

$$\frac{R_R}{R_T} = \frac{\cos\theta_T}{\cos\theta_R} = \frac{1 - e\sin\theta_T}{1 + e\sin\theta_R} \tag{E.17}$$

Combining (E.2b), (E.8), and (E.16a) yields

$$\frac{R_R}{R_T} = \frac{\cos\theta_T}{\cos\theta_R} = \frac{1 - e^2}{1 + e^2 + 2e\sin\theta_R} \tag{E.18}$$

Combining (E.2b), (E.13), and (E.16a) yields

$$\frac{R_T}{R_R} = \frac{\cos\theta_R}{\cos\theta_T} = \frac{1 - e^2}{1 + e^2 - 2e\sin\theta_T} \tag{E.19}$$

E.4 ANGULAR RATES

Solving (E.18) for θ_T yields

$$\theta_T = \cos^{-1}\left(\frac{(1 - e^2)\cos\theta_R}{1 + e^2 + 2e\sin\theta_R}\right) \tag{E.20}$$

Differentiating θ_T with respect to θ_R yields

$$\frac{d\theta_T}{d\theta_R} = \frac{(1 - e^2)[(1 + e^2 + 2e\sin\theta_R)\sin\theta_R + 2e\cos^2\theta_R]}{\left\{1 - \left[\frac{(1 - e^2)\cos\theta_R}{1 + e^2 + 2e\sin\theta_R}\right]^2\right\}^{1/2}(1 + e^2 + 2e\sin\theta_R)^2} \tag{E.21}$$

$$= \frac{(1 - e^2)(2e + e^2\sin\theta_R + \sin\theta_R)}{[(1 + e^2 + 2e\sin\theta_R)^2 - (1 - e^2)^2\cos^2\theta_R]^{1/2}(1 + e^2 + 2e\sin\theta_R)} \tag{E.22}$$

$$= \frac{(1 - e^2)(2e + e^2\sin\theta_R + \sin\theta_R)}{[(1 + e^2)\sin\theta_R + 2e](1 + e^2 + 2e\sin\theta_R)} \tag{E.23}$$

$$= \frac{1 - e^2}{1 + e^2 + 2e\sin\theta_R} \tag{E.24}$$

Equations (E.24) and (E.18) are identical. Thus,

$$\frac{R_R}{R_T} = \frac{\cos\theta_T}{\cos\theta_R} = \frac{d\theta_T}{d\theta_R} \tag{E.25}$$

E.5 PSEUDOBEAMWIDTH AND BEAM POINTING ANGLE

The pseudoreceiving beamwidth is defined as the angle, $\Delta\theta_{Rp}$, at the receiving site subtended by the intersection of the transmitting beam edges and the isorange contour (ellipse), points A and B in Figure E.1. Thus,

$$\Delta\theta_{Rp} = \theta_{R2} - \theta_{R1} \tag{E.26}$$

The pseudoreceiving beam pointing angle, θ_{Rp}, is defined as the bisector of the pseudoreceiving beam. Thus,

$$\theta_{Rp} = (\theta_{R1} + \theta_{R2})/2 \tag{E.27}$$

Exact expressions for θ_{R1} and θ_{R2} are developed as follows. Because points A, B, and the target lie on a single isorange contour in Figure E.1,

$$R_{T1} + R_{R1} = R_{T2} + R_{T2} = R_T + R_R \tag{E.28}$$

Thus, from (E.8):

$$R_{R1} = \frac{L(1 - e^2)}{2e(1 + e\sin\theta_{R1})} \tag{E.29}$$

and from (E.14),

$$R_{R1} = \frac{L(e^2 + 1 - 2e\sin\theta_{T1})}{2e(1 - e\sin\theta_{T1})} \tag{E.30}$$

Combining (E.29) and (E.30) and solving for θ_{R1} yields

$$\theta_{R1} = \sin^{-1}\left[\frac{\sin\theta_{T1}(1 + e^2) - 2e}{1 + e^2 - 2e\sin\theta_{T1}}\right] \tag{E.31}$$

Similarly,

$$\theta_{R2} = \sin^{-1}\left[\frac{\sin\theta_{T2}(1 + e^2) - 2e}{1 + e^2 - 2e\sin\theta_{T2}}\right] \tag{E.32}$$

Now

$$\theta_{T1} = \theta_T - \Delta\theta_T/2 \tag{E.33}$$

and

$$\theta_{T2} = \theta_T + \Delta\theta_T/2 \tag{E.34}$$

Thus, for a given isorange contour of eccentricity e, a transmitting beam pointing angle θ_T, and a transmitting beamwidth $\Delta\theta_T$, the pseudoreceiving beamwidth $\Delta\theta_{Rp}$ and pseudoreceiving beam pointing angle θ_{Rp} can be calculated exactly by (E.33) and (E.34), (E.31) and (E.32), and finally (E.26) and (E.27).

For the opposite case of $(\theta_R, \Delta\theta_R)$ and $(\theta_{Tp}$ and $\Delta\theta_{Tp})$, the following expressions apply:

$$\Delta\theta_{Tp} = \theta_{T2} - \theta_{T1} \tag{E.35}$$

$$\theta_{Tp} = (\theta_{T1} + \theta_{T2})/2 \tag{E.36}$$

$$\theta_{T1} = \sin^{-1}\left[\frac{\sin\theta_{R1}(1 + e^2) + 2e}{1 + e^2 + 2e\sin\theta_{R1}}\right] \tag{E.37}$$

$$\theta_{T2} = \sin^{-1}\left(\frac{\sin\theta_{R2}(1 + e^2) + 2e}{1 + e^2 + 2e\sin\theta_{R2}}\right) \tag{E.38}$$

$$\theta_{R1} = \theta_R - \Delta\theta_R/2 \tag{E.39}$$

$$\theta_{R2} = \theta_R + \Delta\theta_R/2 \tag{E.40}$$

E.6 APPROXIMATIONS TO PSEUDOBEAMWIDTH AND BEAM POINTING ANGLE

An approximation to the pseudoreceiving (or transmitting) beamwidth $\Delta\theta_{Rp}$ (or $\Delta\theta_{Tp}$) can be used in some geometries. The approximation for $\Delta\theta_{Rp}$ is developed by using Figure E.2, where an isorange contour is approximated by the tangent, and respective transmitting and receiving beam rays are assumed parallel. Triangles $\overline{ABC}$ and $\overline{ABD}$ are congruent; thus, $\overline{AD}$, the cross-range dimension of the transmitting beam, $R_T\Delta\theta_T'$, equals $\overline{BC}$, the cross-range dimension of the pseudoreceiving beam, $R_R\Delta\theta_{Rp}'$, or

$$R_T\Delta\theta_T' \approx R_R\Delta\theta_{Rp}' \tag{E.41}$$

where the primes denote approximations. In other words, the cross-range dimension of the transmitting and pseudoreceiving beams are approximately equal at any point on an isorange contour. The opposite case of $\Delta\theta_R$ and $\Delta\theta_{Tp}$ also holds. Note that (E.41) and (E.25) have the same form; thus, for vanishingly small beamwidths, the approximation is exact for all geometries.

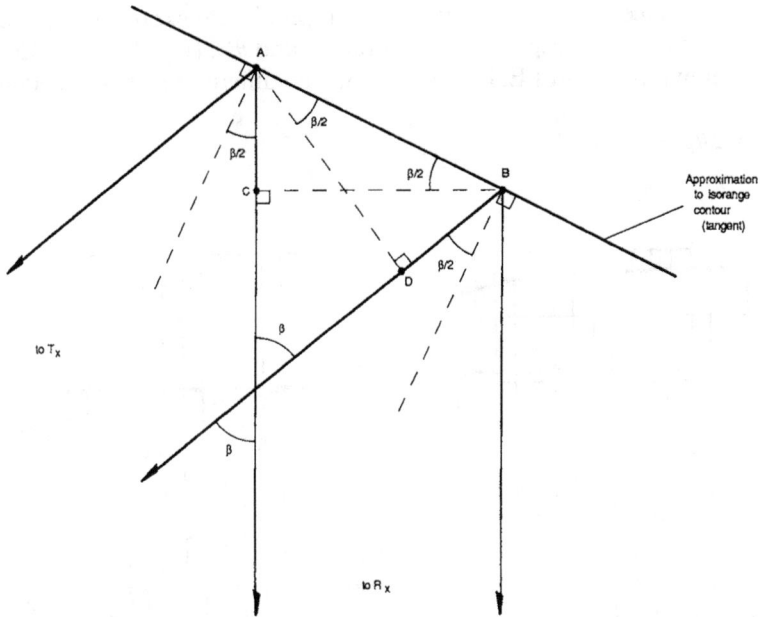

Figure E.2 Geometry showing equal cross-range and pseudocross-range dimensions for parallel beam rays and a tangent approximation to the isorange contour.

Solving (E.41) for $\Delta\theta'_{Rp}$ yields:

$$\Delta\theta'_{Rp} \approx (R_T/R_R)\Delta\theta'_T \tag{E.42}$$

Substituting (E.19) into (E.42) yields

$$\Delta\theta'_{Rp} \approx \frac{(1 - e^2)\Delta\theta'_T}{1 + e^2 - 2e\,\sin\theta_T} \tag{E.43}$$

In similar, but not identical geometries, the pseudoreceiving beam pointing angle θ_{Rp} can be approximated by θ_R. Thus, from (E.16a),

$$\theta'_{Rp} \approx \theta_R = \pm\cos^{-1}[(R_T/R_R)\cos\theta_T] \tag{E.44}$$

Substituting (E.19) into (E.44) yields

$$\theta'_{Rp} \approx \pm\cos^{-1}\left[\frac{(1 - e^2)\cos\theta_T}{1 + e^2 - 2e\,\sin\theta_T}\right] \tag{E.45}$$

The differences between the exact and approximate expressions for $\Delta\theta_{Rp}$ and $\Delta\theta'_{Rp}$, (E.26) and (E.43), respectively, and for θ_{Rp} and θ'_{Rp}, (E.27) and (E.45), respectively, are shown in Figures E.3 through E.8. The differences are defined as

$$\epsilon_\Delta = \Delta\theta_{Rp} - \Delta\theta'_{Rp} \tag{E.46}$$

$$\epsilon_\theta = \theta_{Rp} - \theta'_{Rp} \tag{E.47}$$

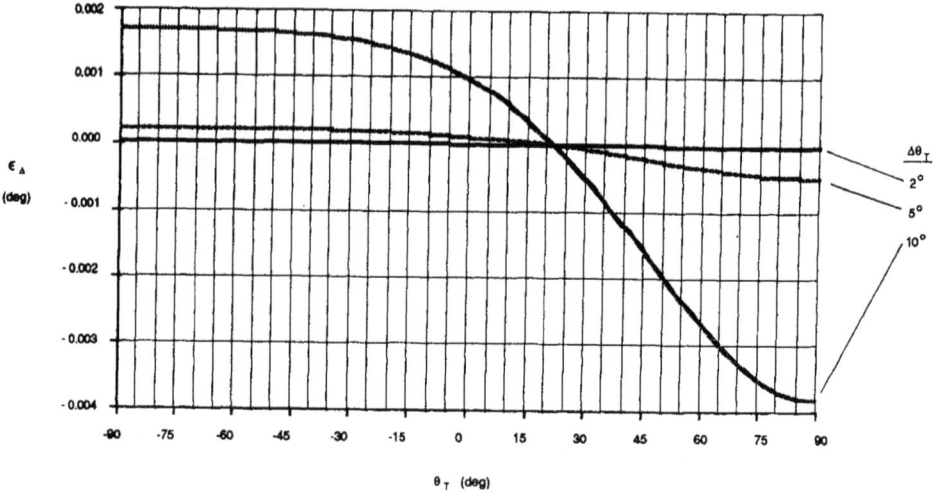

Figure E.3 Pseudoreceiving beamwidth difference (exact minus approximate) for $e = 0.1$.

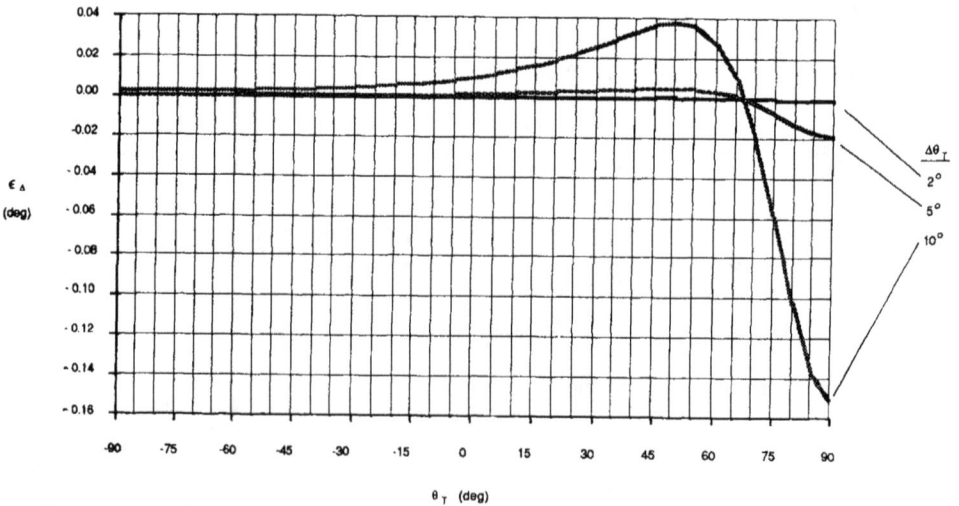

Figure E.4 Pseudoreceiving beamwidth difference (exact minus approximate) for $e = 0.5$.

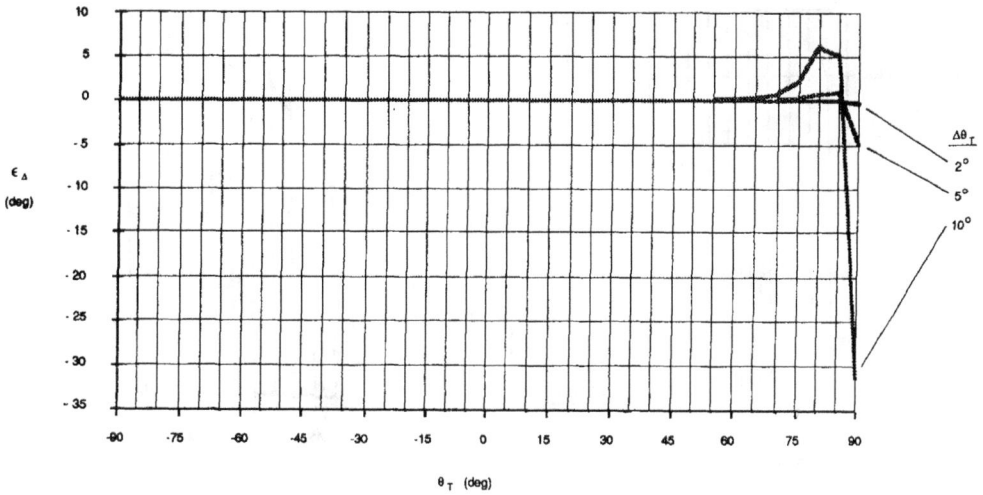

Figure E.5 Pseudoreceiving beamwidth difference (exact minus approximate) for $e = 0.9$.

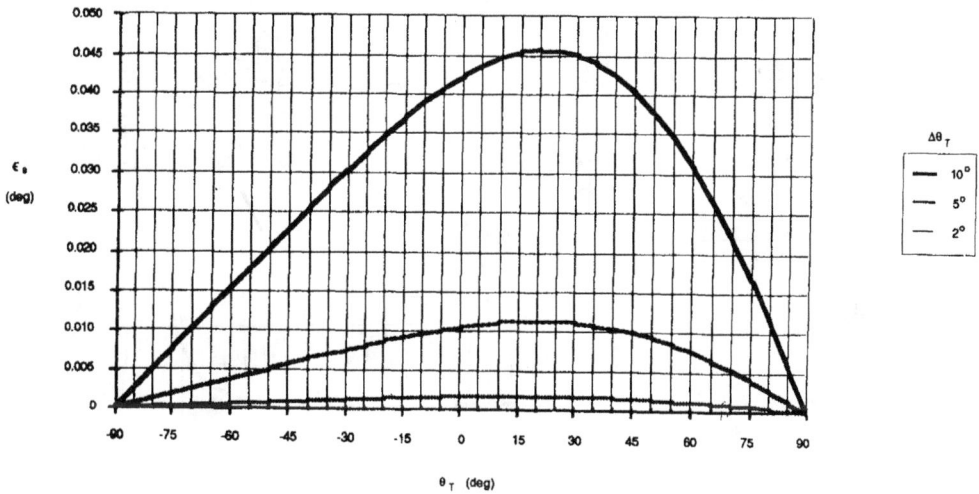

Figure E.6 Pseudoreceiving beam pointing angle difference (exact minus approximate) for $e = 0.1$.

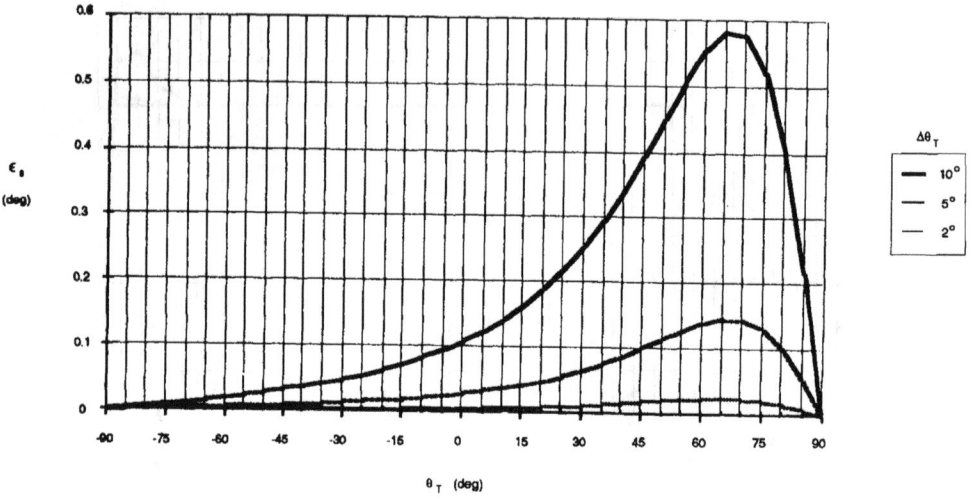

Figure E.7 Pseudoreceiving beam pointing angle difference (exact minus approximate) for $e = 0.5$.

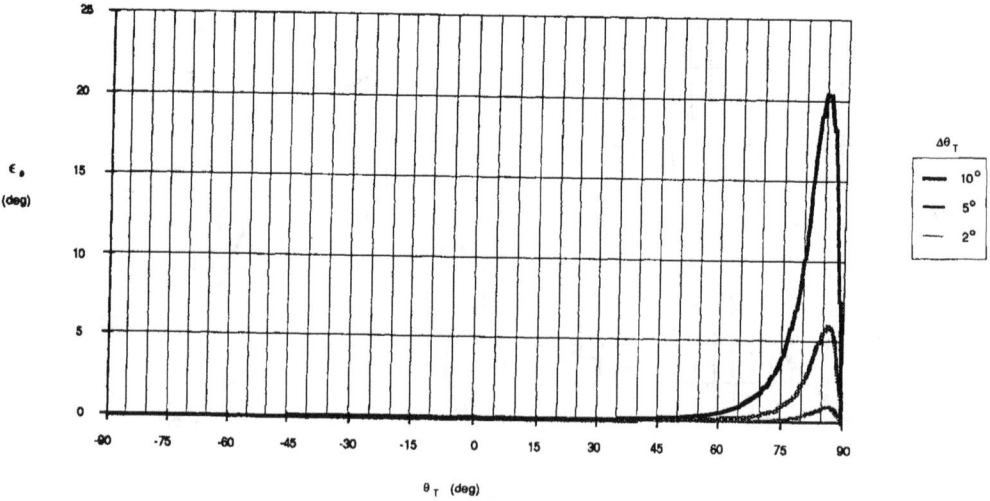

Figure E.8 Pseudoreceiving beam pointing angle difference (exact minus approximate) for $e = 0.9$.

For all ellipse eccentricities and all θ_T, ϵ_Δ and ϵ_θ increase as $\Delta\theta_T$ increases, as shown in Figures E.3 through E.8. Also, for a given ellipse eccentricity, all ϵ_Δ curves cross $\epsilon_\Delta = 0$ at approximately the same point on θ_T. For example, for $e = 0.5$ $\theta_T = 67.09°$, $67.07°$, and $66.99°$ for $\Delta\theta_T = 2°$, $5°$, and $10°$, respectively. The $\epsilon_\Delta = 0$ crossover point monotonically increases as e increases, reaching $\theta_T \approx 86.3°$ at $e = 0.9$, and $\theta_T \approx 87.9°$ at $e = 0.95$ (not shown in the figures).

For the same ellipse eccentricities, ϵ_θ is a maximum at the $\epsilon_\Delta = 0$ crossover point, as shown in Figures E.6, E.7, and E.8. However, ϵ_θ always remains positive. That is, the exact expression for θ_{Rp} leads the approximate expression, θ'_{Rp}, in the clockwise sense of the North-referenced coordinate system for $-90° \leq \theta_T \leq +90°$.

The value of θ_T for the ϵ_Δ crossover (and ϵ_θ maximum) point can in theory be obtained by setting $\epsilon_\Delta = 0$ and solving (E.46) for θ_T in terms of e. However, the expressions for $\Delta\theta_{Rp}$ and $\Delta\theta'_{Rp}$ in (E.46) are particularly messy, which discourages the analytical solution.

A fourth-order polynomial approximation to the crossover or maximum point for $0.1 \leq e \leq 0.95$ is given by

$$\theta_T = \cos^{-1}(1.0842 - 1.5254e - 0.2570e^2 + 1.4189e^3 - 0.7179e^4) \quad \text{(E.48)}$$

The maximum error in this approximation is $\pm 0.3°$, which is of the order of the spread in the crossover points for $2° \leq \theta_T \leq 10°$.

Note that on the extended baseline where $\theta_T = \pm 90°$, $\epsilon_\theta = 0$; thus, $\theta_R = \theta_{Rp} = \theta'_{Rp} = \pm 90°$ exactly, from (E.44) and, with some manipulation, (E.27). However, when $\theta_T = +90°$, ϵ_Δ becomes very large, approaching many tens of degrees for $e \geq 0.9$. This large error is caused by the sharp curvature of the highly eccentric ellipse near the baseline, resulting in a poor estimate of R_T/R_R in (E.42). Therefore, the exact expression for $\Delta\theta_{Rp}$, (E.27), should be used for $e > 0.9$ and θ_T near $+90°$.

As a rule of thumb, the approximations to $\Delta\theta_{Rp}$ and θ_{Rp} will generate errors less than $0.1°$ for $\Delta\theta_T \leq 10°$ under the following conditions:

θ_T	e
$\geq 0°$	≤ 0.2
$< 0°$	all e

with the differences in θ_{Rp} generally dominating.

Appendix F
ORTHOGONAL CONIC SECTION THEOREMS

F.1 INTRODUCTION

In Chapter 3 it was asserted that (1) at any point on an ellipse the bisector of the bistatic angle is orthogonal to the tangent to the ellipse and (2) the tangents of concentric hyperbolas are orthogonal to tangents of concentric ellipses at their points of intersection, when the hyperbolas and ellipses share common foci. These assertions are frequently made in the bistatic radar literature, but to the author's knowledge, their proofs are either not documented or not conveniently available. This appendix provides proofs to these two orthogonal conic section theorems.

F.2 ORTHOGONAL BISECTOR—TANGENT THEOREM

Figure F.1 shows the bisector—tangent geometry. The condition for orthogonality is

$$m_\beta = -(m_e)^{-1} \tag{F.1}$$

where m_β is the slope of the bistatic bisector and m_e is the slope of the ellipse tangent. Now

$$m_\beta = \tan\beta_L \tag{F.2}$$

and

$$m_e = \left(\frac{dy}{dx}\right)_e \tag{F.3}$$

312

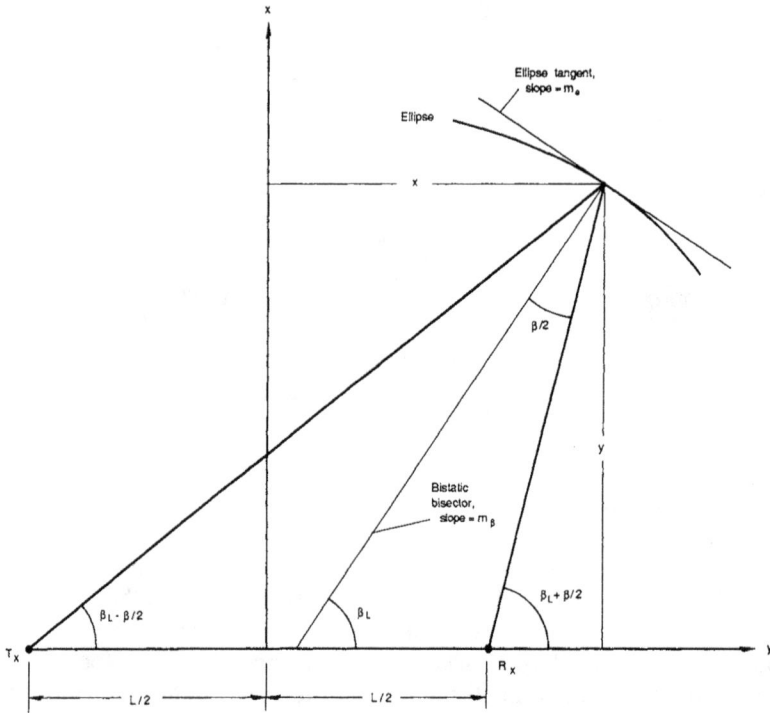

Figure F.1 Ellipse tangent—bistatic bisector geometry.

where β_L is the angle between the bistatic bisector and the baseline, as shown in Figure F.1, and $(dy/dx)_e$ is the derivative of the equation for an ellipse in rectilinear coordinates, defined in Figure F.1. The equation for an ellipse is

$$\frac{x^2}{a^2} + \frac{y^2}{b^2} = 1 \tag{F.4}$$

Solving (F.4) for y and differentiating yields

$$m_e = \frac{dy}{dx} = -(b/a)[a^2/x^2 - 1]^{-1/2} \tag{F.5}$$

The next step is to express (F.2) in terms of the parameters of (F.5), (x, a, b), where a is the semimajor axis of the ellipse and b is the semiminor axis of the ellipse, such that

$$a^2 - b^2 = L^2/4 \tag{F.6}$$

and L is the distance between ellipse foci, or the baseline.

The term $\tan\beta_L$ is developed using the geometry of Figure F.1 as follows:

$$\tan(\beta_L - \beta/2) = \frac{y}{x + L/2} \tag{F.7}$$

$$\tan(\beta_L + \beta/2) = \frac{y}{x - L/2} \tag{F.8}$$

Expanding (F.7) and (F.8) and eliminating the $\beta/2$ term yields

$$\frac{1 - \tan^2\beta_L}{2\tan\beta_L} = \tan(2\beta_L) = \frac{2xy}{x^2 - y^2 - L^2/4} \tag{F.9}$$

Therefore,

$$m_\beta = \tan\beta_L = \tan\left[(\tfrac{1}{2})\tan^{-1}\left(\frac{2xy}{x^2 - y^2 - L^2/4}\right)\right] \tag{F.10}$$

Solving (F.4) for y and substituting it and the expression for $L^2/4$ from (F.6) into (F.10) yields

$$m_\beta = \tan\beta_L = \tan\left\{(\tfrac{1}{2})\tan^{-1}\left[\frac{2bx\sqrt{1 - x^2/a^2}}{x^2(1 + b^2/a^2) - a^2}\right]\right\} \tag{F.11}$$

Substituting (F.11) and (F.5) into (F.1) and manipulating terms yields

$$\tan^{-1}\left[\frac{2bx\sqrt{1 - x^2/a^2}}{x^2(1 + b^2/a^2) - a^2}\right] = 2\tan^{-1}[(a/b)\sqrt{a^2/x^2 - 1}] \tag{F.12}$$

Now the arguments of (F.12) can be equated through the following relationship

$$\tan^{-1}A = 2\tan^{-1}B = C \tag{F.13}$$

$$B = \tan(C/2) \tag{F.14}$$

$$A = \tan C \tag{F.15a}$$

$$= \frac{2\tan(C/2)}{1 - \tan^2(C/2)} \tag{F.15b}$$

Therefore,

$$A = \frac{2B}{1 - B^2} \tag{F.16}$$

Substituting the arguments from (F.12) into (F.16) yields

$$\frac{2bx\sqrt{1 - x^2/a^2}}{x^2(1 + b^2/a^2) - a^2} = \frac{(2a/b)\sqrt{a^2/x^2 - 1}}{1 - (a^2/b^2)(a^2/x^2 - 1)} \tag{F.17}$$

Multiplying the numerator and denominator on the right side of (F.17) by x^2b^2/a^2 yields the left side. Thus, (F.1) is satisfied.

F.3   ORTHOGONAL ELLIPSE—HYPERBOLA THEOREM

Figure F.2 shows the ellipse—hyperbola geometry. The ellipse has foci, T_x and R_x, with separation L; the hyperbola has foci, T'_x and R'_x, with separation L'. The condition for orthogonality at their point of intersection is

$$m_e = -(m_h)^{-1} \tag{F.18}$$

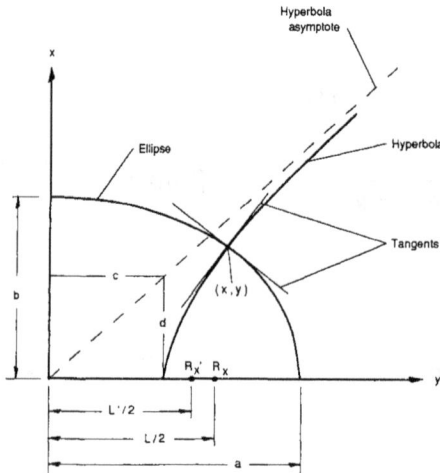

Figure F.2   Ellipse—hyperbola geometry.

where

$$m_e = \left(\frac{dy}{dx}\right)_e = \text{slope of the ellipse tangent,}$$

$$m_h = \left(\frac{dy}{dx}\right)_h = \text{slope of the hyperbola tangent.}$$

The expression for m_e, again in rectilinear coordinates, is given by (F.5). The expression for m_h is developed as follows. The equation for a hyperbola in the same rectilinear coordinate system is

$$\frac{x^2}{c^2} - \frac{y^2}{d^2} = 1 \tag{F.19}$$

where c is the semitransverse axis and d is the semiconjugate axis of the hyperbola, such that

$$c^2 + d^2 = L'^2/4 \tag{F.20}$$

Solving (F.19) for y and differentiating yields

$$m_h = \left(\frac{dy}{dx}\right)_h = (d/c)(1 - c^2x^2)^{-1/2} \tag{F.21}$$

Substituting (F.21) and (F.5) into (F.18), and rearranging terms, yields

$$\left(\frac{bd}{ac}\right)^2 = (a^2/x^2 - 1)(1 - c^2/x^2) \tag{F.22}$$

The point of intersection between ellipse and hyperbola is obtained by combining (F.4) and (F.19):

$$x = \left[\frac{a^2c^2(b^2 + d^2)}{a^2d^2 + b^2c^2}\right]^{1/2} \tag{F.23}$$

Combining (F.22) and (F.23) yields

$$a^2 - b^2 = c^2 + d^2 \tag{F.24}$$

which is the condition for orthogonality. However, $a^2 - b^2 = L^2/4$ and $c^2 + d^2 = L'^2/4$. Thus, $L = L'$; that is, orthogonal ellipses and hyperbolas share a common baseline and common foci. Also, because $a^2 - b^2 = L^2/4$ and $c^2 + d^2 = L^2/4$ define a family of concentric ellipses and hyperbolas, respectively, all with a common baseline, these ellipses and hyperbolas are orthogonal at their points of intersection.

REFERENCES AND BIBLIOGRAPHY

1. Skolnik, M.I., *Introduction to Radar Systems,* McGraw-Hill, New York, 1980.
2. *IEEE Standard Radar Definitions,* IEEE Std. 686-1982, October 15, 1982.
3. Heimiller, R.C., J.E. Belyea, and P.G. Tomlinson, "Distributed Array Radar," *IEEE Trans.,* AES-19, 831–839, 1983.
4. Steinberg, B.D., *Principles of Aperture and Array System Design—Including Random and Adaptive Arrays,* John Wiley and Sons, New York, 1976.
5. Steinberg, B.D., and E. Yadin, "Distributed Airborne Array Concepts," *IEEE Trans. Aerospace and Electronic Systems,* AES-18, 219–226, 1982.
6. Steinberg, B.D., "High Angular Microwave Resolution from Distorted Arrays," *Proc. Int. Computing Conf.,* Vol. 23, 1980.
7. Easton, R.L., and J.J. Fleming, "The Navy Space Surveillance System," *Proc. IRE,* 48, 663–669, 1960.
8. Mengel, J.T., "Tracking the Earth Satellite, and Data Transmission by Radio," *Proc. IRE,* 44, 755–760, June 1956.
9. Merters, L.E., and R.H. Tabeling, "Tracking Instrumentation and Accuracy on the Eastern Test Range," *IEEE Trans.,* SET-11, 14–23, March 1965.
10. Scavullo, J.J., and F.J. Paul, *Aerospace Ranges: Instrumentation,* Van Nostrand, Princeton, NJ, 1965.
11. Steinberg, B.D., *et al.,* "First Experimental Results for the Valley Forge Radio Camera Program," *Proc. IEEE,* 67, 1370–1371, September 1979.
12. Steinberg, B.D., "Radar Imaging from a Distributed Array: The Radio Camera Algorithm and Experiments," *IEEE Trans. Antennas and Propagation,* AP-29, 740–748, September 1981.
13. Salah, J.E., and J.E. Morriello, "Development of a Multistatic Measurement System," *Proc. IEEE Int. Radar Conf.,* pp. 88–93, 1980.
14. "Multistatic Mode Raises Radar Accuracy," *Aviation Week and Space Technology,* 62–69, July 14, 1980.
15. Skolnik, M.I., "An Analysis of Bistatic Radar," *IRE Trans. Aerospace and Navigational Electronics,* 19–27, March 1961.
16. Caspers, J.M., "Bistatic and Multistatic Radar," Chapter 36, in Skolnik, M.I., ed., *Radar Handbook,* 1st Ed., McGraw-Hill, New York, 1970.
17. Ewing, E.F., "The Applicability of Bistatic Radar to Short Range Surveillance," *IEEE Int. Conf. Radar '77,* London, Conf. Pub. No. 155, 53–58, 1977.
18. Ewing, E.F., and L.W. Dicken, "Some Applications of Bistatic and Multi-Bistatic Radars," *Proc. Int. Radar Conf.,* Paris, pp. 222–231, 1978.
19. Farina, A., and E. Hanle, "Position Accuracy in Netted Monostatic and Bistatic Radar," *IEEE Trans. Aerospace and Electronic Systems,* AES-19 (4), 513–520, July 1983.

20. Hanle, E., "Survey of Bistatic and Multistatic Radar," *IEE Proc.,* 133(7), Pt. F, 587–595, December 1986.
21. U.S. Patent 1,981,884, "System for Detecting Objects by Radio," issued to A.H. Taylor, L.C. Young, and L.A. Hyland, November 27, 1934.
22. Williams, A.F., in *The Study of Radar,* "Research Science and Its Application in Industry," 6, 434–440, Butterworth Scientific Publications, London, 1953.
23. Watson-Watt, R., *The Pulse of Radar,* Dial Press, New York, 1959.
24. Skolnik, M.I., "Fifty Years of Radar," *Proc. IEEE,* 73(2), 182–197, February 1985.
25. Guerlac, H.E., *Radar in World War II,* Vols. I and II, Tomask/American Institute of Physics, New York, 1987.
26. Price, A., *The History of U.S. Electronic Warfare,* Vol. 1, The Association of Old Crows, Alexandria, VA, 1984.
27. Barton, D.K., "Historical Perspective on Radar," *Microwave Journal,* 23(8), 21, August 1980.
28. Summers, J.E., and Browning, D.J., "An Introduction to Airborne Bistatic Radar," *IEE Colloquium Ground and Airborne Multistatic Radar,* London, 2/1–2/5, 1981.
29. Eon, L.G., "An Investigation of the Techniques Designed to Provide Early Warning Radar Fence for the Air Defense of Canada," Defense Research Board (Canada) Rep. No. TELS 100, December 1, 1952.
30. Sloane, E.A., J. Salerno, E.S. Candidas, and M.I. Skolnik, "A Bistatic CW Radar," Technical Rep. 82, MIT Lincoln Laboratory, Lexington, MA, June 6, 1955, AD 76454.
31. Skolnik, M.I., J. Salerno, and E.S. Candidas, "Prediction of Bistatic CW Radar Performance," presented at *Symp. Radar Detection Theory,* ONR Symp. Rep. ACR-10, 267–278, March 1–2, 1956, Washington, D.C.
32. Private communication from M.I. Skolnik, NRL, September 1986.
33. Siegel, K.M., *et al.,* "Bistatic Radar Cross Sections of Surfaces of Revolution," *Journal of Applied Physics,* 26(3), 297–305, March 1955.
34. Siegel, K.M., "Bistatic Radars and Forward Scattering," *Proc. Nat. Conf. Aeronautics and Electronics,* 286–290, May 12–14, 1958.
35. Schultz, F.V., *et al.,* "Measurement of the Radar Cross-Section of a Man," *Proc. IRE,* 46, 476–481, February 1958.
36. Crispin, J.W., Jr., *et al.,* "A Theoretical Method for the Calculation of Radar Cross Section of Aircraft and Missiles," University of Michigan Radiation Laboratory Rep. 2591-1-H, July 1959.
37. Hiatt, R.E., *et al.,* "Forward Scattering by Coated Objects Illuminated by Short Wavelength Radar," *Proc. IRE,* 48, 1630–1635, September 1960.
38. Garbacz, R.J., and D.L. Moffett, "An Experimental Study of Bistatic Scattering from Some Small, Absorber-Coated, Metal Shapes," *Proc. IRE,* 49, 1184–1192, July 1961.
39. Andreasen, M.G., "Scattering from Bodies of Revolution," *IEEE Trans. Antennas and Propagation,* AP-13, 303–310, March 1965.
40. Mullin, C.R., *et al.,* "A Numerical Technique for the Determination of the Scattering Cross Sections of Infinite Cylinders of Arbitrary Geometric Cross Section," *IEEE Trans. Antennas and Propagation,* AP-13, 141–149, January 1965.
41. Kell, R.E., "On the Derivation of Bistatic RCS from Monostatic Measurements," *Proc. IEEE,* 53, 983–988, August 1965.
42. Cost, S.T., "Measurements of the Bistatic Echo Area of Terrain of X-band," Ohio State University Antenna Laboratory Rep. 1822-2, May 1965.
43. Pidgeon, V.W., "Bistatic Cross Section of the Sea," *IEEE Trans. Antennas and Propagation,* AP-14(3), 405–406, May 1966.
44. Lefevre, R.J., "Bistatic Radar: New Application for an Old Technique," *WESCON Conf. Record,* San Francisco, 1–20, 1979.

45. Fleming, F.L., and N.J. Willis, "Sanctuary Radar," *Proc. Military Microwaves '80 Conf.*, London, 103–108, October 22–24, 1980.

46. Forrest, J.R., and J.G. Schoenenberger, "Totally Independent Bistatic Radar Receiver with Real-Time Microprocessor Scan Correction," *IEEE Int. Radar Conf.*, 380–386, 1980.

47. Pell, C., *et al.*, "An Experimental Bistatic Radar Trials System," *IEE Colloquium Ground and Airborne Multistatic Radar*, London, 6/1–6/12, 1981.

48. Schoenenberger, J.G., and J.R. Forrest, "Principles of Independent Receivers for Use with Co-operative Radar Transmitters," *The Radio and Electronic Engineer*, 52(2), 93–101, February 1982.

49. Soame, T.A., and D.M. Gould, "Description of an Experimental Bistatic Radar System," *IEE Int. Conf. Radar '87*, Conf. Pub. No. 281, 12–16, 1987.

50. Dunsmore, M.R.B., "Bistatic Radars for Air Defense," *IEE Int. Conf. Radar '87*, Conf. Pub. No. 281, 7–11, 1987.

51. Lorti, D.C., and M. Balser, "Simulated Performance of a Tactical Bistatic Radar System," *IEEE Eascon '77 Record*, IEEE Pub. 77 CH1255-9, Arlington, VA, 4-4A–4-40, 1977.

52. "Tactical Bistatic Radar Demonstrated," *Defense Electronics*, 12, 78–82, 1980.

53. Auterman, J.L., "Phase Stability Requirements for a Bistatic SAR," *Proc. 1984 IEEE Nat. Radar Conf.*, Atlanta, GA, 48–52, March 1984.

54. "Bistatic Radars Hold Promise for Future Systems," *Microwave Systems News*, 119–136, October 1984.

55. Griffiths, H.D., *et al.*, "Television-Based Bistatic Radar," *IEE Proc.*, 133(7), Pt. F, 649–657, December 1986.

56. Tomiyasu, K., "Bistatic Synthetic Aperture Radar Using Two Satellites," *IEEE EASCON '78 Record*, Arlington, VA, 106–110, 1978.

57. Lee, P.K., and T.F. Coffey, "Space-Based Bistatic Radar: Opportunity for Future Tactical Air Surveillance," *IEEE Int. Radar Conf.*, Washington, DC, 322–329, 1985.

58. Hsu, Y.S., and D.C. Lorti, "Spaceborne Bistatic Radar—An Overview," *IEE Proc.*, 133(7), Pt. F, 642–648, December 1986.

59. Anthony, S., *et al.*, "Calibration Considerations in a Large Bistatic Angle Airborne Radar System for Ground Clutter Measurements," *Proc. 1988 IEEE Nat. Radar Conf.*, Ann Arbor, MI, 230–234, April 1988.

60. Walker, B.C., and M.W. Callahan, "A Bistatic Pulse-Doppler Intruder-Detection Radar," *IEEE Int. Radar Conf.*, 130–134, 1985.

61. Dawson, C.H., "Inactive Doppler Acquisition Systems," *Trans. AIEE*, 81, 568–571, January 1963.

62. Detlefsen, J., "Application of Multistatic Radar Principles to Short Range Imaging," *IEEE Proc.*, 133(7), Pt. F, December 1986.

63. Nicholson, A.M., and G.F. Ross, "A New Radar Concept for Short Range Application," *IEEE Int. Radar Conf.*, 1975.

64. Tyler, G.L., "The Bistatic Continuous-Wave Radar Method for the Study of Planetary Surfaces," *J. Geophys. Res.*, 71(6), 1559–1567, March 15, 1966.

65. Tyler, G.L., *et al.*, "Bistatic Radar Detection of Lunar Scattering Centers with Lunar Orbiter 1," *Science*, 157, 193–195, July 1967.

66. Pavel'yev, A.G., *et al.*, "The Study of Venus by Means of the Bistatic Radar Method," *Radio Engineering and Electronic Physics (USSR)*, 23, October 1978.

67. Zebker, H.Z., and G.L. Tyler, "Thickness of Saturn's Rings Inferred from Voyager 1 Observations of Microwave Scatter," *Science*, 113, 396–398, January 1984.

68. Tang, C.H., *et al.*, "Measurements of Electrical Properties of the Martian Surface," *J. Geophys. Res.*, 82, 4305–4315, September, 1977.

69. Zhou Zheng-Ou, *et al.*, "A Bistatic Radar for Geological Probing," *Microwave Journal*, pp. 257–263, May 1984.

70. Peterson, A.M., *et al.*, "Bistatic Radar Observation of Long Period, Directional Ocean-Wave Spectra with Loran-A," *Science*, 170, 158–161, October 1970.

71. Doviak, R.J., *et al.*, "Bistatic Radar Detection of High Altitude Clear Air Atmospheric Targets," *Radio Science*, 7, 993–1003, November 1972.

72. Wright, J.W., and R.I. Kressman, "First Bistatic Oblique Incidence Ionograms Between Digital Ionosondes," *Radio Science*, 18, 608–614, July-August 1983.

73. Jackson, M.C., "The Geometry of Bistatic Radar Systems," *IEE Proc.*, 133(7) Pt. F, 604–612, December 1986.

74. Davies, D.E.N., "Use of Bistatic Radar Techniques to Improve Resolution in the Vertical Plane," *IEE Electronics Letters*, 4(9), 170–171, May 3, 1968.

75. McCall, E.G., "Bistatic Clutter in a Moving Receiver System," *RCA Review*, 518–540, September 1969.

76. Crowder, H.A., "Ground Clutter Isodops for Coherent Bistatic Radar," *IRE Nat. Convention Record*, P. 5, New York, pp. 88–94, 1959.

77. Dana, R.A., and D.L. Knepp, "The Impact of Strong Scintillation on Space Based Radar Design, I: Coherent Detection," *IEEE Trans. Aerospace and Electronic Systems*, AES-19(4), July 1983.

78. Pyati, V.P., "The Role of Circular Polarization in Bistatic Radars for Mitigation of Interference Due to Rain," *IEEE Trans. Antennas and Propagation*, AP-32(3), 295–296, March 1984.

79. McCue, J.J.G., "Suppression of Range Sidelobes in Bistatic Radars," *Proc. IEEE*, 68(3), 422–423, March 1980.

80. Buchner, M.R., "A Multistatic Track Filter with Optimal Measurement Selection," *Proc. IEEE Radar Conf.*, London, pp. 72–75, 1977.

81. Farina, A., "Tracking Function in Bistatic and Multistatic Radar Systems," *IEE Proc.*, 133(7), Pt. F, 630–637, December 1986.

82. Retzer, G., "Some Basic Comments on Multistatic Radar Concepts and Techniques," *IEE Colloquium Ground and Airborne Multistatic Radar*, London, 3/1–3/3, 1981.

83. Hoisington, D.B., and C.E. Carroll, "Improved Sweep Waveform Generator for Bistatic Radar," U.S. Naval Postgraduate School, Monterey, CA, August 1975.

84. Kuschel, H., "Bistatic Radar Coverage—A Quantification of System and Environmental Interferences," *IEE Int. Radar Conf. '87*, Conf. Pub. No. 281, pp. 17–21, 1987.

85. Barrick, D.E., "Normalization of Bistatic Radar Return," Ohio State University Research Foundation Rep. 1388-13, January 15, 1964.

86. Peake, W.H., and S.T. Cost, "The Bistatic Echo Area of Terrain of 10 GHz," *IEEE WESCON Rec.*, Session 22/2, 1–10, 1968.

87. Weiner, M.M., and P.D. Kaplan, "Bistatic Surface Clutter Resolution Area at Small Grazing Angles," MITRE Corporation, Bedford, MA, RADC-TR-82-289, AD A123660, November 1982.

88. Moyer, L.R., C.J. Morgan, and D.A. Rugger, "An Exact Expression for the Resolution Cell Area in a Special Case of Bistatic Radar Systems," *IEEE Trans. Aerospace and Electronic Systems*, Correspondence, July 1989.

89. Lorti, D.C., and J.J. Bowman, "Will Tactical Aircraft Use Bistatic Radar?" *Microwave Systems News*, 8(9), 49–54, September 1978.

90. Private communication from L.R. Moyer, Technology Service Corporation, February 1988.

91. Kock, W.I., "Related Experiments with Sound Waves and Electromagnetic Waves," *Proc. IRE*, 47, 1200–1201, July 1959.

92. Siegel, K.M., *et al.*, "RCS Calculation of Simple Shapes—Bistatic," Chapter 5 in *Methods of Radar Cross-Section Analysis*, Academic Press, New York, 1968.

93. Weil, H., *et al.,* "Scattering of Electromagnetic Waves by Spheres," University of Michigan Radiation Laboratory Studies in Radar Cross Sections X, Rep. 2255-20-T, Contract AF 30(602)-1070, July 1956.
94. King, R.W.P., and T.T. Wu, "The Scattering and Diffraction of Waves," Harvard University Press, Cambridge, MA, 1959.
95. Goodrich, R.F., *et al.,* "Diffraction and Scattering by Regular Bodies—I: The Sphere," University of Michigan Dept. of Electrical Engineering, Rep. 3648-1-T, 1961.
96. Matsuo, M., *et al.,* "Bistatic Radar Cross Section Measurement by Pendulum Method," *IEEE Trans. Antennas and Propagation,* AP-18(1), 83–88, January 1970.
97. Ewell, G.W., and S.P. Zehner, "Bistatic Radar Cross Section of Ship Targets," *IEEE Journal Oceanic Engineering,* OE-5(4), 211–215, October 1980.
98. "Radar Cross-Section Measurements," General Motors, Delco Electronics Division, Rep. R81-152, Santa Barbara, CA, 1981.
99. Basalov, A.A., and V.I. Ostravityanov, *Statistical Theory of Extended Radar Targets,* Artech House, Norwood, MA, 1984.
100. Paddison, F.C., *et al.,* "Large Bistatic Angle Radar Cross Section of a Right Circular Cylinder," *Electromagnetics,* 5, 63–77, 1985, The Johns Hopkins University Applied Physics Laboratory, Laurel, MD, 20707.
101. Glaser, J.I., "Bistatic RCS of Complex Objects Near Forward Scatter," *IEEE Trans. Aerospace and Electronic Systems,* AES-21(1), 70–78, January 1985.
102. Cha, Chung-Chi, *et al.,* "An RCS Analysis of Generic Airborne Vehicles Dependence on Frequency and Bistatic Angle," *Proc. IEEE Nat. Radar Conf.,* Ann Arbor, MI, pp. 214–219, April 20, 1988.
103. Private communication from M.M. Weiner, MITRE Corporation, April 1988.
104. Pierson, W.A., *et al.,* "The Effect of Coupling on Monostatic-Bistatic Equivalence," *Proc. IEEE Letters,* pp. 84–86, January 1971.
105. Barton, D.K., *Modern Radar System Analysis,* Artech House, Norwood, MA, 1988, 121–123.
106. Burk, G.J., and A.J. Foggio, "Numerical Electromagnetic Code (NEC)-Method of Moments," Naval Ocean Systems Center, San Diego, CA, 1981.
107. Pidgeon, V.W., "Bistatic Cross Section of the Sea for Beaufort 5 Sea," *Science and Technology,* 17, 447–448, 1968, American Astronautical Society, San Diego, CA.
108. Domville, A.R., "The Bistatic Reflection From Land and Sea of X-band Radio Waves, Part I," GEC Electronics Ltd., Stanmore, England, Memorandum SLM 1802, July 1967.
109. Domville, A.R., "The Bistatic Reflection From Land and Sea of X-band Radio Waves, Part II," GEC Electronics Ltd., Stanmore, England, Memorandum SLM 2116, July 1968.
110. Domville, A.R., "The Bistatic Reflection From Land and Sea of X-band Radio Waves, Part II—Supplement," GEC-AEI Electronics Ltd., Stanmore, England, Memorandum SLM 2116 (Supplement), July 1969.
111. Larson, R.W., *et al.,* "Bistatic Clutter Data Measurements Program," Environmental Research Institute of Michigan, RADC-TR-77-389, AD-A049037, November 1977.
112. Larson, R.W., *et al.,* "Bistatic Clutter Measurements," *IEEE Trans. Antennas and Propagation,* AP-26(6), 801–804, November 1978.
113. Ewell, G.W., and S.P. Zehner, "Bistatic Sea Clutter Return Near Grazing Incidence," *IEE Conf. Radar '82,* Conf. Pub. No. 216, 188–192, October 1982.
114. Ewell, G.W., "Bistatic Radar Cross Section Measurements," Chapter 7 in *Techniques of Radar Reflectivity Measurement,* N.C. Currie, ed., 2nd Ed., Artech House, Norwood, MA, 1989.
115. Ulaby, F.T., *et al.,* "Millimeter-wave Bistatic Scattering from Ground and Vegetation Targets," *IEEE Trans. Geoscience and Remote Sensing,* GE-26(3), May 1988.
116. Nathanson, F.E., *Radar Design Principles,* McGraw-Hill, New York, 1969.

117. Vander Schurr, R.E., and P.G. Tomlinson, "Bistatic Clutter Analysis," Decision-Science Applications, Inc., RADC-TR-79-70, April 1979.

118. Sauermann, G.O., and P.C. Waterman, "Scattering Modeling: Investigation of Scattering by Rough Surfaces," MITRE Rep. MTR-2762, AFAL TR-73-334, January 1974.

119. Zornig, J.G., *et al.*, "Bistatic Surface Scattering Strength at Short Wavelengths," Rep. CS-9, Yale University, Department of Engineering and Applied Science, AD-A041316, June 1977.

120. Bramley, E.N., and S.M. Cherry, "Investigation of Microwave Scattering by Tall Buildings," *Proc. IEE,* 120(8), 833–842, August 1973.

121. Brindly, A.E., *et al.*, "A Joint Army/Air Force Investigation of Reflection Coefficient at C and Ku Bands for Vertical, Horizontal and Circular System Polarizations," IIT Research Institute, Chicago, IL, Final Rep. TR-76-67, AD-A031403, July 1976.

122. Tang, C.H., *et al.*, "Bistatic Radar Measurements of Electrical Properties of the Martian Surface," *J. Geophys. Res.* (USA), 82(28), 4305–4315, September 1977.

123. Barton, D.K., "Land Clutter Models for Radar Design and Analysis, *Proc. IEEE,* 73(2), 198–204, February 1985.

124. Beckmann, P., and A. Spizzichino, *The Scattering of EM Waves from Rough Surfaces,* Pergamon, New York, 1963. Republished by Artech House, Norwood, MA, 1988.

125. Private communication from F.E. Nathanson, May 1988.

126. Hanle, E., "Pulse Chasing with Bistatic Radar-Combined Space-Time Filtering," in *Signal Processing II: Theories and Applications,* H.W. Schussler, ed., Elsevier Science Publishers B.V., North-Holland, Amsterdam, 665–668.

127. Schoenenberger, J.G., and J.R. Forrest, "Principles of Independent Receivers for Use with Cooperative Radar Transmitters," *The Radio and Electronic Engineer,* 52(2), 93–101, February 1982.

128. Frank, J., and J. Ruze, "Beam Steering Increments for a Phased Array," *IEEE Trans. Antennas and Propagation,* AP-15(6), 820–821, November 1967.

129. Freedman, N., "Bistatic Radar System Configuration and Evaluation," Final Rep. ER76-4414, 1976, Independent Development Project 76D-220, Raytheon Corporation, December 30, 1976.

130. Bovey, C.K., and C.P. Horne, "Synchronization Aspects for Bistatic Radars," *IEE Proc. Int. Conf. Radar '87,* Conf. Pub. No. 281, pp. 22–25, 1987.

131. Schoenenberger, J.G., *et al.*, "Design and Implementation of a UHF Band Bistatic Radar Receiver," *IEE Colloquium Ground and Airborne Multistatic Radar,* London, 7/1–7/3, 1981.

132. Griffiths, H.D., and S.M. Carter, "Provision of Moving Target Indication in an Independent Bistatic Radar Receiver," *The Radio and Electronic Engineer,* 54(7/8), 336–342, July/August 1984.

133. Kirk, J.C., Jr., "Bistatic SAR Motion Compensation," *IEEE Int. Radar Conf.,* 360–365, 1985.

134. Costas, J.P., "Synchronous Communications," *Proc. IRE,* 44, 1713–1718, December 1956.

135. Retzer, G., "A Concept for Signal Processing in Bistatic Radar," *IEEE Int. Radar Conf.,* 288–293, 1980.

136. Blake, L.V., *Radar Range-Performance Analysis,* Lexington Books, Lexington, MA, 1980. Republished by Artech House, Norwood, MA, 1986.

137. James and James, *Mathematics Dictionary,* Van Nostrand, New York, 1949.

138. Milne, K., "Principles and Concepts of Multistatic Surveillance Radars," *IEE Proc. Int. Conf. Radar '77,* London, 46–52, October 1977.

139. Hanle, E., "Distance Considerations for Multistatic Radar," *Proc. IEEE Int. Radar Conf.,* 100–105, 1980.

140. Farina, A., "Multistatic Tracking and Comparison with Netted Monostatic Systems," *IEE Proc. Int. Conf. Radar '82,* Conf. Pub. No. 216, London, 183–187, 1982.

141. Pinnell, S.E.A., "Stealth Aircraft," *Aviation Week and Space Technology,* Letter to Editor, 114, May 4, 1981.

142. Hanle, E., "Influence of the Bistatic Cross Section of Aircraft on Multistatic Radar," *IEE Colloquium Ground and Airborne Multistatic Radar,* London, 5/1–5/4, 1981.

143. Freedman, N., "Bistatic Radar System Configuration and Evaluation," Final Rep. ER76-4414, 1976, Independent Development Project 76D-220, Raytheon Corporation, December 30, 1976. (*See* [129].)

144. Soame, T.A., and D.M. Gould, "Description of an Experimental Bistatic Radar System," *IEE Proc. Int. Conf. Radar '87,* Conf. Pub. No. 281, 12–16. (*See* [49].)

145. Cheston, T.C., and J. Frank, "Array Antennas," Chapter 11 in *Radar Handbook,* M.I. Skolnik, ed., 1st Ed., McGraw-Hill, New York, 1970.

146. Chernoff, R., "Large Active Retrodirective Arrays for Space Application," *IEEE Trans. Antennas and Propagation,* AP-27, 489–496, July 1979.

147. Davies, D.E.N., *et al.,* "A High Resolution Radar Incorporating a Mechanical Scanning Transmitter and a Static Multi-Beam Receiver," *IEE Proc. Int. Conf. Radar '77,* Conf. Pub. No. 155, London, 1977, 59–62.

148. Long, M.W., *Radar Reflectivity of Land and Sea,* Lexington Books, Lexington, MA, 1975, 79–84. Republished by Artech House, Norwood, MA, 1983.

149. Pell, C., "Some Systems Aspects of Multistatic Radar for Long Range Air Defense," *Military Microwave Conf.,* London, 85–90, 1984.

150. Private communication from A. Gold, RDA, November 1988.

151. Fisher, N.I., *et al., Statistical Analysis of Spherical Data,* Cambridge University Press, Cambridge, UK, 1987, p. 32.

152. Gradshteyn, I.S., and I.M. Ryzhik, Section 3.671 in *Table of Integrals Series and Products,* A. Jeffrey, ed., Academic Press, New York, 1980.

153. Stinson, G.W., *Introduction to Airborne Radar,* Hughes Aircraft Company, El Segundo, CA, 1983.

154. Barton, D.K., *Modern Radar System Analysis,* Artech House, Norwood, MA, 1988.

155. Barton, D.K., *Radar System Analysis,* Artech House, Norwood, MA, 1976.

156. Englund, C.R., *et al.,* "Some Results of a Study of Ultra-Short-Wave Transmission Phenomena," *Proc. IRE,* 21(3), 464–492, March 1933.

157. Bowen, E.G., *Radar Days,* IOP Publishing Ltd., Bristol, UK, 1987.

158. Price, A., *Instruments of Darkness: The History of Electronic Warfare,* Charles Scribner's Sons, New York, 1978.

159. Swords, S.S., *Technical History of the Beginnings of RADAR,* IEE History of Technology Series, Vol. 6, Peter Peregrinus, London, 1986.

160. Private communication from Robert G. Bulk, TSC, June 1989.

161. Fink, D.G., *Radar Engineering,* McGraw-Hill, New York, 1947.

162. Page, R.M., *The Origin of Radar,* Anchor Books, Doubleday, New York, 1962.

163. Secor, H.W., "Tesla's Views on Electricity and the War," *Electrical Experimenter,* 5(4), 270ff., 1917.

164. Allison, D.K., "New Eye for the Navy: The Origin of Radar at the Naval Research Laboratory," NRL Rep. 8466, Naval Research Laboratory, Washington, DC, 1981.

165. Howeth, L.S., "History of Communications-Electronics in the United States Navy," U.S. Government Printing Office, Washington, DC, Chapter XXXVIII, 1963.

166. Gebhard, L.A., "Evolution of Naval Radio-Electronics and Contributions of the Naval Research Laboratory," NRL Rep. 8300, Naval Research Laboratory, Washington, DC, 1979.

167. Tiberio, U., "Some Historical Data Concerning the First Italian Naval Radar," *IEEE Trans. Aerospace and Electronic Systems,* AES-5, 733–735, 1979.

168. Wilkinson, R.I., "Short Survey of Japanese Radar-II," *Electrical Engineering,* 65, 455–463, 1946.

169. Nancarrow, F.E., A.H. Mumford, P.C. Carter, and H.T. Mitchell, "Interference by Aeroplanes," GPO Radio Rep. 233, Part V, June 3, 1932.

170. Watson-Watt, R., *Three Steps to Victory*, Oldhams Press, Ltd., London, 1957.
171. Johnson, B., *The Secret War*, Methuen, Inc., New York, 1978.
172. Cooper, H.W., "Radar: History of a Need-Fostered System," *Microwave Journal*, May 1989.
173. Fawcette, J., "Bistatic Radar May Find a 'Sanctuary' in Space," *Electronic Warfare/Defense Electronics*, 84–88, January 1978.
174. Fawcette, J., "Vulnerable Radars Seek a Safe Sanctuary," *Microwave Systems News*, 45–50, April 1980.
175. Elson, B.M., "Bistatic Airborne Use Studied," *Aviation Week and Space Technology*, 57–58, August 13, 1979.
176. Buchbinder, M., F. Esposito, and H. Perini, "Sanctuary Receive Antenna System," Antennas and Propagation Society, 1980 International Symposium, IEEE and USNC/URSI Meeting, Quebec, Canada, June 1980.
177. Bailey, J.S., G.A., Gray and N.J. Willis, "Sanctuary Signal Processing Requirements," Circuits, Systems and Computers, 11th Asilomar Conference, Pacific Grove, CA, November 7–9, 1977.
178. Lewis, E.A., *et al.*, "Hyperbolic Direction Finding with Sferics of Transatlantic Origin," *Journal Geophysical Research*, 5, 1879–1905, July 1960.
179. Espeland, R.H., "Experimental Evaluation of SCORDES Resolution Capabilities," Naval Ordnance Laboratory Rep. 652, May 1966.
180. Preston, G.W., "The Theory of Stellar Radar," Rand Corp. Mem. RM-3167-Pr, May 1962.
181. Patton, R.B., "Orbit Determination from Single Pass Doppler Observations," *IRE Trans. Military Electronics*, 336–344, April–July 1960.
182. deBey, L.G., "Tracking in Space by DOPLOC," *IRE Trans. Military Electronics*, 332–385, April–July 1960.
183. *IEEE Standard Radar Definitions*, IEEE Std. 686-1982, October 15, 1982.
184. Gumble, Bruce, "Air Force Upgrading Defenses at NORAD," *Defense Electronics*, 86–108, August 1985.
185. Klass, P.J., "Navy Improves Accuracy, Detection Range," *Aviation Week and Space Technology*, 56–61, August 16, 1965.
186. Ridenour, L.N., *Radar System Engineering*, M.I.T. Rad. Lab. Series, Vol. 1, McGraw-Hill, New York, 1947.
187. Private communication from James S. Miller, Technology Service Corporation, 1989.
188. Locke, A.S., *Principles of Guided Missile Design*, Van Nostrand, Princeton, NJ, 1955.
189. James, D.A., *Radar Homing Guidance for Tactical Missiles*, Halsted Press, New York, 1986.
190. Fenster, W., "The Application, Design and Performance of Over-the-Horizon Radars," *IEE Proc. Int. Conf. Radar '77*, Conf. Pub. No. 155, pp. 36–40, 1977.
191. Simpson, T.J., "The Air Height Surveillance Radar and Use of Its Height Date in a Semi-Automatic Air Traffic Control System," *IRE Int. Conf. Record*, 8, Pt. 8, 113–123, 1960.
192. Van Brunt, L.B., *Applied ECM*, Vols. 1 and 2, EW Engineering, Inc., Dunn Loring, VA, 1978.
193. Schleher, D.C., *Introduction to Electronic Warfare*, Artech House, Norwood, MA, 1986.
194. Boyd, J.A., *et al.*, *Electronic Countermeasures*, Peninsula Publishing, Los Altos, CA, 1978.
195. Johnston, S.L., ed., *Radar Electronic Counter-Countermeasures*, Artech House, Norwood, MA, 1979.
196. Maksimov, M.V., *et al.*, *Radar Anti-Jamming Techniques*, Artech House, Norwood, MA, 1979.
197. Leonov, A.I., and K.I. Fomichev, *Monopulse Radar*, Artech House, Norwood, MA, 1986.
198. John P. Murray, Chapter 29 in *Radar Handbook*, M.I. Skolnik, ed., McGraw-Hill, New York, 1970.
199. Cantafio, L.J., ed., *Space-Based Radar Handbook*, Chapter 5, Artech House, Norwood, MA, 1989.
200. Johnston, S.L., "Guided Missile ECM/ECCM," *Microwave Journal*, Int. Ed., 21(9), 20–26, September 1978.

201. Grant and Collins, "Introduction to Electronic Warfare," *IEE Proc.,* 129(3), Pt. F, 113–130, June 1982.
202. Barton, D.K., and H.R. Ward, *Handbook of Radar Measurement,* Artech House, Norwood, MA, 1984.
203. Cline, J.F., "Multilateration Error Ellipsoids," *IEEE Trans. Aerospace and Electrical Systems,* AES-14(4), Correspondence, 665–667, July 1978.
204. Lee, H.B., "A Novel Procedure for Assessing the Accuracy of Hyperbolic Multilateration Systems," *IEEE Trans. Aerospace and Electrical Systems,* AES-11(1), 2–29, January 1975.
205. Torrieri, D.J., "Statistical Theory of Passive Location Systems," *IEEE Trans. Aerospace and Electrical Systems,* AES-20(2), 183–198, March 1984.
206. Johnson, R.L., "Elliptical Error Statistics for Radiolocation Analysis," *IEEE Trans. Aerospace and Electrical Systems,* AES-14(4), Correspondence, 663–665, July 1978.
207. Booker, H.G., "Slot Aerials and Their Relation to Complementary Wire Aerials (Babinet's Principle)," *IEE Journal,* 93(4), Part IIIa (Radiolocation), 620–626, 1946.
208. Private communication from Daniel C. Lorti, Xontech, 1985.
209. Farina, A., and F.A. Studer, *Radar Data Processing, Vol. 1–Introduction and Tracking,* Research Studies Press Ltd., UK, 1985.
210. Farina, A., and F.A. Studer, *Radar Data Processing, Vol. 2—Advanced Topics and Applications,* Research Studies Press Ltd., UK, 1986.
211. Atkinson, R., "Project Senior C.J., the Story Behind the B-2 Bomber," *Washington Post,* October 8, 1989.
212. Adams, J.A., "How to Design an 'Invisible' Aircraft," *IEEE Spectrum,* 24–31, April 1988.
213. Kolosov, A.A., *et al., Over-the-Horizon Radar,* Artech House, Norwood, MA, 1987.
214. Greenwood, T., "Reconnaissance and Arms Control," *Scientific American,* 228(2), 22, Feb. 1973.
215. Steeg, G.F., "Stealth and the Straight Through Repeater," *Journal of Electronic Defense,* 53, May 1988.
216. Ivanov, A., "Radar Guidance of Missiles," Chapter 19 in *Radar Handbook,* 2nd Ed., M.I. Skolnik, ed., McGraw-Hill, New York, 1990.
217. Stein, I., "Bistatic Radar Applications in Passive Systems," *Journal of Electronic Defense,* 13(3), 55–61, March 1990.
218. Ruvin, A.E., and L. Weinberg, "Digital Multiple Beamforming Techniques for Radars," *IEEE Eascon '78 Rec.,* 152–163.
219. Swartzlander, E.E., and J.M. McKay, "A Digital Beamforming Processor," *Real Time Signal Processing, III, SPIE Proc.,* 241, 232–237, 1980.
220. Holzer, R., and G. Leopold, "Call for R&D Effort is Latest Twist in Radar-Homing Missile Debate," *Defense News,* 5(14), 3, April 2, 1990.
221. Peebles, P.Z., Jr., "Bistatic Radar Cross Sections of Chaff," *IEEE Trans. Aerospace and Electronic Systems* 20(2), 128–140, March 1984.
222. Peebles, P.Z., Jr., "Bistatic Radar Cross Sections of Horizontally Oriented Chaff," *IEEE Trans. Aerospace and Electronic Systems,* AES-20(6), 798–809, November 1984.
223. Garbacz, R.J., *et al.,* "Advanced Radar Reflector Studies," Report AFAL-TR-75-219, DTIC ADB013005, ElectroScience Laboratory, Ohio State University, Columbus, OH, December 1975.
224. Dedrick, K.G., A.R. Hessing and G.L. Johnson, "Bistatic Radar Scattering by Randomly Oriented Wires," *IEEE Trans. Antennas and Propagation,* AP-26(3), 420–426, May 1978.
225. Baumgarten, D., "Comparison of Optimum and Suboptimum Receiver Performance for Multistatic Radar Configurations," *IEE Colloquium on Ground and Airborne Multistatic Radar,* London, 4/1–4/3, 1981.
226. D'Addio, E., A. Farina, M. Longo, and E. Conte, "Optimum and Sub-Optimum Processors for Multistatic Radar Systems," *Rivista Tecnica Selenia,* 8(2), 21–28, 1982.

227. Conte, E., E. D'Addio, A. Farina, and M. Longo, "Multistatic Radar Detection and Comparison of Optimum and Suboptimum Receivers," *IEE Proc.* 130(6), Pt. F, 484–494, October 1983.

228. D'Addio, E., and A. Farina, "Overview of Detection Theory in Multistatic Radar," *IEE Proc.* 133(7), Pt. F, 613–623, December 1986.

229. Private communication from R.G. Martin, Westinghouse Corporation, December 1989, patent pending.

230. Strebeigh, F., "How England hung the 'curtain' that held Hitler at bay," *Smithsonian,* 21(4), 120–129, July 1990.

231. Nathanson, F.E., *Radar Design Principles,* 2nd Ed., McGraw-Hill, New York, 1990.

232. Bendor, G.A., and T.W. Gedra, "Single-Pass Fine-Resolution SAR Autofocus," *IEEE NAECON Rec.,* 482–488, 1983.

233. Matthewson, P., B. Wardrop, and D.M. Gould, "An Adaptive Bistatic Radar Demonstrator," *Military Microwaves Conf. Proc.,* London, 471–476, July 1990.

234. Dunsmore, M.R.B., "Bistatic Radars," *Alta Frequenza,* LVIII (2), 53–79, March–April 1989.

INDEX

Active cancellations, 224
Active homing missile, 189, 213
Active loading, 225
Adaptive beamforming, 195
Aircraft security radar, 53, 216
Air traffic control radar, 3
Alerting, 72
ALTAIR radar, 55
Antenna scanning modulation, 257
Antenna taper loss, 253
Anti radiation missiles, 9, 10, 11, 12, 15, 42, 205
AN/FPS-23, 8, 20, 21, 24, 37, 55
Area surveillance, 245
Augmented decoys, 224
Autofocus, 236, 263
Azuza, 38

Babinet's Principle, 150
Barrage Electromagnetique, 30
Barrage noise jammer, 176
Baseline, 59
Beam mismatch loss, 256
Beam scan-on-scan, 43, 44, 245
Beam scan-on-scan loss, 246
Beam scan-on-scan remedies, 245, 253
Beat frequency, 18, 20, 21, 22, 24, 25, 30, 32, 33, 53, 272, 273
Bistatic alerting and cueing, 50, 204
Bistatic angle, 3, 59
Bistatic bisector, 60, 311
Bistatic maximum range product, 68
    under noise jamming, 179
Bistatic plane, 59
Bistatic radar,
    constant, 70
    issue, 7, 8
Bistatic range cell, 78, 277
Bistatic reflectivity measurements, 157
Bistatic surveillance range equation, 247
Bistatic triangle, 7, 59

Brewster effect, 166, 170
Broadcast noise jamming, 176
Burnthrough, 179, 187

Cesium oscillator, 259
Chain Home Radar, 9, 15, 28, 34, 42
Characteristic ellipse, 81
Clutter,
    scattering components, 169
    skewing, 122, 258
    spreading, 124, 258
    tuning, 10, 15, 43, 46, 234
Coherent radar net, 5
Communication satellite transmitter, 53
Common beam volume, 245, 255
Common coverage area, 108, 257, 259
Conic distortion, 65, 103
Constant bistatic range ratio contours, 256
Constant β contours, 63, 211, 228
Constant γ clutter model, 162, 165
Continuous wave radar, 3
Cosecant-squared antenna pattern, 251
Cosite operating region, 72, 96, 106, 115, 198, 250, 255
Costas loop, 261
Coordinate systems,
    baseline, referenced, 101
    clutter, centered, 160
    elliptic, 104
    north, referenced, 59
    other systems, 66
    polar, 70
    true north, referenced, 102
Coordinated scan bistatic radar, 257
Counter-retro system, 222
Counter-stealth, 10, 12, 205, 218
Crosseye, 225
Cross-polarization jamming, 186, 190
Crossover range, 184

Cusp, 71
CW wave interference, 20, 24, 267, 274

Differential doppler, 5
Digital beamforming, 225, 256
Direct breakthrough, 259
Direct phase synchronization, 261
Direct path excision, 7, 210
Direct time synchronization, 258
Displaced phase center aperture, 244
Distant early warning line, 37
Distributed array radar, 56, 191
Doppler
    acquisition system, 56
    location equation, 88
    phase lock (DOPLOC) system, 41, 89

Eccentricity, 61
Effective system noise temperature, 178
Effective σ_B^0, 165
Electronic support measures, 5
ELINT, 214
Ellipse tangent, 60, 62, 311
Emitter locators, 5, 86
Extended baseline, 59
Equivalence theorem, 145, 163
Equivalent monostatic operation, 64, 73
Equivalent monostatic range, 68

Flutter, 8, 20, 21, 24, 37, 55, 273, 275
Floodlight beams, 245
Forward cover extension, 202
Forward-looking SAR, 10, 234
Forward-scatter
    fence, 8, 9, 12, 15, 20, 33, 37, 73, 83, 90, 218, 222
    OTH radar fence, 8, 9, 37
    RCS, 10, 21, 215, 218, 270
Frequency agility, 191

Glistening region, 164

Hitchhiking, 10, 11, 15, 36, 49, 73, 90
Homodyne receiver, 267
Hybrid radar, 5, 13, 184, 224
Hyperbolic asymptote, 62
Hyperbola tangent, 62, 314

Impulse waveform, 270
Indirect phase synchronization, 261
Indirect time synchronization, 259

Integrated sidelobe ratio, 236, 261
Integration efficiency, 249
Interference phenomenon, 18, 22, 24
Interferometer, 39, 56, 89
Ionospheric experiments, 18
Isodoppler contour, 114, 122
Isorange contour, 59, 112, 277
Iso β contour, 63
Isorange resolution, 138, 234

Klein Heidelberg, 8, 9, 34, 49, 199

Leapfrog, 256
Lemniscate, 71
Linear arrays, 192
Local oscillator, 258
Loran C, 259
Low probability of intercept, 205, 226

Maille en Z, 31, 90, 219, 222
Mainbeam footprint, 111
Mainlobe jamming, 177
McGill fence, 37
Minimum detectable velocity, 221, 241
Minitrack, 40, 41
Missile launch alert, 72
Mistram, 38
Monitoring, 72, 209
Motion compensation, 264
Moving target indication, 10, 238, 257
Multiple simultaneous beams, 203, 233, 245, 248
Multistatic measurement system, 4, 8, 9, 12, 55, 191, 198
Multistatic radars, 1, 4, 6, 8, 13, 17, 32, 38, 62, 100, 191

Navstar GPS, 259
Nonmilitary bistatic radars, 57
Normalized reflectivity parameters, 162

Ogive, 63
Ordir, 38
Oval of Cassini, 11, 66, 67, 70, 72, 80, 106, 115, 181, 200, 228, 269, 295
Over-the-horizon backscatter radar, 3
Over-the-shoulder geometry, 50, 72, 88, 96, 203, 245, 291

Parasitic radar, 8
Passive missile seeker, 213
Passive situation awareness, 11, 71, 206

Plan position indicator, 104
Planar array, 192
Plato, 38
Phased array antenna, 65, 103, 194, 253, 255
Phase noise, 261
Phase priming, 260
Power spectral density, 176
Power management, 187
Propagation time delay factor, 104
Pseudobeam pointing angle, 126, 303
Pseudobeam width, 126, 303
Pulse chasing, 15, 43, 115, 190, 203, 212, 233,
    245, 254

Quadratic phase errors, 261
Quartz oscillator, 259, 262

Radal, 41
Radar net, 4, 111, 191, 224
Radar warning receivers, 206, 213, 216
Radio camera, 4
Rapid, 32
Receiver-centered operating region, 73, 106, 116,
    198, 202, 206
Receiver isolation, 3
Repeater, 185
Retrodirective
    bistatic angle, 226
    jammer, 10, 13, 42, 177, 183, 224, 230
Retroreflector, 49, 225
Reven, 32
Rhubarb, 32
Rotman lens, 184, 225, 257
RSRE Bistatic Radar Programme, 257
Rubidium oscillator, 259
RUS-1, 33

Sanctuary, 10, 43, 50, 235, 244, 249
Satellite transmitter, 53
Scattering coefficient, 157
Scattering regions, 162
    bistatic, 163
    low grazing, 163
    specular ridge, 164
    very low grazing, 163, 164
Self cohering, 195
Self screening, 184, 187
Semiactive homing missiles, 8, 12, 15, 36, 72,
    189, 190, 212
Semiconjugate axis, 62
Semimajor axis, 60

Semiminor axis, 60
Semitransverse axis, 62
Sidelobe cancellation, 7
Sidelobe jamming, 177
Sinusoidal phase errors, 261
Situation awareness, 206
Sky wave interference, 20, 24, 265, 274
Smart jamming, 183
Space Surveillance System (SPASUR), 4, 8, 9,
    12, 39, 89, 191
Spatial decorrelation, 191
Spatial excision of clutter, 258
Spot noise jamming, 176, 183, 230
Spy buster, 57
STALO phase noise, 271
Standoff jamming, 181, 230
Stealth targets, 10, 198, 202, 204, 207, 218
Straight-through repeater, 185
Studied ship effect, 34
Step scan, 245
Surveillance radar range equation, 246
Synthetic aperture radar, 10, 46, 234, 261

Tailored flood beam, 250
Tangent to isorange contour, 60, 62, 311
TARSARC, 195
Television flutter effect, 21
Television transmitters, 50
Theta-theta location, 89, 100
Time-difference-of-arrival, 4, 5, 55, 62, 90, 100
Time-of-arrival, 4
Time-multiplexed beams, 245
Transmitter-centered operating region, 73, 106,
    116, 198, 209, 212
Transmitter configurations, 73
TRADEX radar, 55
Transponder, 185
Triangulation, 5, 183, 185, 232
Tripwire fences, 73, 90
Tropospheric scatter, 259
Two frequency range measurement, 89
Type A radar, 33
Type B radar, 33

Udop, 38

Wave interference, 20, 24, 267, 275